Himalayan Perceptions

In the 1970s and 1980s many institutions, agencies, and scholars believed that the Himalayan region was facing imminent environmental disaster. They argued that rapid growth in population had caused extensive deforestation, which in turn had led to massive landsliding, soil erosion and widespread flooding downstream in northeast India and Bangladesh. This series of assumptions was first challenged in the book *The Himalayan Dilemma* (1989: Ives and Messerli, Routledge). Nevertheless, the environmental crisis paradigm still exerts considerable support and continues to be promoted by the news media.

Himalayan Perceptions discusses the evolving attitudes toward environmental change, the confusion of misunderstanding, vested interests, and institutional unwillingness to base development policy on sound scientific knowledge. It brings together and analyses the extensive amount of new research published since 1989 and totally refutes the entire construct of the environmental crisis paradigm. This is followed by examination of recent social and economic developments in the region and it identifies widespread oppression of poor ethnic minorities, which leads to civil unrest, guerrilla activities and warfare, as the primary cause for the instability that pervades the entire region. It is argued that the development controversy is further confounded by exaggerated reporting, even falsification, by news media, environmental publications, and agency reports alike.

Highly illustrated with numerous photographs and detailed examples, *Himalayan Perceptions* will prove an invaluable resource for all those interested in the Himalayan region.

Jack D. Ives is Senior Advisor on Sustainable Mountain Development to The United Nations University, Tokyo, and Honorary Research Professor, Carleton University, Canada.

Routledge Studies in Physical Geography and Environment

This series provides a platform for books which break new ground in the understanding of the physical environment. Individual titles will focus on developments within the main subdisciplines of physical geography and explore the physical characteristics of regions and countries. Titles will also explore the human/environment interface.

Himalayan Perceptions

Environmental change and the well-being
of mountain peoples

Jack D. Ives

Routledge
Taylor & Francis Group

LONDON AND NEW YORK

Published 2014 by Routledge
First published 2004
by Routledge
2 Park Square, Milton Park, Abingdon, Oxfordshire OX14 4RN

Simultaneously published in the USA and Canada
by Routledge
711 Third Avenue, New York, NY 10017
First issued in paperback 2014

Routledge is an imprint of the Taylor & Francis Group, an informa business

© 2004 Jack D. Ives

Typeset in Galliard by
Keystroke, Jacaranda Lodge, Wolverhampton

British Library Cataloguing in Publication Data
A catalogue record for this book is available from the British Library

Library of Congress Cataloging in Publication Data
Ives, Jack D.
   Himalayan Perceptions : environmental change and the well-being of mountain peoples /
Jack D. Ives.
       p. cm. – (Routledge studies in physical geography and environment ; 6)
   Includes bibliographical references and index.
   ISBN 978-1-138-86713-0 (pbk)
   1. Human ecology–Himalaya Mountains Region. 2. Indigenous peoples–Himalaya
Mountains Region–Ecology. 3. Mountain people–Social conditions.
4. Deforestation–Himalaya Mountains Region. 5. Culture and tourism–Himalaya
Mountains Region. 6. Environmental degradation–Himalaya Mountains Region.
7. Environmental policy–Himalaya Mountains Region. 8. Himalaya Mountains Region–Social
conditions. 9. Himalaya Mountains Region–Environmental conditions. I. Title. II. Series.

   GF696.H55I945 2004
   304.2′095496–dc22   2003027488

ISBN 978-0-415-31798-6 (hbk)
ISBN 978-1-138-86713-0 (pbk)

To Pauline – for fifty years

If the doors of perception were cleansed everything would appear to man as it is. . .
William Blake

Contents

Figures

Foreword

The United Nations University (UNU) has been involved in Himalayan research from its formative years. This commitment, supported by each of my predecessors as Rector and myself over a quarter century, came about by a curious chance. When former UNU Vice-Rector, Walther Manshard, asked Jack Ives in 1977 to serve as coordinator of the then new UNU project on 'Highland–Lowland Interactive Systems', he accepted on condition that the project, still to be defined, include an assessment of mountain hazards facing development of the Himalayan region. This condition was met with enthusiasm and set in motion an enterprise that extended eastward from the Nepal Himalaya to the Hengduan Mountains of southwest China and the highlands of northern Thailand and northwestward into the Pamir Mountains of Tajikistan. It rapidly expanded from consideration of mountain hazards to embrace a much broader study of the relationships between development and the well-being of the mountain people. It also led to a most effective partnership between Ives and Bruno Messerli and their colleagues and students, and a training programme for young scholars from many of the countries of the Himalayan region.

The first phase of research led to the Mohonk Mountain Conference on the 'Himalaya–Ganges Problem' in May 1986. This facilitated collaboration with the East–West Center (Honolulu), the Nepal–Australia Forestry Project, the Chinese Academy of Sciences, the National Planning Commission (Nepal), and many scholars from within the region and from Europe and North America.

The Mohonk Mountain Conference and additional research prompted Ives and Messerli to co-author the book *The Himalayan Dilemma*, published in 1989. This effort has been credited as the first successful attempt to overthrow one of the prevailing environmental paradigms of the period – the assumption of imminent Himalayan environmental collapse.

The success, however, has remained primarily academic. Post-1989 research, deriving directly from the project or undertaken independently by other institutions and individuals, has repeatedly reinforced the somewhat preliminary conclusions of the 1989 book, thus calling for an up-dated version.

Despite the academic success, decision-makers, development assistance agencies, and government institutions have scarcely altered their missions relating to development of the region. Policies continue to be implemented based on little solid scientific evidence. Furthermore, they are often inimical to the well-being of the poor mountain people of the region.

The persistent acceptance of Himalayan catastrophe theory, widely supported by the popular news media, deflects political and administrative attention away from the real problems that face the region.

In writing this book, Jack Ives has revisited the Theory of Himalayan Environmental Degradation by analysing the large amount of relevant research that has been completed since 1989. He has, through this effort, succeeded in laying to rest, once and for all, the regrettable misrepresentations that have been made about the Himalayan environmental situation. His analysis provides the foundations for identification of what he believes to be the real problems facing the region – their many facets focus on a single theme – the mistreatment of the mountain people, coupled with unconscionable competition for resources and territory. He portrays conflict in all its forms as elements of the regional threat: conventional warfare, military stand-offs, guerrilla activity, terrorism, insurgency, expropriation of land for mega-projects that provide minimal benefits for local people, casual discrimination and neglect.

The book concludes with a proposal for a regional conference, on the pattern of the Rio de Janeiro Earth Summit of 1992, that may hold out the prospect for collaboration rather than competition and for a fuller measure of equitable access to the undoubtedly enormous resources of the region.

This volume is of immense value in bringing light to a long-standing area of misunderstanding and misrepresentation. Its findings will contribute to a rethinking of the policies and approaches of decision-makers and government agencies concerned with the Himalayan region. Such work is a vital part of the UNU's mission to focus on complex problems and issues that require sustained effort. Through such efforts the University seeks to bring together theory and practice to ensure that our research informs not only academics, but also policy-makers and practitioners. The human-centred approach that Jack Ives has always kept at the heart of his research will continue to be an example to others working on issues such as the Himalayan dilemma which are at the intersection of development and human well-being.

Prof. Dr. Hans J.A. van Ginkel
Rector
United Nations University

Preface

Perceptions of environmental change affecting the Himalayan region have undergone extensive revision over the last 30 years. During the first half of this period it had been widely assumed that environmental collapse was imminent due to exponential increase in pressure on the natural resources driven by rapid population growth and deepening poverty. Although the poverty–population equation continues to attract serious debate, more recent research has totally demolished the earlier paradigm of impending environmental disaster, aided by the very passage of time during which the heralded disaster has not occurred. This has opened the way for a more realistic appraisal of the actual dynamics of change in the region.

The publication of *The Himalayan Dilemma* (Ives and Messerli 1989) 15 years ago derived from an international conference on the 'Himalaya–Ganges Problem' held at Mohonk Mountain House, New York State, in May 1986. The conference had been called to investigate the validity of the prevailing Himalayan environmental paradigm of the 1970s and 1980s that came to be known as the *Theory of Himalayan Environmental Degradation*. In brief, the Theory proposed that increased flooding on the Ganges and Brahmaputra lowlands was the result of extensive deforestation in the Himalaya. The deforestation was presumed to result from a rapid growth in the mountain subsistence farming populations dependent on the forests for fodder and fuel and for conversion to terraced agriculture. As steep mountain slopes were denuded of forest cover, it followed that the heavy monsoon rains caused accelerated soil erosion, numerous landslides, and increased runoff and sediment transfer onto the plains. This was assumed to induce progressive increase in flooding of Gangetic India and Bangladesh that was putting at risk the lives of several hundred million people.

The 1986 deliberations were frequently controversial but a consensus was reached to the effect that the Theory lacked scientific substantiation. This was reflected in the 1989 book; it was understood, however, that much more empirical research was required before absolute refutation of the Theory could be claimed. Nevertheless, the academic response to the book was generally positive and it is still quoted in almost every scholarly publication on the Himalayan region. Forsyth (1996) credited the Mohonk Conference with achieving the first major environmental paradigm shift and, along with Thompson (1995), referred to the unfolding discourse as *The Mohonk Process*.

My personal concern in 1989 was the lack of rigorous and focused research that was needed to substantiate the many issues that had been raised. In practice, *The Himalayan Dilemma*, while effectively contesting many unproven assumptions that collectively formed the Theory, was an attempt to prove a series of negatives. And despite the subsequent academic support, it had little impact on environmental policies. Regional authorities, for example, to this day maintain embargoes on logging in the mountains with the justification that extensive deforestation is causing serious flooding and major dislocations downstream.

Since 1989, and partly as an outcome of the Mohonk Conference, a vast amount of related environmental research has been undertaken; its publication, however, has been scattered widely throughout the literature. The present book, therefore, attempts to bring together and analyse the more recent studies in the context of the earlier work that led up to the 1989 publication. It presents a final rejection of the earlier environmental paradigm; this becomes the more important considering the inappropriate environmental and developmental policy decisions to which the region is still subjected. Furthermore, the inept and sustained focus of much of the government legislation has served to paint the poor mountain minority people as the prime cause of environmental degradation and so deflect attention from the real problems.

Himalayan Perceptions has two primary aims: one is to follow through on the academic discourse, to examine the results of the post-1989 research, and thus to up-date *The Himalayan Dilemma*; the second is to assess the problems that threaten the stability of the region as the new century unfolds. As a corollary to this, some of the reasons why scholarly research has had little, or no, inherent impact on environmental policy making are discussed. In particular, the perpetration of disaster scenarios by the news media is explored because it is believed that this is one of the reasons why the public at large still accepts the notion of impending environmental catastrophe.

The region discussed here extends well beyond the limits of the Himalaya *sensu stricto* (the 2,500-kilometre arc from Nanga Parbat, above the middle Indus Gorge in the northwest, to Namche Barwa, above the Yarlungtsangpo–Brahmaputra Gorge in the east). Coverage is extended to include the Karakorum, Hindu Kush, and Pamir mountains in the northwest, and the Hengduan Mountains of Yunnan, the mountains of northern Thailand, and the Chittagong Hill Tracts, in the southeast. The United Nations University mountain research project, from its initiation in 1978, has investigated test areas throughout this broader region and the present study represents a contribution that concludes the quarter century of UNU effort.

This book attempts to analyse the manner in which the perceptions of the Himalayan region have evolved over the last three decades. It explores how the simplistic environmental alarm arose and why it held sway for so long. Without doubt, in the 1970s and 1980s, the environmental problems assumed to be threatening the region were causing widespread concern and affected the way in which international aid was moulded. Over the last 15 years it has become increasingly clear that the more dominant causes of instability are socio-economic, administrative, political, and the spread of violence and terrorism. The continued debilitating poverty is regarded, at least in part, as a consequence of mismanagement in its broadest sense. Therefore, in addition to assessing how the environmental discourse has played out since 1989, issues involving poverty, oppression of the mountain peoples, unequal access to resources, insurgency, and military conflict are presented. The importance of tourism is also addressed because it is a major force that has both positive and negative aspects and is now menaced by the growing political tensions in the region.

I have tried to write in the spirit of the United Nations General Assembly of 1997 (Rio-Plus-Five) held to evaluate the progress achieved in the five years following the 1992 Rio de Janeiro Earth Summit (UNCED), and the UN declaration of 2002 as the International Year of Mountains (IYM). Since the primary goal of IYM is 'sustainable mountain development', it is considered that prospects for achieving this goal, at least within the Himalayan region, will be limited by the degree to which the problems can be correctly defined. If progress has been made towards producing a more accurate definition then the writing of this book will have been well worthwhile.

There are 11 chapters. Chapter 1 provides an overview of how the Himalayan region has been perceived over the last 30 years and of how research has progressively influenced, or failed to influence, efforts to obtain regional 'sustainable development'. A brief outline of the physical extent of the region comprises Chapter 2. Chapters 3, 4, and 5 provide a detailed discussion of the research that contests the core premises of the *Theory of Himalayan Environmental Degradation* – the questions of deforestation, soil erosion and downstream flooding. The major mountain hazards (physical) that pose a challenge to sustainable development are the concern of Chapter 6. This is followed (Chapters 7–11) by a series of overviews of the current crises and opportunities that are part of the greater Himalayan region. They frame the main conclusion that the Theory was not only a fallacy but also an unfortunate impediment to identification of the real obstacles to sustainable development. These include administrative incompetence, corruption, greed, oppression of mountain minority peoples, political in-fighting, and even military and political competition for control of resources and strategic locations. The well-being of the 70–90 million mountain people has been largely neglected and so they are left with little alternative but to exert increasing pressure on whatever natural resources are accessible, whether legally or illegally.

I have tried to make each chapter as self-contained as possible. This has led to a considerable amount of repetition. However, I believe this approach will be most beneficial for the reader who has not had direct experience of the Himalaya. It would also be advisable to have a high-quality atlas available while reading the text. None of the topics has received an exhaustive treatment. Rather, by selecting a series of issues, I have tried to keep the task within reasonable limits, yet also to ensure a broad view of this vast and complex mountain region and the challenges facing its diverse mountain peoples, who deserve far better treatment than they have so far received. Without their direct involvement sustainable mountain development will remain a bureaucratic pipedream.

Jack D. Ives
Senior Advisor, Environment and Sustainable Development Programme,
The United Nations University, Tokyo, Japan
Honorary Research Professor, Department of Geography and Environmental Studies, Carleton University, Ottawa, Canada
(e-mail: jackives@pigeon.carleton.ca)

Acknowledgements

Many individuals and institutions have contributed to the writing of this book. The actual writing, of course, is only a small part of the total energy that has been expended. The underlying fieldwork by many researchers has been extensive and, at times, physically and mentally exacting. I have received the direct contributions of several dozen colleagues and former students, many of whom have worked under the auspices of the United Nations University (UNU) mountain research project, closely allied with the International Geographical Union's Commission on Mountain Geoecology. In all, associates from more than 20 universities and institutions have been involved, many of them for a quarter century.

Above all, I am indebted to my good friend Bruno Messerli, of the Institute of Geography, University of Berne, Switzerland; this indeed has been a 25 year partnership of shared mountain experiences and mutual support all over the world. Béatrice Messerli also, has joined Bruno in making their Zimmerwald residence a veritable home-from-home.

Much of the earlier fieldwork was undertaken by a distinguished group of colleagues under the umbrella of the United Nations University mountain research project. They are fully acknowledged in *The Himalayan Dilemma*. My own fieldwork subsequent to 1989 was generously assisted by the UNU/Yunnan Academy of Social Sciences team – He Yaohua, Du Juan, Feng Xiao, Mu Liqin, Yang Fuquan, Zuo Ting, Janet Momsen, Seth Sicroff, Peggy Swain, and Lindsey Swope; the Tajikistan team – Yuri Badenkov, Steve Cunha and Irina Merzliakova; the Lake Sarez team – Don Alford, Jörg Hanisch, Gerald Le Claire, Bruno Periotto, Goulsara Pulatova, Carl-Olaf Söder, Svetlana Vinnichenko, Teiji Watanabe and Attilio Zaninetti.

Extensive use has been made of the continuing research in the Khumbu and the Arun-Barun National Park and Conservation Area by Alton Byers, in the Khumbu in the Kangchenjunga Conservation Area by Teiji Watanabe, and in the Arun-Barun National Park and Conservation Area by Robert Zomer.

Early drafts of the following chapters, or parts of chapters, have been critically read by Roger Bilham (Chapter 6), Piers Blaikie (Chapters 1 and 3), Alton Byers (Chapter 7), Dinesh Dhakal (Chapter 8), John Gerrard (Chapters 1, 3, and 4), Larry Hamilton, (Chapters 1 and 3), Thomas Hofer (Chapter 5), Bruno Messerli (Chapters 1 and 5), Don Messerschmidt (Chapter 8), Mike Thompson (Chapters 1, 3, and 4), and Justin Zackey (Chapter 7). Seth Sicroff has provided invaluable assistance by reviewing and suggesting revisions to Chapters 1 and 7.

The following have graciously made available unpublished manuscripts: Roger Bilham, Piers Blaikie, Alton Byers, David Griffin, Thomas Hofer, Hermann Kreutzmann, Yam Malla, Bruno Messerli, Udo Schickhoff, and Teiji Watanabe. All of those mentioned above, together with Brot Coburn, Tim Forsyth, Monique Fort, Dipak Gyawali, Aghar Haroon,

Don Messerschmidt, Pradeep Mool, Martin Price and Greta Rana, have assisted over the years with incidental comments, information and ideas vital to the compilation of the book. Nina Gurung, ICIMOD Librarian, has been most generous in providing a large amount of library and bibliographic support. The Mountain Institute and the Mountain Forum and all their staff members have assisted in numerous ways, always with enthusiasm and care.

The following are thanked for making available photographs: Alton Byers Figures 3.1–3.4, 7.2; K. Kawaguchi 6.4; Bruno Messerli 7.1; and Stan Stevens 10.1. All the unacknowledged photographs are my own. Thom Millest is thanked for his production of the line drawings.

The book itself is a synthesis of the work of many scholars. In such a situation it is difficult to separate out one's own contribution from those of colleagues. Heated debate often sparks new thoughts, sometimes after a long lapse of time. I can best explain that I am trying to portray a team effort. The mistakes are all my own. The original inspiration derives in large part from the Mohonk Mountain conferences of 1982 and 1986; this book is merely the latest result of the 'Mohonk Process'. It also represents my final contribution to the mountain project of the United Nations University for which I served as co-ordinator over a quarter century. My involvement, beginning in 1978, was made possible by the sometime Vice-Rector of UNU, Walter Manshard, and the late James M. Harrison, sometime chair of the UNU Advisory Committee on Renewable Natural Resources. Many members, and former members, of the staff of UNU, but especially Lee MacDonald, Juha Uitto, and Libor Jansky, have been of immeasurable help. UNU Rector, Professor Hans van Ginkel, has been supportive throughout the later years of the mountain project.

Financial, material, and diplomatic support has been provided by UNU, IGU, the United States National Science Foundation, the Ford Foundation, the National Geographic Society, the Swiss Development Coorporation, ICIMOD, and the Academies of Science of China, the Soviet Union, Russia, Tajikistan, and the Nepalese National Planning Commission. The award of a John Simon Guggenheim Memorial Fellowship that facilitated a year's Visiting Professorship tenable at the University of Berne, Switzerland (1976–77) precipitated a turning point in my career that led to the research in the mountains of South and Central Asia. The universities of Colorado at Boulder, California at Davis, and Carleton University, Ottawa, have each made substantial contributions.

My editors at Routledge, Andrew Mould, Melanie Attridge and Tracy Morgan, have been most helpful throughout the writing and production process; I greatly appreciate their care and constant good advice.

In my retirement years I can claim that academic life, with all its pros and cons, has been a fulfilling experience. In proper academic style, it is necessary to state that the individuals mentioned above have contributed enormously in terms of ideas, new insights, disagreements, and vigorous debate. To have been able to teach mountain geography to several hundred graduate and undergraduate students and, with the passage of time, to realize that this has extended through three generations, has been both an exhilarating and a learning experience. To imply that there is a consensus would be absurd – there remain many alternate views and differences of opinion, although I retain the impression that there is broad agreement on the general theme of the book among active Himalayan scholars. But above all, the ability to work and travel throughout extensive areas of the mountains of Asia has been both a privilege and one of life's great adventures.

There remains two major acknowledgements. First, my deepest thanks to the mountain people of this wide and diverse region. Their encouragement has surpassed belief; their friendship, spontaneous joy, frequent deep understanding of the dynamic environmental and socio-economic problems that beset them, have convinced me that Western-based scientific effort can only be enlightened and enhanced as it is realized what *they* have to contribute. Their hospitality and humanity can be encapsulated in a single example – that of an elderly widow in the Pamir who grumbled that I thwarted her attempt to sacrifice one of her two remaining chickens so that she could prepare a suitable lunch. These mountain, frequently minority, people are the 'salt-of-our-earth', but so often its victims.

Finally, without the help of Pauline Ives, all of the above would count for little: steadfast support, sustenance, criticism, editorial assistance, forbearance, and encouragement over 50 years. Our partnership led to our most important contribution: Nadine, Tony, Colin, and Peter Ives; each chose a career of academic research and teaching. I thank them for friendship and support, vigorous discussion, and for our ventures together in the mountains.

1 The myth of Himalayan environmental degradation

New scientific truth does not triumph by converting its opponents and making them see the
light. It triumphs because its opponents eventually die.

(Max Planck)

Introduction

The Himalaya and the mountain ranges and plateaus to the north, west, and east have
remained a focus for myth and drama that have inspired Western thought for centuries. The
birth and spread of great religions, conquest, adventure, and trade have all played their part,
from the times of Alexander the Great, Akbar, and Tamerlane to the present day. The struggle
for control of various sections of the Silk Road emerged through the period of the British
Raj to influence the founding of many independent nation-states during the mid-twentieth
century. This latter period saw the evolution of the Great Game[1] (the competition between
the British Raj and the empires of Czarist Russia and China for control of Central Asia
that began early in the nineteenth century) and the concomitant struggle to effect accurate
survey. The survey, that led to the demarcation of frontiers across some of the world's most
severe and inaccessible landscapes was, of course, an integral part of Britain's strategy in
the deadly competition of the Great Game. The Game has left a legacy of present-day geopolitical anomalies and an array of contested territories that remain persistent flashpoints for
military confrontation: the Wakhan Corridor; Afghanistan itself; the Durand Line, originally
an attempt to define British control in the northwest; the McMahon Line, a similar attempt
in the northeast.

As the world's highest mountains and deepest gorges were identified, exploration and
mountaineering expeditions proliferated. The withdrawal of the colonial powers from
the region both overlapped and followed this phase with the creation of independent states;
those of Central Asia that became established in the last decades can be regarded as the latest
precipitates from the same phenomenon of over-extended imperialism – a centrifugal process
that is by no means complete. It is likely that further evolution is only temporarily arrested
by the long-standing stalemate and armed confrontation in Kashmir, the uncertainties that
engulf Afghanistan, and the ambiguity that prevails in northeastern India.

Just as Westerners have imposed their own cartographic and political constructs on the
Himalayan region, they have also imposed their alien intellectual constructs. Neither the
yearning escapism of the myth of Shangri-la, nor the condescending reductionism of development policies represent useful characterizations of this highly variegated region. While
the region's complexity gradually prevailed over the undifferentiated blank spaces on maps

of Central Asia, the continued existence of 'forbidden kingdoms', fostered by acute inaccessibility (both physical and political), left intact the popular fantasies of hidden valleys and sacred peaks. By the mid-nineteenth century the demands of empire led to the first large-scale exploitation of the luxuriant mountain forests of the Central and Western Indian Himalaya. This prompted the founding of the Forest Department in 1864, the Forest Research Institute at Dehra Dun, and the first institutionalized efforts to achieve sustained timber yield based on scientific forestry. The ensuing expropriation of forest resources on which mountain people depended led inevitably to serious unrest. Such deprivation and unrest have spread and persisted, a trend that is a central focus of this treatise.

Attempts, successful and otherwise, to control and manage the mountain forests have taken diverse forms and have been both cause and effect of many conflicts. There were several early cautionary reports of forest depletion, and even deliberate burning of mountain forests by the disenchanted subsistence farmers, as reported by Tucker (1987) and Guha (1989). However, the alarms drew only local and regional attention. Worldwide concern for the environment and for inequitable development originated much later with the United Nations Conference on the Human Environment held in Stockholm in 1972. While the world's mountains received no direct consideration in Stockholm, the global environment and the North–South dialogue became priorities on the world political agenda. Concurrently, the International Biological Programme (IBP: 1968–72) introduced computer model-ling as a tool for investigation of complex biological phenomena. UNESCO's Man and the Biosphere (MAB) Programme contributed an important course-correction: the programme emphasized the intrinsic linkage between environment and human utilization of natural resources and, more broadly, the absolute necessity of bridging the functional chasm between the natural and the human sciences. One of its twelve individual projects, 'MAB-6: Study of the Impact of Human Activities on Mountain Ecosystems' undertook an examination of the accelerating demands for industrial and amenity resources and of the resultant pressure on natural and cultural systems in mountainous regions. MAB-6 provided the point of departure for what became a global approach to scholarly and applied mountain research; its first institutional link was with the mountain commission of the International Geographical Union (IGU).

The accompanying surge in international academic and popular interest in mountains occurred at a time when mountain environments were comparatively little understood and their importance to world social and economic development hardly realized. However, when a number of Himalayanists joined discussion with the newly formed UNESCO MAB-6 Working Group in Munich in 1974 by invitation of the German Foundation for International Development (GTZ), an alarm was sounded over the perceived imminent environmental collapse of the Himalaya. Erik Eckholm, *New York Times* science editor, was invited to Munich by the conference organizing committee specifically to transpose the conference findings for widespread popular distribution.

It must be emphasized that the Munich conference alarm did not develop out of thin air. There had been earlier expressions of concern from the Himalayan region (Kaith 1960) and in the late 1960s and early 1970s development agency staff had been reporting to their headquarters that there was cause for alarm, although publication appeared after 1974 (Mauch 1976; Rieger [1978/79] 1981). Similarly, Indian Himalayanists followed a parallel path (Lall and Moddie 1981; Centre for Science and Environment, India 1982, 1985).

The Munich conference on 'The Development of Mountain Environment' served to focus much of the pre-existing concern regarding the projected deterioration of the Himalayan environment and its assumed causes (Müller-Hohenstein 1974). The conference not only motivated a considerable number of academic scholars and development planners, but actually spear-headed a global campaign. The furore must be credited largely to one participant: Erik Eckholm; his paper in *Science* (1975) and his book *Losing Ground* (1976) had a major impact. However, his *New York Times* pulpit was probably even more influential in rallying popular interest.

A virtual flood of publications sought to address perceived crisis in the Himalaya–Ganges region. It engulfed the news media and scientific journals. Sandra Nichol's documentary movie *The Fragile Mountain* (1982) caused a great stir. In addition, the bandwagon was heaped with monographs, primarily of Indian and Nepalese provenance (Poffenberger 1980; Lall and Moddie 1981; Bandyopadhyay *et al.* 1985; J.S. Singh 1985; T.V. Singh and Kaur 1985; and S.C. Joshi 1986). Among Western academics, the perceived threat of imminent disaster held sway for more than a decade (Bishop 1978, 1990; Blaikie *et al.* 1980; Karan and Iijima 1985; Myers 1986; Allan 1987). Even today it is still taken seriously in several sectors.

Among planners and policy makers a new paradigm was born and quickly achieved dominance; through this simplistic lens all the problems of the Himalayan region were perceived as a result of population growth and consequent deforestation. There can be no doubt that the paradigm had a powerful impact on development and foreign aid policy formulated by both international and bilateral agencies. The United Nations Environment Programme (UNEP), for example, contracted with the International Institute for Applied Systems Analysis (IIASA) for a detailed analysis of the assumed Himalayan–Ganges crisis. Ironically, the resultant report became the basis of a book, *Uncertainty on a Himalayan Scale* (Thompson *et al.* 1986), that contributed to exposing the fallacies under-pinning this dangerously constructed approach to development. Even as late as 1989 there was a concerted move to designate funds and engineering schemes aimed at channelling the Brahmaputra in an attempt to reduce the annual monsoon flooding in northeastern India and Bangladesh. This would have expended resources far in excess of the Gross Domestic Product (GDP) of the region concerned bringing funding, engineering expertise, and contractual arrangements from Japan, Germany, The Netherlands, the United States, and the World Bank. Fortunately, good sense prevailed (Rogers *et al.* 1989 – see below p. 21), although major river diversion schemes involving the Ganges and the Brahmaputra are currently under active discussion between India and Bangladesh.

What was so startling at the time was that the poor mountain subsistence farmer had become a scapegoat, summarily convicted of causing the deforestation that was initiating a cascade of environmental catastrophes. It was in view of the far-reaching implications of this paradigm that, based on our subsequent field experience in Nepal, Bruno Messerli and I undertook to write *The Himalayan Dilemma* (1989), and it is because of the persistence and exemplarity of the paradigm that I offer this follow-up account. But before going further, it is necessary to reiterate the environmental threat as it was perceived in the mid- to late 1970s and early 1980s.

Origin and misuse of the *Theory of Himalayan Environmental Degradation*

At the outset, I must acknowledge responsibility for introducing the expression, the 'Theory of Himalayan Environmental Degradation' (Ives 1984, 1986). The phrase was an expedient in my early attempt to examine what seemed the most compelling themes in a liturgy of dire predictions of environmental catastrophe, ranging from total deforestation of steep mountain slopes to plumes of sediment far into the Bay of Bengal. The use of the word *theory* may be questioned. Nevertheless, the expression, either in its original form or slightly modified, has been widely quoted and used to this day (Schickhoff 1993, 1998; Guthman 1997; Stellrecht 1998; Zurick and Karan 1999; Funnell and Parish 2001; Parish 2002; Smadja 2003; Blaikie and Muldavin 2004).

The construct as presented is an amalgam of ostensibly scientific findings, casual observations, assumptions, opinions, and moral imperatives. The early focus of these pronouncements was Nepal, primarily because its accessibility resulted in a rapidly accumulating data base and also a concentration of foreign aid projects.

In the synthesis, the large number of propositions were reduced to an eight-point scenario characterized as *The Theory of Himalayan Environmental Degradation* (hereinafter, 'the Theory'). This was presented for debate at the opening of the Mohonk Mountain Conference on 'The Himalaya–Ganges Problem' in May 1986 (Ives 1987a). The eight points were as follows:

1   After about 1950, the entire Himalayan region began to experience a spurt in population growth. The 1981 Nepalese census, for instance, reported a population in excess of 16 million, indicating a growth rate of 2.6 per cent per annum and a doubling time of 27 years. This was assumed to be representative of much of the region.

2   The population explosion was underway among a community that was more than 90 per cent rural, impoverished, and dependent on subsistence farming. In the context of a country that was primarily dependent on fuel wood for energy, with presumably expanding herds of domestic animals that required grazing, fodder, and bedding materials from the forests, and with demonstrable expansion of agricultural lands, this population growth was causing extreme intensification of stress on forest resources.

3   The increasing pressure on the mountain forests was leading to extensive deforestation, with authoritative predictions that all accessible forest in Nepal would disappear by the year 2000. Similar losses were predicted for much of the Indian Himalaya.

4   The process of deforestation was stripping steep mountain slopes of their protective vegetation cover and causing a catastrophic increase in soil erosion and landsliding.

5   This, in turn, was leading to increased surface runoff during the summer monsoons and an increase in disastrous flooding and siltation on the plains of the Ganges and Brahmaputra, together with accentuated water shortages during the dry season.

6   The increased sediment transfer was extending the delta of the combined Ganges and Brahmaputra and causing new islands to form in the Bay of Bengal.

7   Continued loss and degradation of agricultural land in the mountains, due to soil erosion and landslides, had forced another round of forest depletion for construction of more agricultural terraces so that more food could be grown. But as the forest margins were pushed back further from the villages the women had to walk progressively further in order to collect fuel; eventually this burden exceeded human capacity, driving farmers to rely more and more on animal dung for fuel.

8   This, in turn, deprived terraced soils of natural fertilizer, crop yields declined, and the weakened soil structure further increased the incidence of landslides. Thus, even more trees were cut on still steeper slopes to make room for additional terraces so that food production might keep pace with the continuing population growth.

The Theory seemed to be couched as a series of eight affirmations – rather than as tendentious hypotheses inviting further analysis – leading several subsequent writers to assume that they represented the views of the authors of the synthesis (i.e. Ives and Messerli 1989), rather than a target deliberately erected in order to instigate productive controversy.

These eight points were represented as a series of vicious circles, each linked to the others, driving an inexorable downward spiral. It was the apparent impossibility of breaking any of these vicious circles that supported the prediction of imminent widespread environmental collapse. Furthermore, the gathering weight of the environmental and economic tragedies would put increasing pressure on the already fragile political balances of the wider Himalayan region.

Many other ominous epicycles were spun out from the eight-point scenario. For example, it had been calculated that in 1981 (Nepal Census) there was, on average, less than one hectare of land per family; now, as the population of Nepal doubled every 27 years, the pressures created by the subdivision of this finite amount of agricultural land would balloon exponentially. Similarly, the added pressure of collecting and carrying fuelwood and fodder and fetching water fell predominantly on the women. They in turn became progressively overworked and undernourished and the next generations of children would begin life more and more deficient in essential nutrients, so that the situation would continue to deteriorate. Domestic animals, essential to the Middle Mountain subsistence mixed farming system as suppliers of fertilizer and draught energy, depended heavily on fodder from the depleted forests, so that their capacity and total numbers also would diminish.

The net results of the various destabilizing processes in the Middle Mountains were predicted to be: diminished crop productivity, both in terms of total national production and as yield per unit area; increase in proportion as well as absolute numbers of that part of the rural population whose nutrient intake was below a minimum acceptable level; absolute deforestation; and progressive mountain desertification. Since the desertification was assumed to be occurring on steep slopes, the associated processes of gullying, soil erosion, and land-sliding would have calamitous downstream effects. Thus were envisaged the rapid siltation of reservoirs, excessive shortening of the useful life of major hydro-electric and irrigation projects, increased flooding on the plains (already an annual disaster for India and Bangladesh), increase in the levels of river beds, and destruction of rich lowland farmland by the spread of sand and gravel as rivers break their banks and change their courses. The worst-case scenario predicted that the terrain of Nepal and that of adjacent areas of the Himalaya, and certainly the very basis of life, the topsoil, would have virtually washed away down the Ganges by the year 2000. Eckholm wrote eloquently that Nepal was exporting the commodity that it could least afford to part with, namely topsoil, to India, in the form that India could least afford to receive it (Eckholm 1976).

And there were further ramifications, such as claims by environmentalists that deforestation was affecting the climate in such a way as to reduce normal annual rainfall amounts. This, of course, would set up another vicious circle to accentuate the effects of the others. A further problem, emphasized by Thompson *et al.* (1986), was that several aid agencies and government institutions were predisposed to adopt the essence of the Theory because it justified their preferred agendas. The Theory, in short, was an intellectually satisfying concept, an environmental 'theory of everything', so plausible that it was widely accepted as fact.

The scenario presented above led to a number of critical implications which further enlarged what could be described as the *perceived* Himalaya–Ganges Problem. It implied that a few million Himalayan hill farmers were responsible for the massive landscape changes that were affecting the lives and property of several hundred million people in Gangetic India and Bangladesh (World Resources Institute 1985). This false perception raised two related points: first, that the downstream countries, as victims of this unwarranted and irresponsible environmental disruption, could have justified reprisals in economic, political, or military terms; and second, that so long as no reprisals were actually taken, Nepalese interests were well-served by the understanding that it was drifting helplessly into environmental and socio-economic chaos, which evidently accounted for its disproportionate share of international and bilateral development aid.

Whether or not the scenario of disaster for Nepal could have been extended along the entire Himalayan system was not discussed further by Ives and Messerli (1989) except for emphasis of several related points. The Kumaun and Garhwal Himalaya appeared to conform to the Theory's generalizations, with two additional components. One was the intensive commercial cutting of mountain forest stands to meet timber demands of the lowland population centres (until checked somewhat by the Chipko Movement). The other component was the extensive development of mountain roads, especially as a military response by India to the border war of 1962 with China. Much of the road construction was substandard and caused a great increase in landslide incidence; the roads also opened up extensive mountain forests to commercial clear-felling.

Proceeding westward into Himachal Pradesh, Jammu and Kashmir, and the Karakorum and Hindu Kush, a series of mountain and highland landscapes with very different climatic regimes occur. Increasing aridity with distance from the influence of the summer monsoons greatly reduced the value of comparison with the Central Himalaya.

Conversely, eastward from Nepal, into Sikkim and the Darjeeling (West Bengal) Himalaya, and Bhutan, there were mountain areas with increasing amounts of summer monsoon precipitation. Bhutan, and probably even the less accessible and less well known Indian state of Arunachal Pradesh, were viewed by adherents of the Theory as exceptional cases where postulated deforestation and environmental disturbance were believed to be moderate or insignificant. However, it was believed that even these regions were probably poised to follow the same road to disaster, along which Sikkim, Nepal, and the Central Himalaya of India were purported to be travelling.

This early attempt to define and explain the *Theory of Himalayan Environmental Degradation* was intended to provide a synthesis of the widely prevailing convictions of the time. Descriptions were also obtained during initial field experience in the Hengduan Mountains of Yunnan and Sichuan (southwestern China) and northern Thailand, where similar assessments of environmental deterioration and its causes were made. The issue of political instability in parts of the region, even ongoing warfare (as in Afghanistan) and military stand-off (as in Kashmir along the Line-of-Control), was also introduced although not developed. Finally, the extreme complexity of the region, topographic, climatic, bio-geographic, historic, demographic, and ethnic, was emphasized. The problem of evaluating the Theory is best illustrated by the following quotation:

> The perceived problem . . . is in the minds of the vested interests – whether the World Bank, the Chipko Movement, different national governments, or the scientists. It is likened to a kaleidoscope which will change its pattern depending upon the way in which it is tilted, or upon the angle of view. This is the essence of Thompson and Warburton's (1985) *Uncertainty on a Himalayan Scale*. The Uncertainty is a large element of the Problem.
>
> (Ives 1987a: 193)

Initial response to the Theory

The foregoing discussion is intended to provide a collective overview of the perception of the threat to the Himalayan and plains environments and its cause that prevailed during the period from approximately 1960 to 1990. Next will follow a synthesis of the response that was evoked during the Mohonk Mountain Conference of May 1986. The full conference proceedings were published (Ives and Ives 1987). The editorial policy of the journal *Mountain Research and Development* had been adapted to explore the discourse as far as possible prior to the conference, thereby augmenting the basis for discussion. As editor, I unashamedly encouraged the review and publication of many papers to ensure that an active discussion was available in a mainline mountain journal. Thus, many papers directly related to the Himalaya–Ganges problem appeared during the period 1982–1989. Some of the most important of these were:

- the series by Thompson and Warburton (1985a, 1985b) and Hatley and Thompson (1985) that became the basis for the book, *Uncertainty on a Himalayan Scale*;
- the series by Mahat *et al.* (1986a, 1986b, 1987a, 1987b) and Griffin *et al.* (1988) arising from the Nepal–Australia Forestry Project and culminating in David Griffin's book *Innocents Abroad in the Forests of Nepal* (1989);
- an early and important individual article by Deepak Bajracharya that was published in 1983;
- there were also the initial results of the UNU–Nepal Mountain Hazards Mapping Project that had sparked this entire process. One of the most important papers was that by Johnson *et al.* (1982); others included the associated papers on the Kakani test area (Caine and Mool 1981, 1982; Kienholz *et al.* 1983, 1984);
- similarly, new insights were being obtained from the early UNU research in the Khumbu Himal, some of the most revealing being those of Byers (1987a, 1987b);
- other contributions resulted from the initial examination of the actual mountain hazards affecting this small but representative sector of the High Himal (Bjønness 1980; Ives 1986; Zimmermann *et al.* 1986; Vuichard and Zimmermann 1986; Brower 1990).

A contemporaneous process of early questioning the orthodoxies of the 1970s and 1980s, especially concerning the relationships between forest clearance and watershed response, had been taking place at the East–West Center, Honolulu. Lawrence S. Hamilton had mounted some of the first challenges to the FAO and World Bank forestry assumptions. He initiated a Watershed Programme at the East–West Center in 1981 that included presentation of a concept that he referred to as the '4 M's': myth, misunderstanding, misinterpretation, and misinformation (Hamilton 1983). Consequently, the East–West Center became a co-sponsor of the Mohonk Mountain Conference (Hamilton 1987). Two important challenges emerged which have pervaded much subsequent thought: (a) the recommendation that the term 'deforestation' be eliminated from the environmental discourse, or else rigorously defined; and (b) the theoretical notion that 'it floods in Bangladesh when it rains in Bangladesh'.

Consideration was also extended far beyond the geographical limits of Nepal as it was realized that similar assumptions of population growth forcing deforestation had been influential as far afield as northern Thailand, southwestern China (Yunnan and Sichuan), and the Central and Western Indian Himalaya, as well as in the Darjeeling area, as previously mentioned.

Many other individual studies were published during the 1980s that had a bearing on the quest to evaluate the Theory: Carson (1985); Ramsay (1985); Haigh (1982a, 1982b, 1984). During the Mohonk Conference, Tucker (1987) and Richards (1987) had provided valuable historical depth that related especially to nineteenth-century forest exploitation in the Central Himalaya of India. While the Conference certainly included contributions that strongly supported the Theory (Narayana 1987; Tejwani 1987), it was nevertheless generally agreed that the Theory had been demolished, or at the very least, seriously challenged. The essence of the response to the Theory appeared as the book *The Himalayan Dilemma* (Ives and Messerli 1989) and its general conclusions were reinforced by a literature review undertaken by Bruijnzeel and Bremmer (1989). Metz (1991), drawing on his own fieldwork in Nepal, produced a valuable overview and assessment of many of the publications cited above. The following summary draws on an overview published in *Land Use Policy* (Ives 1989a).

The title of the *Land Use Policy* paper posed the key question: 'Deforestation in the Himalayas – The cause of increased flooding in Bangladesh and northern India?' As illustrated above, this apparently simple question had already been answered in the affirmative over the previous decades by many scholars, decision-makers, politicians, foreign aid and United Nations donor agencies, and the news media. Although the question had appeared simple, the answer was to prove extremely complex. As the debates during and after the Mohonk Conference demonstrated, the dramatic assumptions that combined to make up the Theory had been based upon surprisingly little reliable information. There were virtually no hard data to support the contention that land-use changes in the Himalaya had had any direct impacts on the plains and delta of the Ganges and Brahmaputra. It was equally uncertain whether there had been recent and significant loss of forest cover in the mountains. In fact, what was truly alarming was that, with so little substantiation, the World Bank (1979), the Asian Development Bank (1982) and others (e.g. Karan and Iijima 1985; Myers 1986) had declared that by about AD 2000 no accessible forest cover would remain in Nepal.

In many of the areas for which reasonably reliable data were available, it was demonstrated that the portrayed reports of widespread deforestation since 1950 were simply insupportable. For Nepal itself, the process of systematic clearing of the forests had begun as long ago as the eighteenth century. In fact, it could be shown that following the militant unification of the region under the House of Gorkha in 1769 state policy was to encourage conversion to arable land. This had been motivated by the state's need to increase fiscal viability: forest land was not taxable but considered wasteland, while agricultural land was the main source of state revenue through taxation. A gift of forest land was also used as a form of payment to high-ranking military personnel, prompting further forest clearance and conversion to an increase in the tax base. In fact, the vast proportion of the arable land in the Middle Mountains (the most densely populated region) had been converted from the forests by the early twentieth century (Mahat *et al.* 1986a). Little reduction in total forest cover (as distinct from deterioration in forest quality) in the last century has been demonstrated to have occurred since then. Another critical point that was raised during the Mohonk Conference was that fodder, not fuelwood, was the primary commodity taken from the forests by the rural people.

While loss of forest cover had certainly occurred since 1950, this was largely restricted to the Terai, the Inner Terai, and the Siwalik Hills, much of the loss being the result of government resettlement policies, spontaneous migration from the Middle Mountains, and commercial as well as illegal cutting. In addition, based on repeated aerial reconnaissance of selected areas, it was demonstrated that landslide incidence and damage was inversely proportionate to population density. This indicated that farmers worked hard to maintain slope stability and that repair of landslide damage was a farm priority. Dung was rarely used for fuel, except by high-altitude communities as part of their traditional practice. In other words, this form of fuel use was ethnically based and was not related to fuelwood scarcity among the majority Hindu population. There were exceptions however. For instance, dung was used for fuel in the Kathmandu Valley in the late winter when there was a surplus after organic fertilization had been completed. Thompson *et al.* (1986) had made a mockery of the statistics derived from actual measurement of fuelwood consumption obtained from internal reports by Donovan (1981). Mahat *et al.* (1986b) had also undertaken a thorough analysis to explain the hazards of the earlier approaches to 'measuring' per capita consumption of fuel: in many instances the investigator assumed that 'fuelwood' referred to cut and chopped timber while the interviewee understood 'fuelwood' to encompass all manner of organic material, including crop residues.

The Mohonk Conference discourse presented to this point had been mainly limited to Nepal. Parallel observations were offered concerning other sectors: Dehra Dun (Richards 1987); Yunnan and Sichuan (Messerli and Ives 1984; Ives 1985). Some of the claims for extensive loss of forest cover on steep slopes, for instance in the Khumbu, were shown to be false by repeat ground photography (Byers 1987a, 1987b).

Hamilton (1987) insisted that 'deforestation' was an emotive term; that its use had caused more confusion than clarification; that deforestation was not inherently *bad*; and that, depending on mode of tree removal and subsequent land use, conversion from forest to terraced agriculture could prove more beneficial in terms of soil conservation than leaving the area under forest. Use of the term 'deforestation' is also technically imprecise and fails to take into account the extensive research on forest quality, canopy, ground cover, and the dynamics of rain-drop impact. Gilmour *et al.* (1987) also produced preliminary data, if only for a single experimental site, indicating that soil erosion and flooding are not invariably consequences of forest clearance, and that reforestation of cleared land does not necessarily result in any reduction in soil erosion and overland flow from monsoon rainfall. Gilmour further demonstrated that extensive tree planting was being successfully undertaken on privately controlled land through the initiative of the supposedly ignorant farmer. This was in contrast to the failure of many government and foreign aid reforestation projects (Gilmour 1988). Through repeat photography (Figure 1.1 and Figure 1.2), it was shown that landslide scars were frequently re-terraced by the local farmers resulting in rapid restoration of agricultural productivity within a few years (Ives 1987b). In some instances landslides were deliberately initiated through the diversion of mountain streams in order to facilitate new terrace construction.

A much larger problem was determination of the relative importance of human land-use interventions as compared to the natural processes: mountain uplift, development of steep slopes, and massive transfer of weathered rock materials from the mountains to the plains over geological time. The Himalaya are recognized as one of the most tectonically active mountain regions of the world and they are incurring some of the highest rates of sediment transfer (erosion). The very presence of the Ganges and Brahmaputra plains, in places underlain by as much as 5,000 metres of sediment derived primarily from the Himalaya, depends upon these dynamic natural processes. Attempts to analyse stream flow and sediment transfer data from gauging stations on the Ganges, Brahmaputra, Tista, and Sun Kosi rivers have yielded inconclusive results because the data sets were incomplete or unreliable. Even so, the data that were available showed little indication of any correlation between changes in land use in the mountains and changes in stream flow and sediment transfer on the plains. Some of the adverse effects of flooding, assumed to be the result of environmental degradation in

Figure 1.1 View taken from the Trisuli Road below Kakani, just outside the Kathmandu Valley, Nepal. A classic debris flow (landslide) had released during the previous monsoon season (1978), several houses had been swept away, the debris was considered extremely unstable, and the disturbance likely to expand. The site was re-photographed at intervals until November 2000 by which time the landscape had changed so much that comparison with 1978 was not possible (previously published as Figure 5.2 in Ives and Messerli 1989; see also Figure 5.3 in the same publication that shows almost complete recovery in October 1987).

the mountains, were certainly caused by human activities on the plains. Clearly, the enormous increase in population and infrastructure on the plains during the course of the twentieth century would have resulted in accelerating human and economic losses with or without intensified flooding. Furthermore, flood losses have almost invariably been registered in terms of the loss of human and domestic animal lives and property – and the more serious the losses that could be demonstrated, the larger would be the amount of international aid.

It was clear to the Mohonk Conference participants that a broad range of apparently simple questions urgently required investigation. What was the cause of rapid population increase? Could it really be proven that flooding on the plains had increased systematically over the last century? If so, could the change be related unambiguously to forest clearing in the mountains? Sufficient was already known by the 1980s so that a tightly focused investigation could have been mounted for a fraction of the cost of a single hydro-electric power project, the effectiveness of which could be eliminated within a few years by siltation or the outbreak of a glacier lake. Unfortunately, many policy makers tend to regard researchers as obsessive compulsives who delay and disrupt efficient decision-making. At the same time, there was

Figure 1.2 October 1997: approximately the same view as Figure 1.1. The two photographs show major changes across the entire mountain side. The tree cover is more extensive. Most significantly, it is very difficult to recognize many of the individual terraces. Except for the terraces at the top right, the landscape has been extensively re-shaped. The local farmers have been effective in conserving the terraces and this has more than counter-balanced natural mass movement. The landscape is essentially stable. The farmers explained that the old landslide scar, which cannot be detected on the photograph but which they can trace precisely, is one of the more productive sections of the hillside. The situation in November 2000 was essentially unchanged.

an urgent need to explore a broad range of interwoven issues underlying the existing instability and potential future crisis: unequal access to resources; mistreatment of mountain minorities; persistent confusion between cause and effect; 'outside' aid; technological, monetary, and political interventions; political fragmentation within the region; a Eurocentric approach to problem solving based upon short-term budgets and a 'target' mentality of bureaucratic rigidity; and lack of attention to the need for appropriate institution building (Griffin 1987, 1989).

Setting aside the question of the reliability and relevance of the available demographic and economic data, as well as the difficulty of analysing aggregate data from one of the most complex regions in the world, the Mohonk Conference participants rejected the validity of the conventional wisdom that an environmental crisis in the Himalayan region was imminent. I was concerned about publishing this statement in 1987 because it bore on the security of several hundred million people. There was absolutely no intention of minimizing the problems facing the region. The danger was seen as the uncertainty and confusion regarding the causes and scope of the problem. Lack of reliable information could lead to the intervention by institutions and agencies whose natural inclination would be to expend vast sums of money on a major technological 'fix' that would not only fail to cure the perceived symptoms but would likely result in significant collateral damage. Most unfortunately, it would malign subsistence farmers. These men and women, who had accumulated extensive environmental knowledge and experience, have much to offer the world at large. The final words arising from the Mohonk deliberations were that the subsistence farmer must be appreciated as an important part of the solution rather than deprecated as a substantial cause of the problem. Without this accord, 'development' would be seriously handicapped, assuming that its goal is to achieve beneficial and sustainable change. Without it, decision-makers would themselves lead the region into a true super-crisis. The causes would be political and administrative rather than environmental, and once again the minority peoples of the high hills and mountains would be front-line victims. On behalf of the Mohonk Conference participants, we therefore called for a global commitment to the preservation of cultural diversity, a commitment without which our avowed concern for biological diversity would be futile and fruitless (Ives and Messerli 1989).

Overview of research in the Himalayan region: 1989–2003

As noted above, numerous critical questions were left unanswered at the close of the Mohonk Conference. What follows is a selective overview of research pertaining to those questions undertaken in the 15 years that have elapsed since publication of *The Himalayan Dilemma*.

Nepal–Australia Forestry Project

This project (subsequently renamed the Nepal–Australia Community Forestry Project) has focused for 25 years on two districts, Sindhu Palchok and Kabhre Palanchok, situated immediately to the east of the Kathmandu Valley, Nepal. The study site encompasses a cross-section from the foothills, across the Mahabharat Lekh and the Middle Mountains, to the High Himal and the Tibetan border. The series of papers by Mahat *et al.* (1986a, 1986b, 1987a, 1987b), Griffin (1987, 1989), and Griffin *et al.* (1988) has been recognized as a major contribution to the 'Mohonk Process'. The pre-1989 work of Gilmour (1988) has already been cited. Subsequent work has both reinforced Gilmour's preliminary findings and has lent additional support to that section of the Mohonk Process dealing with forestry and rural land use in the Middle Mountains of Nepal. Thus, by the early 1990s Gilmour and Nurse (1991) and Gilmour and Fisher (1991) demonstrated that between 1972 and 1989 there had been an increase in tree cover on land controlled by the local farmers in contrast to forests on common land and those directly controlled by the government. Furthermore, this refor-estation had been spontaneous, without government or foreign aid assistance. Nevertheless, continued investigations have yielded important qualifications. Progress is reported in the transfer of responsibility for forests from direct government control to the villagers and in the establishment of Forestry User Groups (FUGs) within the villages.[2] Jackson *et al.* (1998) have demonstrated that between 1972 and 1992 community forestry activities at lower altitudes had a beneficial impact on the local environment and stability of the land-use system. For instance, shrublands and grasslands were converted into more productive forest land. In contrast, however, forest cover on the upper slopes (above about 2,500 metres) was being rapidly lost. This is a significant qualification of the earlier work of Gilmour and his associates and is related directly to the degree of local control of the higher land. This in turn is related to distance from the village centres.

Malla and Griffin (1999) report on the changing role of the forests in Nepal's Middle Mountains. Their study area is representative of Kabhre Palanchok District and involves a transect, extending from Dhulikhel, a small urban centre on the main road to Kathmandu, to Budhakhani, a remote village in the Mahabharat Lekh more than a day's walk from the motor road. This cross-section of the Middle Mountains embraces a continuum from a remote site with a traditional near-subsistence economy to an area substantially influenced by the rapidly growing market economy of the late-twentieth century. Malla and Griffin demonstrate that, in proportion to the proximity to influences of the market economy, there have been major changes in the farming systems and in the use of forest resources. At the Dhulikhel extreme there has been a rapid and accelerating development of local manufacturing and establishment of small businesses that has led to a change in farm economy: increased reliance on cash crops as well as off-farm wages; a shift to fewer livestock, stall feeding, and emphasis on buffalo and goat (for sale of milk and meat); and a change in forest product use from predominantly fodder and leaf-litter to mainly timber and firewood. The conversion from fodder to timber and firewood also reflects a change from use of the public forests to dependence on private lands because commercial use of public forests is prohibited. Changes also include importation of wood from the Terai and progressive abandonment of marginal *bari* terraces and their conversion to shrub and forest. Malla and Griffin underline the need for the Nepalese government to recognize the flexibility of Middle Mountain farming systems rather than to assume an unchanging traditional subsistence status. In turn, this indicates the need for policy to respond accordingly and, in particular, to meet the requirements of the rapidly expanding domestic market.

Overall, because of the continued inflexibility of government forest policy, these studies demonstrate the serious lag between government and aid agency attitudes and the rapidly changing economic situation as it affects the rural farming scene in the Middle Mountains. Although conclusions drawn from studies in Kabhre District cannot necessarily be extrapolated to the entire Nepal Middle Mountains, they are likely applicable to the environs of the growing number of small towns with road connections. However, this work of the Nepal–Australia Community Forestry Project was completed before the impacts of the Maoist Insurgency began to be felt and so does not fully reflect conditions prevailing today (see Chapter 9 pp. 198–201).

Highland–lowland interactions

A key issue in evaluating the Theory is whether or not erosion and increased runoff in the Middle Mountains (caused by human activities such as deforestation) has been producing increased flooding and siltation on the plains. Since 1989, a considerable number of highly sophisticated studies of the geomorphology of agricultural terraces and of degraded and unused land have focused on the impact of deforestation.

Schreier and colleagues, attached to Canadian, Swiss, and Nepalese institutions, have studied a number of small watersheds in the Nepalese Himalaya (Schreier and Wymann von Dach 1996). The Jhikhu Khola watershed in the Middle Mountains became the main focus of the Schreier team. Detailed field experiments were undertaken over a period of more than nine years. Instrumentation has been maintained until the time of this writing. Work has included the collection and analysis of thousands of soil samples, determination of the impacts of farming activities, and application of GIS modelling. One question of central importance was how the episodic torrential rainstorms of the summer monsoon affect runoff and sediment dynamics. The researchers were able to conclude that at the local scale human intervention plays a significant role in redistributing the losses incurred through cultivation on steep slopes. Wymann (1991) had already demonstrated that soil nutrient content is inversely proportional to altitude in the Jhikhu Khola: the lowest rice paddies (*khet*) have a high nutrient status due to the progressive downslope movement of sediments. One corollary to this is that the farmers, with their elaborate indigenous system of retaining and redistributing sediments, maintain an important degree of control over their fields. Only in extreme rainfall events was it possible to obtain some measurable response beyond the limits of the small watershed. Schreier and Wymann von Dach (1996: 82) concluded that as their research extended downslope from a first order stream system to the complexity of the Himalayan foothills the scale of the processes became unmanageable and the effects of human impact are eclipsed by the magnitude of the natural processes (see Chapter 4 pp. 99–100).

Independently, and in association with a project initiated by the Royal Geographical Society, Wu and Thornes (1995) undertook a detailed geomorphological study of terrace irrigation in the Likhu Khola catchment in the Middle Mountains north of Kathmandu. They determined that terracing does not change the hydrological behaviour of the hill slopes and that the effects of human impacts are positive rather than negative (Figure 1.3). They demonstrated that individual terrace failures during torrential monsoon downpours do not contribute any sediment to the downslope nor augment lower slope stream flow. The mechanisms of the large-scale terraced slope failures, which involve dozens of terraces, are much more complicated. Even on such a scale of failure, however, farmers who manage terraces and irrigation canals on land naturally prone to erosion and failure can repair most of the damage and reduce sediment yield and overland flow.

Gerrard and Gardner (1999, 2000a, 2000c, 2002) and Gardner and Gerrard (2001, 2002, 2003), in a series of papers detailing discrete aspects of extensive field investigations in 1992, 1993 and 1994, have also added extensively to our knowledge of the geomorphological dynamics of farming on steep slopes in the Middle Mountains of Nepal. Five sub-catchments of the Likhu Khola Valley were instrumented. They investigated the nature, cause, and incidence of landsliding; the relationship between runoff and land degradation on cultivated rainfed (*bari*) terraces, irrigated (*khet*) terraces, and on uncultivated land; and the flow regime characteristics of Himalayan river basins in Nepal. They focused on these processes because they underlie the entire Theory: that deforestation on steep slopes will lead to the disruption of the normal hydrological cycle and that modification of the slopes will cause increased runoff and sediment transfer during the summer monsoon.

The conclusions of Gerrard and Gardner are complex. Furthermore, they caution against the value of extrapolation to a wider area because of the enormous variation in soils, hydrology, slope aspect and micro-climate, and land-management practices. They conclude that large-scale landsliding over geological time has been a major contribution to the landscape as it appears today, but that by far the largest number of present-day events are small, shallow slippages easily repaired by the local farmers. They set an average rate of surface lowering for the whole Likhu Khola Valley at 3.25 mm/yr which is consistent with the results of other studies in comparable areas. Gardner and Gerrard show that large landslides do occur in their field area but are very rare during a normal monsoon season and are probably not caused by farming activities. An additional component, beyond the time frame of the Gardner and Gerrard studies, is the rare seismic event that may cause extensive failure over a limited area.

In conjunction with the work of Wu and Thornes (1995), Gardner and Gerrard (2003) were able to determine that there is less cause for concern about erosion on agricultural terraces than hitherto assumed. They also admonish that 'it is also time to reconsider the

Figure 1.3 Sindhu Palchok District, Middle Mountains near Chautara, Nepal, previously published as Figure 1.4 in Ives and Messerli (1989). The photograph is repeated here because it demonstrates how these remarkable sets of *khet* terraces represent hundreds of individual small reservoirs capable of retaining water during heavy monsoon rains. The small notches in the bunds of the foreground terraces represent controlled outlets for surplus water that descends progressively through the entire system to the valley floor. In effect, these flights of terraces are highly efficient soil conservation systems and not the cause of landslides, as has often been claimed (see Chapter 4).

common perceptions in an environment where farmers, for the most part, exercise great care and judgement in managing their limited land resources'. These results are discussed in detail in Chapter 4 (pp. 95–9).

Further west in Nepal, Thapa and Paudel (2002) have studied farmland degradation in two watersheds above Pokhara. One research area (Phewa Tal) has received considerable external assistance through the Department of Soil Conservation and Watershed Management, while the other (Yamdi–Mardi) has received no such benefits. The assessment of land degradation in both watersheds was based on extensive interviews with farmers, and revealed declines in crop productivity. Farmers were questioned about their perceptions of the impacts of the various cropping systems and types of agricultural terraces (valley *khet*, hillslope *khet*, and rainfed hillslope *bari*). Phewa Tal has benefited from considerable inputs of subsidized chemical fertilizer in comparison to the Yamdi–Mardi watershed. The authors conclude that both watersheds are undergoing increased degradation in the form of loss of nutrients and reduced crop yields, acceleration of landsliding, and soil erosion. While the Phewa Tal land degradation was considered slightly less severe than that experienced in the Yamdi–Mardi watershed, Thapa and Paudel regard the Phewa Tal Watershed Management Project as a partial failure.

The research methods applied by Thapa and Paudel differed markedly from those employed in the studies of Likhu Khola and Jhikhu Khola watersheds. Nevertheless, their conclusions are somewhat contradictory in terms of their assessed increase in surface erosion in recent decades. This illustrates the problem of comparison of even small watershed studies within Nepal itself; how much greater is the difficulty of comparison across the broader Himalayan region?

Awasthi *et al.* (2002) detail land-use and land-cover change in the same two watersheds as Thapa and Paudel (2002). Their study is based primarily on analysis of two sets of air photographs, taken in 1978 and 1996, which they converted to Digital Elevation Models. They show that between 1978 and 1996 some thin forests have been converted into dense forests and degraded forests have improved to immature forests, indicating an overall increase in forest cover and quality. This general statement disguises internal exchanges between different land-use types: shifting cultivation at higher elevations has been abandoned and low-lying shrublands have been converted to rice fields. The actual amount of forest cover has increased in protected community forests and some degraded lands have been reforested. At the same time, low-lying forests have been reduced by agricultural expansion and the harvest of trees for timber and fuelwood. Awasthi *et al.* (2002) maintain that both watersheds had 'undergone severe erosion during the past and are susceptible to surface erosion and soil degradation'. However, apart from the air photograph interpretation, there are no actual field measurements to support these conclusions.

Fox (1993) and Metz (1997) have both undertaken separate detailed replicate studies on forest cover at the village level in different areas of the Nepal Middle Mountains. Fox concluded that, over a ten-year period, forest cover had increased in his field area and that '[d]espite an annual [human] population growth rate of 2.5 %, forests were found to be in much better condition in 1990 than they were in 1980'. Metz introduced a new consideration. He raised the issue of natural plant succession as a hitherto ignored factor in the 'deforestation discussion'. In a study of seven relatively undisturbed forest stands, he demonstrated that the only major canopy dominant that appeared to be regenerating with sufficient vigour to maintain its current population was *Tsuga dumosa*. He hypothesized that other species such as oaks (*Quercus* spp) required more severe disturbance (intense ground fires or large mass movements, for example) before effective reproduction could occur. In other words, the deforestation debate has rarely departed from the limited consideration of anthropogenic influences on forest stability. Natural, or biological, processes, so far, have received inadequate attention (see Chapter 3 pp. 47–8).

Khumbu Himal

In part, because the UNU Mountain Hazards Mapping project had recently been extended into this small but representative area of the High Himal, the Khumbu was a prominent topic of discussion at the Mohonk Conference (Zimmermann *et al.* 1986; Byers 1986, 1987a, 1987b; Vuichard and Zimmermann 1987). Byers has continued his research in this area up to the present. After 1989 work on the impacts of tourism also began to attract much more attention. One of the more recent reports is a special publication supported by the Swiss Foundation for Alpine Research to mark the International Year of Mountains (Nepal *et al.* 2002). And Ortner's (2000) analysis of mountaineering and its effects in the Khumbu caps a long series of anthropological studies of the Sherpas.

From all of these studies a series of observations has emerged that consolidate the perceptions discussed during the Mohonk Conference. The most obvious conclusion is that the Theory was either insupportable or at least overly simplistic. Nevertheless, Byers has shown also that extensive destruction of juniper shrubs is occurring in the upper treeline/subalpine belt. This is due, in part, to acquisition of fuelwood for trekking tourists and mountaineering expeditions and to collection of juniper for religious observances. The disturbance of the vegetation cover in specific areas is causing substantial soil erosion. Similar conclusions to those of Byers have emerged from research by Zomer *et al.* (2001) in the Barun–Arun area to the east of Khumbu (see Chapters 3 and 7).

While there is room for disagreement over the positive and negative aspects of the substantial cultural and economic changes that have occurred and are occurring in the Khumbu, the spectacular increase in affluence of the Sherpas is resulting in a resurgence of financial support for the monasteries and traditional practices. It also belies the prediction that environmental disturbance will lead to increased hardship for women since they now can hire lowlanders to collect firewood for them.

An equally important aspect of the UNU work in the Khumbu has been recognition that the precipitous outbreak of rapidly forming lakes on the surface of, or in front of, receding glaciers is a significant hazard to life and property. The outbreak of the glacier lake Dig Tsho in 1985 destroyed a nearly complete Austrian Aid hydro-electric project. The examination of this disaster led to a reconnaissance of other potentially dangerous lakes. Detailed studies were undertaken on the Imja Glacier, where a surface lake that had formed within the preceding 30 years was initially considered a potential threat to the trekking routes leading to the Everest Base Camp (Watanabe *et al.* 1994, 1995). Additional studies were undertaken as planners realized that such glacial lake outburst floods (*jökulhlaup*, or GLOFS) posed a threat not just to local bridges and mini-hydro plants but also to expensive infrastructure, such as the 60 MW Khimti hydro power installation in the Rolwaling Valley to the east of Khumbu (Mool *et al.* 2001a). The problem of glacial lake outburst floods is discussed in detail in Chapter 6 (pp. 126–33).

This focus on the threat of catastrophic events, however, has recently exceeded the bounds of propriety with over-dramatized prediction that, worldwide, hundreds of millions of people will be killed and billions of dollars in damage will be incurred during the twenty-first century as glaciers continue to recede and ice-melt lakes break out (Pearce 2002). As the super-crisis of Himalayan environmental degradation is exposed as unsupported sensationalism, this new and possibly more spectacular hazard of glacial lake outburst floods is challenging the comparatively sedate processes of deforestation and erosion for the news media spotlight (see Chapter 10).

Mountain research in the broader Himalayan region

Researchers at the G.B. Pant Institute for Himalayan Environment and Development, together with several of the Indian Hill universities, have published increasingly well-documented results that indicate the need for modification of several of the 'Mohonk' generalizations. It has been demonstrated that road construction, the accelerated out-migration of young males, unequal treatment of women, and increased pressures from tourism, are all contributing to a worsening of the environmental conditions in the Garhwal and Kumaun Himalaya. In addition, studies by Hoon (1996) and Chakravarty-Kaul (1998) have demonstrated serious problems of degradation at high altitude – up to and above the treeline in the Central Indian Himalaya. Both maintain that the problems result from poorly thought-out government policies and prejudicial treatment of the transhumant Gaddis, Bakrwals, and the Bhotiyas of Kumaun Himalaya. Present policies are disrupting the long-standing synergism between the transhumant herders and the settled subsistence farmers described by Uhlig (1995). The changes are detrimental to both groups as well as to the stability of the alpine pastures, the winter grazing zones at lower altitudes, and vegetation along the transfer routes.

Negi *et al.* (1997) have argued for applying a historic perspective before attempting to depict the prevailing environmental and socio-economic situation in the Central Indian Himalaya. Building on the earlier works of Tucker (1987), Richards (1987), and Guha (1989), they indicate that present-day dynamics are best understood in terms of patterns of land use and extraction of forest resources that were established by the middle of the nineteenth century in former British Garhwal and in the then autonomous state of Tehri Garhwal. They also claim that, following Independence in 1947, the central governmental authorities have relied increasingly on exploitation of the forests to provide revenue. Herein lies the main driving force for accelerated environmental pressures and growing social tensions. Their study also provides insight into the origins and early progress of the Chipko Movement and its consequences.

Rawat and Rawat (1994a, 1994b) have shown that accelerated erosion associated with different cover types and changing land-use patterns presents an extremely complicated picture. This supplements the earlier work of Bartarya and Valdiya (1989) and Valdiya and Bartarya (1991). Further west, in Himachal Pradesh, extensive damage to the alpine pastures is believed due to poor management and excessive livestock pressure (Sharma and Minhas 1993), corroborated by conclusions drawn by Hoon (1996) and Chakravarty-Kaul (1998).

Two important publications (Berkes *et al.* 1998; Duffield *et al.* 1998) ably explain the relatively more stable environmental and socio-economic conditions of the Kulu Valley, Himachal Pradesh, compared with other sectors of the Central Indian Himalaya. Nevertheless, there are problems relating to both diversity and sustainability of resource use, problems that have been created by the replacement of traditional agriculture with orchards and other cash crops and the heavy use of subsidized pesticides. According to these researchers, ongoing changes within the Kulu Valley support the claim of Shiva (1993: 7) that development policies and prescriptions imported from the West 'may be turning this culturally and biologically diverse world into a monoculture'. Even more dramatic is what Shiva refers to as 'monocultures of the mind' that are undermining the very social institutions that have been a major factor in securing sustainable resource use on a local scale.

The western Himalaya, Karakorum, and Hindu Kush were geographically and environmentally beyond the effective reach of the original United Nations University study on highland–lowland interactive systems. However, they have found their way into most discussions of Himalayan environmental degradation. In the past 15 years there has been a broadening and intensification of research on the environmental problems of this vast mountain region. Hewitt and his research team based at Wilfrid Laurier University, Ontario, Canada (Hewitt 1985, 1997; MacDonald 1996; MacDonald and Butz 1998) have made important contributions. Another major source of information is the 'Culture Area Karakorum' project, a joint Pakistan–German research effort led by Stellrecht (Kreutzmann 1991, 1993, 1994, 1995, 2000; Schickhoff 1995, 1998, 2004; Nusser and Clemens 1996; Stellrecht 1997, 1998). This project tightly integrates the physical and human mountain sciences and, with many publications still in progress, is significantly increasing understanding of this complex area. Shroder (1998) also has made important contributions to the geo-morphology and geophysics of the mountains of northern Pakistan and the Himalaya in general. Any attempt to summarize these research activities would be premature because of the large number of pending publications. However, it can be emphasized that the complexity of the northern Pakistan region and the diverse nature of its problems are becoming more and more apparent.

In the northwestern sector, excluding for the moment the special cases of Afghanistan and Tajikistan, the extensive work of the Aga Khan Rural Support Programme is especially important. Widespread and sensitive efforts have been made to help the local people adapt to the encroachments of outside pressures that are increasingly disrupting traditional ways of life of the many different ethnic groups. Kreutzmann (1991, 1993, 1995, 2003) has taken a strong lead in identifying the many issues that face the peoples and landscapes of northern Pakistan and adjacent areas. The major driving forces are regional and world politics and military pressures. The Karakorum Highway, since its completion in 1978, has greatly increased accessibility. Thus it has played a crucial role, both in terms of its facilitation of foreign intrusions and local out-migration and in the maintenance that is required to counter mountain hazards that confront it (Hewitt 1993).

The effects of these recent changes on the traditional roles of the mountain women are also of considerable importance. Felmy (1996) and Azhar-Hewitt (1998, 1999) have acquired valuable insights from their studies of the impacts of development on the position of women in hitherto isolated Muslim (Ismaili) communities. Hunza, Gilgit, and Chitral, names to conjure with during the Great Game, have been open to foreign trekkers and mountaineers and the illegal timber trade since completion of the Highway, yet this once hidden redoubt of the Himalaya remains a male-dominated environment. The writings of Felmy, following her long sojourn in a Hunza village, and of Azhar-Hewitt whose frequent visits resulted in close communication with many village women, have led to a much fuller understanding of some of the negative impacts of the best conceived efforts toward 'development'.

The Pamir are subjected to powerful tectonic forces and there are numerous severe mountain hazard problems that will inhibit effective surface communications. Thus, extreme relief, unstable slopes, glaciers, and ice- and landslide-dammed lakes combine to create a major challenge to development (Alford *et al.* 2000). However, the overarching problems of Central Asia emanate from Afghanistan and from the broader issues of warfare, terrorism, drugs, water politics, gender, and poverty, characteristic of the entire region (for a valuable review of the current political situation in Central Asia, see the article by Lambert in *The Economist*: 26 July 2003).

Since research in Afghanistan has been practically impossible following the Soviet invasion of 1981, any account of the prevailing environmental and socio-economic factors has been omitted from this overview. Tajikistan is a less extreme centre of conflict, although still politically very unstable. It is ironic that the brutality of Stalin and the Red Army left this spectacular mountain region as close to pristine wilderness as any mountain region in Central Asia (Badenkov pers. comm. 15 October 1990). Many of the mountain Tajiks were forced out of Gorno-Badakhshan (Pamir) to provide cheap labour in the cotton fields of the southwestern lowlands when the overriding need for gun cotton became evident in the late 1930s. After the Second World War the Soviet Union's continuing demand for cotton ensured that this forced migration became long term. With the relaxation of control from Moscow (Gorbachev's policy of *perestroika*) in 1987, however, there arose the beginnings of mass re-migration as many of the mountain Tajiks returned to their formerly abandoned subsistence villages (*kishlaks*). The lack of a co-ordinated land-use policy posed a threat of serious environmental pressure, but this scenario has been overwhelmed by widespread civil unrest and armed conflict (Badenkov 1992, 1997, 1998; Cunha 1994, 1995). With the partial resolution of hostilities in now independent Tajikistan a degree of optimism has returned (Breu and Hurni 2003). The Aga Khan is moving to establish a regional university at Khorog in the Pamir that will specialize in professional training for sustainable mountain development and will attract students from all the neighbouring countries, including northwestern China.

Moving now to the far eastern perimeter of the mountain system, it is reiterated that from the earliest years of the UNU project attention had been focused on the Hengduan Mountains of western Sichuan and northwestern Yunnan, China. Some of the preliminary findings were introduced at the Mohonk Conference (Messerli and Ives 1984; Ives 1985) when it became apparent that we had encountered a Chinese variant of the Theory. Early reconnaissance tended to reinforce the primary results derived from Nepal (Ives and Messerli 1989: 53–9). Subsequent, and more extensive field investigation in the Lijiang area of northwestern Yunnan in 1991, 1993, 1994, and 1995 has revealed unsuspected layers of complication (Ives and He 1996; Swope *et al.* 1997; Sicroff and Ives 2001). Our early findings, which took issue with the prevailing environmental appraisal of the 1970s and 1980s, have been sustained. There was little factual support for assertions that massive deforestation after about 1950 (in this case attributed to the policies of Mao Zedung and the Gang of Four) had led to soil erosion and intensified downstream flooding and siltation. More recent studies, however, indicate a complex pattern of natural and human/political dynamics; over the last 70 years innumerable small mountain areas have been responding in diverse fashions to varying modalities of resource exploitation and environmental pressures. These have been driven largely by China's fluctuating internal politics.

From about 1990 onward there has been a massive influx of tourists into southwestern China, both domestic and foreign, promoted by provincial and central governmental policy. A single example is provided here. Lijiang Town has changed from an isolated traditional market town closed to the outside world in 1982, to a rapidly expanding modern town with over 200 hotels, an airport, and first class motor roads, in 2002; this is almost beyond comprehension. It raises the question: can the potential advantages of mass tourism be managed so that they outweigh the severe threats of environmental and cultural disruption (Swope 1995: Swope *et al.* 1997; Sicroff and Ives 2001)? Furthermore, it is predicted that prospects for equitable development are further confounded by recent moves to outlaw logging in Yunnan and other upper watershed regions of China where the majority of the very poor ethnic minorities are heavily dependent on access to forest resources (see Chapters 7 and 8).

The final area for consideration here is the hill region of northwestern Thailand, scene of the first stages of the UNU highland-lowland interactive systems project (Ives *et al.* 1980). Work by Forsyth (1994, 1996, 1998), Fox *et al.* (1995), Roder (1997), Schmidt-Vogt (1998), and Renaud *et al.* (1998) in northern Thailand, Laos, and Vietnam add further detail and reinforce the main Mohonk conclusion that the Theory is not supportable. However, they also demonstrate that, as more studies are undertaken, the complexity of the entire region defies simple analysis. A recent overview is provided by the book *Montane Mainland Southeast Asia in Transition* (Rerkasem 1996). Yet again, imposition of logging bans by the central government, on the assumption that the mountain subsistence minorities are responsible for downstream siltation and flooding due to deforestation, further impedes the prospects for sustainable mountain development. This entire situation, of course, is driven by powerful business interests, forest contractors and local elites, and poachers, as well as local indigenous felling activity (see Chapter 3 pp. 60–7).

Flooding on the Ganges and Brahmaputra plains

The 'Eastern Waters Study'

On 30 October 1988 United States President Reagan signed into law an Act of Congress (H. R. 5389) entitled *The Bangladesh Disaster Assistance Act*. It required that within six months of enactment the President report to Congress:

> on the efforts by the international community and the governments of the region to develop regional programs for the Ganges basin and the Brahmaputra basin that are designed (1) to ensure an equitable and predictable supply of water in the dry season, and (2) to provide better flood control mechanisms to mitigate in the mid-term, and to prevent in the long-term, floods as severe as the 1988 floods in Bangladesh.

In addition, short-term relief efforts were authorized.

In preparation for the President's report to Congress, US AID recruited a study team under the leadership of Peter Rogers and including Peter Lydon and David Seckler. Funds were also provided for support staff and for the involvement of specialist consultants for input to the different sub-sectors of the report. This study was initiated when the politically favoured approach to prevention of flooding in Bangladesh – that is, a combination of river channel training and reforestation in the Himalaya – was also unofficially endorsed by the World Bank, and aid agencies in Japan, The Netherlands, and other potential donors, as well as the government of Bangladesh.

Rogers *et al.* (1989) found no statistically reliable evidence that the physical extent and severity of the flooding had increased over the 100 years for which data were available. They cautioned that:

> deforestation of the Himalaya was not likely to have a significant effect on the extent of the floods in the plains and the delta below . . . generalizations that changes in the mountain tree cover are responsible for siltation and floods in the densely populated plains below are not justified. [and that] conventional flood mitigation approaches of upstream storage in mountain zones and embankments on the main rivers raise basic questions regarding technological and economic feasibility. They are also approaches that lead to the largest environmental disruptions.

It was shown that loss of crop production due to flooding in the more seriously affected districts was compensated for by improved soil moisture conditions during the following dry season. In effect, 1987 and 1988 provided bumper crop yields for Bangladesh as a whole because dry season production more than offset flood losses and because there were higher than average yields during the wet season in areas that had experienced only light flooding.

The emphasis of the Eastern Waters Study was on the Bangladesh floods of 1987 and 1988 and the UNU study to this point had focused on the mountains, but the two research efforts were mutually supportive in terms of their major findings. One crucial result was that the pressure for a major lowland technological 'fix' that would consume vast financial and other resources appears to have faded. Rogers *et al.* (1989) did draw attention to the alternate opportunities for flood mitigation through surface and underground water storage. They proposed that such an undertaking also would have attendant dry season benefits in terms of the potential for extension and reliability of the existing irrigated agriculture (see Chapter 5).

Summary of former perceptions of environmental degradation

The discussion so far has not incorporated the overarching threats of warfare, terrorism, guerrilla activity, and political unrest. Nor has the question concerning excess summer monsoon precipitation and the actual physical transfer of eroded material from the mountains onto the plains been discussed in any detail. These topics have been omitted to preserve a degree of clarity, or manageability, in an already complicated introduction.

I believe that, despite a complicated and far-ranging discussion, fraught with internal ambiguities and apparent contradictions, the overall refutation of the Theory of Himalayan Environmental Degradation (Ives 1984, 1987a; Thompson *et al.* 1986; Hamilton 1987; Ives and Messerli 1989) has been finally validated. The paradigm shift away from the Theory, as indicated by Forsyth (1998) and Blaikie and Muldavin (2004), has been fully implemented within academia. Nevertheless, a series of difficulties has surfaced, relating to the differences among research methodologies and to the interpretations of the various sets of results. Yet another aspect is that conditions prevailing in the first part of the twenty-first century are somewhat different from those that characterized the region from the 1960s through the 1980s. In addition to the major changes on the world scene and the accelerating penetration of globalization into the entire region, is the realization that the mountain farmers were never fixed in their ways as had been imagined. There have been continuing adjustments, family by family, village by village, and region by region, to the changing opportunities and pressures in the struggle to survive in a difficult environment. In many areas, cash-cropping and off-farm wage earning have developed extensively and have either strongly modified or eliminated traditional subsistence farming. The extent of remittances from non-agricultural work and seasonal out-migrations and their environmental implications have been greatly underestimated (Blaikie and Coppard 1998). Thus, cash from off-farm labour now substitutes for forest-derived nutrients by facilitating purchase of chemical fertilizer. It also has furthered a shift in labour allocation from the arduous processes of harvesting forest products. The increasing spread of modern communications, including e-mail and the Internet, into all parts of the region has had a profound effect. These developments, moreover, may have beneficial impacts on the mountain resource base; they may also be promoting increased awareness of inequities in hitherto isolated rural areas that leads to unrest and political instability.

Although detailed and exacting studies have been undertaken, differences in research methods sometimes make the comparison of results ambiguous. There has been a heavy emphasis on analysis of cover-type change and farm practices and on instrumented geomorphic studies of small watersheds. Little effort, with the exception of Alford (1992a, 1992b) and Hofer (1993), has been expended on analysis of the actual transfer of water and sediment through the mountains and out onto the plains. Also after 1989, social and political science studies have made a much stronger entry into the Himalayan arena (Ives 1996; Jodha 1997; Rhoades 1997; Banskota *et al.* 2000; Blaikie and Sadeque 2000; Blaikie and Muldavin 2004).

Certain ambiguities that require further investigation have been identified. For instance, while Wu and Thornes (1995) have shown that individual terrace failures do not contribute sediment to the downslope, they also state that in the case of large-scale slope failures involving dozens of terraces the situation is much more complicated. Does this mean that during major storm events (for instance, those with a recurrence interval of 25, 50, or 100 years) there is, or may be, a large impact from first-order to third-order streams, and so out onto the plains? This cannot be answered unequivocally at the moment. Similarly, Gardner and Gerrard (2003) conclude that large landslides do not occur during 'normal' monsoons; but what about monsoons of higher than average intensities and total precipitation amounts? Several workers have indicated that the vast majority of slope failures are small ones, such that the local farmers can easily effect repairs. Yet, as these increase in frequency (if they do increase in frequency), the farmers have to expend more of their available energy on repair. What happens when they can no longer keep pace? Several researchers have indicated that abandonment of many of the more steeply sloping *bari* terraces is occurring. It is not known, however, except on an individual case assessment, whether this is environmentally good or bad, or if it is bad today, will it be good next year, or in 10 or 20 years, as vegetation re-establishes. It is well known that in the Alps, for instance, abandonment of traditionally farmed land on steep slopes produced a cycle of accelerated erosion with serious downstream effects in some places, followed by increasing stability as natural revegetation occurred.

These are all important, if apparently secondary points. They serve to demonstrate, however, that both mountain landscapes and mountain societies are complex, flexible, and that there will always be change. This should warn even more dramatically of the dangers inherent in generalization. If the current outpourings of media melodrama are added, such as predictions of millions of deaths from the outbreak of glacier- and landslide-dammed lakes, with a different set of examples and circumstances to those that originally prompted Thompson *et al.* (1986) to effective irony, we are still left with the question 'what would we like the facts to be?'

There remain the overarching issues. What causes flooding in Gangetic India and Bangladesh? How will the various parts of the region withstand the current impacts of warfare, threats of war, and political unrest that in turn promote large-scale movements of poor people? Is sustainable mountain development feasible in all or in some parts of the region? How do we define 'sustainable mountain development', 'degradation', 'deforestation'?

Above all, it seems that much of the apparent complexity and uncertainty, or the conflicting conclusions, would be better handled if the question of *scale* were addressed more effectively. This topic was originally introduced in the Himalayan context by Bruno Messerli during the Mohonk Conference (Ives *et al.* 1987) and was elaborated by Andreas Lauterburg (Ives and Messerli 1989: 131–2). An attempt will be made to illustrate, even to answer, some of these questions in the following chapters.

This opening chapter is not intended solely as a summary of the book. Rather it has been constructed as a narrative of the changing perceptions of the condition of Himalayan environment over the last half century to set the stage for detailed discussion of the specific topics in the chapters that follow. It is argued there is a general consensus, at least within academia, that there is little support for the notion that uncontrollable environmental degradation, from the mountains to the Bay of Bengal, poses an imminent super-crisis. The World Bank's (1979) year of reckoning (AD 2000) has passed and Nepal's mountains and forests are relatively intact. Without question, changes are occurring rapidly and there are many instances of serious but local degradation off-set by local improvements in other areas. Nevertheless, some of the grave threats have received only incidental attention. Thus, international competition, warfare, guerrilla activity, unequal access to natural resources, pervasive mistreatment of mountain minority peoples, continued inequality imposed upon women, and widespread corruption could well combine and propel the region into something approaching super-crisis.

These themes will be taken up in the body of the book. It is emphasized that the sheer volume of post-1989 publication defies an exhaustive synthesis in the available format. Finally, and to repeat, many of the vested interests within the region's governments and central institutions and the donor agencies, sometimes aided by the news media intent upon melodrama, either ignore or disagree with the conclusions being presented here. The causes underlying this dissonance will also be examined in later chapters.

2 The Himalayan region

An overview

In a thousand ages of the Gods
I could not tell thee of the glories of the Himalaya
(Puranas)

Setting the scene

To help meet the overall objectives of this book this chapter presents an overview of the greater Himalayan region. The description is by no means complete nor uniform in any geographical sense and because the central theme of the book is an analysis of the Theory of Himalayan Environmental Degradation, most of the discussion will be centred on the Himalayan range *sensu stricto*. Thus, the great mountain arc that extends for 2,500 kilometres from Nanga Parbat (8,125 m) and the Indus Gorge in the northwest to Namche Barwa (7,756 m) and the Yarlungtsangpo–Brahmaputra Gorge in the east will receive the most detailed attention. Furthermore, within this 2,500-kilometre extent, the Nepal Himalaya will loom large because that is the core area from which the Theory emerged in the first place. It is also the sector that has received the most thorough attention from scholars, development agency personnel, bureaucrats, and politicians who have sought to support or refute the Theory subsequent to publication of this book's forerunner, *The Himalayan Dilemma* (Ives and Messerli 1989).

Ives and Messerli's original treatment received some qualification in several of the series of book reviews because of over-emphasis on Nepal. The present book will likely demonstrate the same slant. Nevertheless, a concentration on Nepal is inevitable for the same reason: concern is focused on the Theory, therefore most attention will fall on Nepal because of its central relevance. However, as already indicated, several adaptations of the Theory have been, and still are being, applied to other extra-Himalayan (*sensu stricto*) regions: for instance, the Hengduan Mountains of Yunnan, southwestern China. And Forsyth (1998) has drawn attention to the furthest southeastern extension that can reasonably be included. The linkage with the mountains of northern Thailand is emphasized in the title of one of his several papers: 'Science, Myth, and Knowledge: Testing the Theory of Himalayan Environmental Degradation in northern Thailand' (Forsyth 1996). This area, therefore, will also receive detailed consideration.

In the same vein, several other scholars have examined the Theory as a component of their studies in northern Pakistan (Schickhoff 1995, 1998, 2004; Nusser and Clemens 1996); Schickhoff refers to the 'Karakorum Dilemma'. This has led to inclusion of the western Himalaya, the Karakorum, and the eastern Hindu Kush. These mountain regions, from a climatological and vegetational perspective, contrast with, rather than compare with, the Nepal Himalaya. The Pamir are taken into account for the sake of continuity. The question might be raised: why stop there and not include the Alai, the Tien Shan, and the other ranges of arid Central Asia? The answer is pragmatic: scarcity of relevant data and lack of personal experience in the field.

The great Tibetan Plateau, of course, is integral to any discussion of the geophysics and geography of this vast mountain region. Nevertheless, it remains peripheral to a discussion of the Theory despite its important role in the emergence of regional political tensions. Therefore, it will only receive passing attention, especially since forests and their utilization, real or supposed, are central to the Theory, so that only the forested deeper southeastern valleys that cut into the Plateau are directly relevant, and these are subsumed in the analysis of Yunnan's environment.

The physical dimensions

Any general atlas plate that covers the great mountain and plateau complex of Central Asia brings into sharp focus several predominant geographic components. These include: the huge Tibetan Plateau (frequently coloured in whites and pale purples); the brown linear forms of its containing mountain ranges; the sharply etched north-south parallel blues that cut into the Plateau's southeastern quadrant (the River Gorge Country, or the Hengduan Mountains); the Tarim Basin terminated in the west by the Pamir Knot; the brown linear forms that spin off from the 'knot' (to the west and south into Afghanistan – the Hindu Kush; to the east and north into Mongolia – the Tien Shan; eastward along the northern rim of the Plateau – the Kun Lun Shan; to the south and east, through northern Pakistan, Kashmir, Nepal, Bhutan, Assam as far as the River Gorge Country – the Karakorum and Himalaya. Equally prominent on the continental-scale atlas plate will be peninsula India jutting southward between the Bay of Bengal and the Arabian Sea deep into the Indian Ocean, almost to the equator (Figure 2.1 and Figure 2.2).

Together, this array of landforms encompasses the world's largest accumulation of high mountain ranges and plateaus and provides the platform upon which the monsoon system functions. The lowlands immediately to the south (plains of the Indus, Ganges (Ganga), Brahmaputra, and Meghna rivers) and to the east (the Sichuan Basin), watered by some of the world's greatest rivers and the sometimes excessive monsoon precipitation, support more than a quarter of humankind including a high proportion of the world's poor.

North and northwest of the Central Asian mountain heartland the lowlands are semi-arid and arid and hence support much lower human densities. Here the rivers draining from the mountains are the main life-support systems as direct precipitation is inadequate.

The overall mountain and plateau heartland, introduced above, has a west-east dimension of about 3,500 km and a north-south extent of 2,400 km. Global position, topography, climates, vegetation patterns, and human history justify the epithet – the most diverse region on earth. But it is not only diverse in the broad regional sense, ranging from high-altitude cold desert and permanent ice and snow to tropical monsoon rainforest and the humid mangrove swamps of the Ganges–Brahmaputra delta. It is diverse from point-to-point in a single mountain valley or on a single slope facet. When the cultural, historical, and political diversity is added, meaningful generalization beyond that on the atlas plate map is unproductive. In a geopolitical context, the time-worn musings of Halford J. Mackinder (1919) warrant repetition:

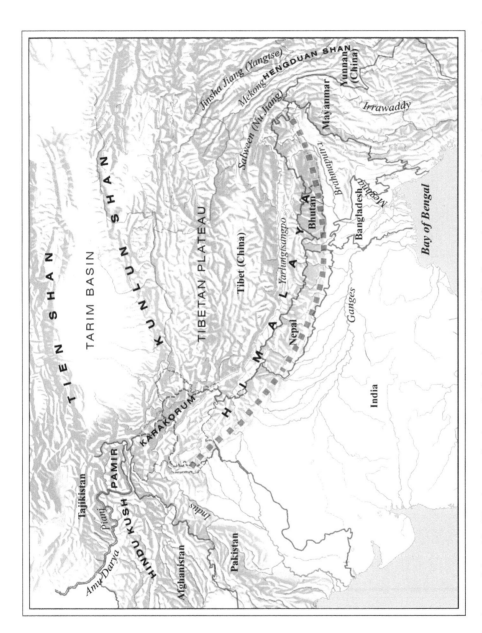

Figure 2.1 Sketch map of the wider Himalayan region (*sensu lato*) showing national frontiers, major rivers, and selected specific mountain regions. Shaded relief base modified after Ives and Messerli (1989, Figure 2.1). The line of small squares denotes the Himalayan Front.

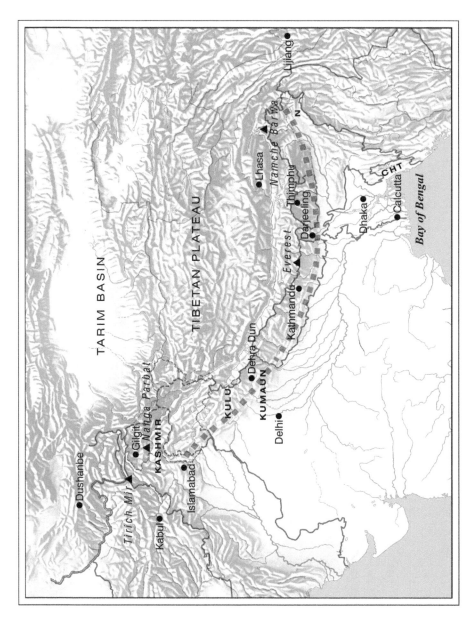

Figure 2.2 Sketch map of the wider Himalayan region (*sensu lato*) showing national frontiers, principal towns and cities, selected mountain peaks and regions. Shaded relief base modified after Ives and Messerli (1989, Figure 2.1). The line of small squares denotes the Himalayan Front, CHT denotes Chittagong Hill Tracts.

Who rules East Europe commands the Heartland:
Who rules the Heartland commands the World-Island:
Who rules the World-Island commands the World.

These words were written in the later stages of the Great Game. While tempered by the contemporary counter-balance of sea power and subsequently by development of post-modern weaponry, access to oil, and technology in general, Mackinder's speculation retains some significance. Thus, if severe and current deforestation of the southern and eastern perimeters of the mountain and plateau heartland is accepted as a truism by power centres in India and China, the assumed consequences of such environmental perception for the hundreds of millions living on the subjacent plains hold considerable geopolitical signifcance. Its acceptance also constitutes a serious liability for many of the 70–90 million mountain dwellers.

This brief overview signifies why attention is focused on selected sections of Asia's mountain heartland and periphery. The pivotal section, as stated above, is the 2,500-kilometre extent of the Himalaya proper; of comparable importance are the Hengduan Mountains and parts of northern Thailand. The Karakorum, Hindu Kush, and Pamir provide an interesting contrast. Finally, the river basins of the south and east are significant as their dense populations are the perceived victims of irresponsible environmental disruption in the mountains.

The Himalaya (*sensu stricto*)

The great Himalayan arc extends from latitude 35° 40′ North, longitude 74° 50′ East in the northwest, to latitude 26° 20′ North, longitude 95° 40′ East in the east. It is a complex of sub-parallel structural units produced by the northward thrust of the Indian tectonic plate beneath the Central Asian plate. The Indian plate is moving northward at a rate of about 45 mm/year (see Chapter 6). As explained by Bilham (2003), its movement is not simply due north but it is turning slightly in an anti-clockwise direction. This largely accounts for the complexities of the structural units along its northwestern boundary (Karakorum, Hindu Kush, Pamir) and along its eastern boundary (Hengduan Mountains and the configuration of the mountains of Myanmar and Thailand). Figure 2.3 provides a simplified cross-section through the Central, or Nepalese, Himalaya. It demonstrates the conventional structural divisions through this section of the Himalaya, from north to south: the Tibetan Plateau; the Tibetan Marginal Mountains; the Inner Himalaya; the Greater Himalaya (Himal); a transition belt; the Middle Mountains; the Mahabharat Lekh; the Siwalik; the Terai; and the Ganges Plain.

The Tibetan Plateau ranges in elevation from about 3,500 metres in the southeast to more than 5,500 metres in the northwest. Several major mountain ranges cut obliquely across the Plateau, the most conspicuous being the Nyainqêntanglha Shan, that approach 7,000 metres in places. The Greater Himalaya (Himal), together with the sub-parallel off-shoot, the Karakorum (sometimes referred to as the Karakorum-Himalaya) contain all 14 of the world's summits that rise above 8,000 metres, eight of which are located in Nepal or on its northern border with Tibet (China). The Himalaya appear as a series of aligned massive blocks separated by lower sections that are often the locations of deep gorges. In the east the great peaks of Everest, Cho Oyu, Makalu, Kangchenjunga, and their associated massifs, form the international border. Further west the height of land diverges from the international border so that the Annapurna and Dhaulagiri massifs lie entirely within Nepal, and the Nanda Devi massif within India.

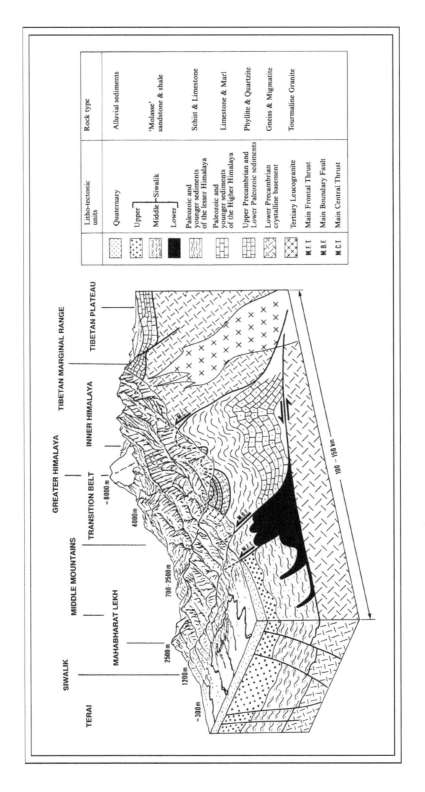

The legend of the cross-section contains the following rock type classifications:

Litho-tectonic units	Rock type
Quaternary	Alluvial sediments
Upper / Middle — Siwalik / Lower	'Molasse' sandstone & shale
Paleozoic and younger sediments of the lesser Himalaya	Schist & Limestone
Paleozoic and younger sediments of the Higher Himalaya	Limestone & Marl
Upper Precambrian and Lower Paleozoic sediments	Phyllite & Quartzite
Lower Precambrian crystalline basement	Gneiss & Migmatite
Tertiary Leucogranite	Tourmaline Granite
M.f.T.	Main Frontal Thrust
M.B.F.	Main Boundary Fault
M.C.T.	Main Central Thrust

Figure 2.3 Schematic cross-section of the Nepalese Himalaya, modified after W.J.H. Ramsay and first published as Figure 2.3 in Ives and Messerli (1989). The cross-section is an idealized transect from the Terai, on the left, to the vicinity of Mount Everest.

The narrowest cross-section of the entire Himalayan system (that portrayed in Figure 2.3) also contains the world's highest mountain (Mount Everest, at 8,850 m., Chinese–Chomolungma; Nepalese–Sagarmatha). Also in this section the Terai on the Nepalese border with India is barely 200 metres above sea level. This produces the greatest relief on earth – 8,500 metres vertical within a horizontal distance of little more than 100 kilometres.

Seen from the height of a commercial jet on a flight from Bangkok to Kathmandu in May–June, the Himalaya appear as a massive white wall rising abruptly from the flat dry plains of pre-monsoon Bangladesh, Bihar, and Uttar Pradesh. The higher massifs and the great peaks are apparent, yet the mountain wall appears complete and unbroken. This is an illusion of distance and season. There are many high passes that within a few weeks will appear brown, as the lower snows melt, and provide access for traders, transhumance herdsmen, and trekkers. Rivers penetrate the great Himalayan wall in very few places. The most spectacular breaches are those made by the Indus and the Yarlungtsangpo–Brahmaputra that form the northwestern and eastern limits of the conventionally defined Himalaya. These great rivers, together with several lesser ones, the Sutlej, Karnali, Kali-Gandaki, and Arun, are classified as antecedent streams (i.e. they were formed before uplift of the mountains) that have been able to maintain their courses by down-cutting despite the rapid uplift of the mountain wall through recent geological time.

The two major antecedent rivers, the Indus and the Yarlungtsangpo–Brahmaputra rise comparatively close together in the vicinity of Mount Kailas on the Plateau. They flow in opposite directions along the northern front of the Himalaya for more than a thousand kilometres before turning abruptly south through deep gorges to debouch onto the southern plains of Pakistan and Assam respectively. In comparison, the Ganges is distinct in that most of its water originates from a series of consequent streams that flow down the southern slope of the mountain chain, supplemented by the southern tributaries that rise on the plateau of peninsula India. Nevertheless, all these rivers are strongly controlled by the geological structure and exhibit a pronounced rectilinear pattern. Collectively, the valleys and gorges through which they flow provide an enormous amount of actual and potential hydro-electric power. This is probably the largest concentration of current and future water energy in the world. It is one of the region's greatest assets and also one of its greatest scourges, if improperly harnessed (see Chapters 6 and 9).

The continuing rapid uplift of the Himalaya (calculations vary between 0.5 and 20 mm/yr), the tectonically disturbed bedrock, the steep slopes, the great differences in altitude over short horizontal distances, and the heavy monsoon precipitation have provided the physical basis for the very high rates of erosion, transport, and deposition of sediments. Thus the plains of the Indus, the Ganges, and the Brahmaputra are built on sediments that, in places, are several thousand metres thick. It is this enormous geophysical–geomorphological system that has functioned over geological time and must be balanced against the real and assumed human-induced soil erosion and downstream sediment transfer that lies at the root of the Theory (see Chapter 4).

The so-called Middle Mountains of Nepal (often referred to as the Middle Hills) have been the locale of the country's highest human population densities, at least until recent out-migration to the Terai. The Middle Mountains merge almost imperceptibly with the Himal. They are bounded on the south by the Mahabharat Lekh that reach heights of about 3,000 m and do not support permanent ice and snow. Tectonic activity has produced a series of deep basins and valleys within the Middle Mountains (such as the Kathmandu Valley), between them and the Mahabharat Lekh, and between the latter and the Siwalik Hills (foothills). These include the Inner Terai and the dun valleys, typified by Dehra Dun Valley.

The Siwalik (also known as the Churia Hills) are by far the youngest of the Himalayan units with large areas underlain by unconsolidated sands and gravels and hence are highly susceptible to erosion. The 'outer' Terai can be divided into two sub-regions. The first lies immediately below the foothills and is formed by a series of extensive coalescing alluvial fans and torrent fans produced by the mountain rivers as they debouch from the outer range and deposit much of their coarse alluvium at and beyond the break of slope. This is known locally as the *barbar* (porous place). Given the coarse and porous nature of the deposits much of the monsoon flood water enters the soil to recharge the vast ground-water table that underlies much of the Ganges Plain. With increasing distance from the outer range, the slope of the *barbar* is reduced and the materials become finer until the surface merges imperceptibly into the Ganges flood plain proper, forming the second sub-region.

Given such a vast east-to-northwest extent and massive south-to-north increase in altitude, it follows that the diversity in regional and micro-climates will be comparably great. Maximum monsoon precipitation occurs immediately north of the Bay of Bengal where the warm moist air masses strike the Meghalaya Hills and the main mountain front. Here in the Assam, Darjeeling, and Bhutan Himalaya, annual precipitation amounts exceed 5,000 mm with up to 80 per cent falling between the end of May and the beginning of October (Cherrapungi, on the Meghalaya Hills and well south of the Himalayan front, receives in excess of 11,000 mm/yr). The influence of the summer monsoon diminishes progressively northwestward, both in terms of total amounts and later dates of onset and earlier dates of withdrawal. Westward to Kashmir and the Western Himalaya of Pakistan monsoon rainfall amounts are usually below 1,500 mm/yr whereas the winter anti-cyclonic circulation assumes an increasing level of importance at high altitudes. The valleys of the upper Indus and its tributaries, and the Hindu Kush and Pamir valleys are semi-arid and arid at low levels (Figure 2.4). Here precipitation is largely confined to the higher altitudes and occurs mostly in the form of snow.

A north-south transect displays a comparable change in precipitation receipts. Taking the standard eastern Nepal cross-section shown in Figure 2.3, the heavy precipitation of the outer hills diminishes with altitude and changes in form from liquid to solid. There is the additional factor of the rain-shadow effect. As each of the major east-west ridges is crossed, its northern slope receives less precipitation than the southern windward slope. In addition, the overall perception of the monsoon, at least to non-residents, has erred seriously until quite recent accurate precipitation observations have become available. For example, D.P. Joshi (1982) recorded a mean annual precipitation average of 1,048 mm (1949–77) for Namche at 3,450 m above sea level (asl) in the Khumbu Himal. Byers's (1986, 1987a) research, also in the Khumbu, which included the first soil erosion studies to be undertaken throughout a complete summer monsoon season, indicated that conditions were far different from those described in the general literature. Total amounts were modest (Khumjung at 3,790 m received only 725 mm during the 1985 summer monsoon, much less than Namche), mornings were frequently clear, with cloud building toward midday and showers generally confined to the afternoons. These observations are of great significance when testing the earlier assumptions of massive soil erosion and sediment transfer on steep slopes denuded of protective vegetation cover (see Chapter 4).

The natural vegetation cover similarly varies greatly in both south to north and east to west directions. The dense mountain forest cover, with extreme biodiversity, characteristic of the eastern Nepal, Sikkim, and Bhutan sectors, gives way progressively to thorn shrub and bare ground in the northwestern valleys where agriculture is dependent on irrigation from carefully controlled canals that lead snow- and glacier-melt down the mountain sides to small

Figure 2.4 Glaciers and ice- and snow-covered peaks of the High Himal in the Mount Everest region. Photograph taken during the Swiss Mount Everest Expedition of 1956 by the late Professor Fritz Müller.

irrigated fields. The south–north transect demonstrates a comparable change in vegetation from the same tropical monsoon rain forest at the lowest elevations through a variety of forest belts to timberline at about 4,000 metres to alpine shrub and meadow to rock, snow and ice (Figure 2.5), and high plateau semi-desert on the northern side of the Himalaya.

Undoubtedly, this great diversity of topographic, climatic, and vegetational conditions has a special relevance to any discussion on the actual or hypothetical consequences of human-induced changes in forest cover. In the extreme case of the northwestern Himalaya, Karakorum and Hindu Kush, total natural forest cover prior to any human impact would have been insignificant in terms of total land area. Thus forest destruction by human intervention, while very important at the micro-scale, would be extremely difficult to detect in terms of its influence on the sediment transferred from a meso-scale watershed. The original dense forest cover of much of the Central and Eastern Himalaya provides an entirely different environment for sustained and intensive human use.

Climate change must also be considered. The progressive global warming of the last 150 years, and especially the last 30 years, has already affected the Himalaya as many of the mountain glaciers have thinned and retreated. This has led to the formation of a large number of glacier-melt lakes, some of which have drained suddenly to cause dangerous downstream flooding, a topic discussed in detail in Chapters 6 and 10. The same process of global warming is expected to cause the numerous mountain vegetation belts to slowly adjust to higher levels, although this will require more time before becoming clearly discernible.

Figure 2.5 Some of the highest yak pastures in the upper Dudh Kosi Valley near Machhermo (4,410 m), Khumbu Himal. The valley leads to Gokyo, a recent destination point for trekkers, beneath the summit of Cho Oyu (8,201 m), March 1979.

The mountains of northern Thailand

The mountains of northern Thailand, situated between latitudes 16° and 20° North, occupy most of Region Five of the country which includes the eight provinces of Mae Hong Son, Chiang Mai, Chiang Rai, Nan, Lamphun, Lampang, Phrae, and Uttaradit. It is a region of north–south trending mountains and hills separated by narrow valleys and areas of inter-montane depressions. The landscape is highly complex and merges across the international borders with similar terrain in Myanmar, Laos, and southern Yunnan, China. The highest point (2,595 m) is Mount Doi Inthanon so that the region's altitude and relief are not comparable with the other mountain areas that are discussed in this book. Nevertheless, relief approaches 2,000 m and the pronounced north–south structural and topographic trend prompts classification as the southern extension of the Hengduan Mountains of Yunnan and Sichuan.

The central part of the region is drained by the headwaters of Thailand's main river system, the Chao Phraya that flows south through Bangkok to empty into the Gulf of Siam. This drainage alignment in relation to Bangkok and the country's main intensive agricultural lowland has drawn into the national political agenda the question of the perceived environmental mismanagement of the mountains. The northwestern section drains into the Salween and Irrawaddy rivers, the northeast into the Mekong.

Northern Thailand experiences a pronounced monsoonal climate with heavy summer rains between early June and October. This is followed by a cool dry season from October to February and a hot dry season that culminates in April/May with temperature maxima exceeding 40° C (Yoshino 1980).

With a range of relief of nearly 2,000 metres (Figure 2.6), the natural vegetation can be divided onto a series of distinct altitudinally controlled ecosystems. The upper slopes, between about 1,000 and 2,600 m, are dominated by a Lower Montane Forest, consisting of oaks, chestnuts, laurels, birches, and other species. A Coniferous Forest ecosystem partly overlaps; it is edaphically controlled and usually confined to the steeper slopes and exposed ridges between 800 and 1,600 m. The dominants are *Pinus merkusii* and *P kesiya*. They have been used extensively as a source for the widespread national reforestation programme in the mountains. The lower slopes support a range of forest types including dry evergreen, dry deciduous, mixed moist deciduous, and dipterocarp forests, leading down to the intensively cultivated river terraces and flood plain.

As with the natural vegetation, now much depleted, the modes of human occupation are also closely influenced by altitude. Despite the rapid growth in total population since about 1950, the uplands are still relatively sparsely populated, predominantly by a large number

Figure 2.6 The mountains of northern Thailand reach a maximum altitude of 2,595 metres. This view, from about 1,200 metres, shows a Lisu swidden in the foreground and partially forested mountain ridges fading eastward in the haze of forest fires. The hill rice has been harvested and a large number of root crops are still to be collected. Very few signs of gullying or rill wash could be detected on the steep slopes of the swidden, April 1978.

of distinct ethnic groups, the so-called hill tribes, compared with the lower slopes and valley floors. These lowland areas support the much larger population of ethnic northern Thai, traditionally paddy rice farmers. Population densities here are as high as 800–1,000/km^2.

The Hengduan Mountains

The Hengduan Mountains form the southeastern margins of the Tibetan Plateau. They are structurally and topographically distinct, although southeasternmost outliers of the Plataeau are isolated between the gorges of the great rivers. These rivers include the upper Yangtze (Jinsha Jiang), the Salween (Nu Jiang), the Irrawaddy (Dulong Jiang), and the Mekong (Lancang). Their upper courses have a generally southeast trend although, on leaving the Plateau proper, they flow almost due south and closely approach each other as they exploit the pronounced Hengduan structural alignment. Here the gorges are extremely deep and steep, some of the deepest in the world. It is their close parallel proximity that has given rise to use of the terms 'River Gorge Country' and 'Transverse Mountains' as regional names. At their closest the outer members of the four gorges are little more than 100 km apart and the intervening ridges rise well over 5,000 m in many places, thus forming an extremely rugged and difficult mountain terrain. The higher massifs, such as the Gongga Shan (Minya Konka) at 7,556 m, the Meili Xue Shan at 6,740 m, and the Yulong Xue Shan at 5,596 m are extensively glacierized and support the most southerly glaciers and ice caps of Eurasia (Figure 2.7).

Figure 2.7 The highest summit of the Yulong Xue Shan (Jade Dragon Snow Mountains, 5,596 m) in northwestern Yunnan, China. This is the sacred mountain of the Naxi people. The glacier to left of centre is one of the most southerly of Eurasia. By 2002 it was accessible by gondola off the far left of the photograph, November 1995.

The Hengduan Mountains have a north–south extent of about 800 km and stretch about 400 km east to west, giving them a total area equal to that of France. Their situation between latitudes 22° and 32° North, combined with their almost Himalayan scale of elevation and monsoon climate, has produced one of the world's most varied and complex vegetation regions.

Within this extensive and diverse region special emphasis is given to the Yulong Xue Shan (Jade Dragon Snow Mountains). Located immediately north of the great bend of the Jinsha Jiang in latitude 27° North, they are reasonably representative of the whole mountain region. Furthermore, this area has been the most extensively studied in the entire region by a succession of scholars, including Joseph Rock (1924, 1947), Handel-Mazzetti (1921), and Kingdon Ward (1973) prior to the Second World War, and the Chinese Academy of Science during the closed Communist era, 1950–1985 (Ngo 1958; Li *et al.* 1982). The combined studies of the Chinese Academy of Science and the Nature Conservancy since 1996 led to the designation of a vast area of northwestern Yunnan as the Three Parallel Rivers National Park and World Heritage Site in 2003. Earlier, one of the region's major towns, Lijiang (specifically, the Old Town – Dayan), the historical capital of the Naxi minority nation, had been designated as a World Heritage site in 1997.

Mountains of northern Pakistan

The mountainous region of northern Pakistan extends between latitudes 33° and 37° North and longitudes 70° and 75° East. It occupies an area in excess of 70,000 km^2 and supports a population of about 800,000 people living in scattered villages throughout the region (Rieck 1997). The much more northerly latitude than, for instance, Nepal or the Indian Eastern Himalaya, is very significant in terms of climate and vegetation patterns and adds greatly to the contrast between the two mountain regions.

The mountains of northern Pakistan are conventionally classified into three major groups: the northwestern Himalaya, the Karakorum, and the Hindu Kush. In addition, there are many lesser mountains and foothills extending south of the Himalaya almost to Islamabad and southward along the Pakistan–Afghanistan border zone into Baluchistan and the coastal zone fronting the Arabian Sea. These latter are not included in the present discussion.

The Pakistan Himalaya trend southeast to northwest from the Indian province of Jammu and Kashmir across the Line-of-Control into Azad-Kashmir to culminate in the great ice pyramid of Nanga Parbat (8,125 m) above the middle Indus Gorge. They are fronted by the Pir Panjal, the northwestern equivalent of the Nepalese Mahabharat Lekh.

Northwest of the Himalaya, and separated from them by the upper Indus Gorge, lies the Karakorum range. It runs parallel with the Himalaya and extends for some 350 km from the vicinity of the politically contested area of the Siachen Glacier along the border with China as far as the Ishkamun River. The Karakorum include the world's second highest peak (K-2: 8,611 m) that rises from the midst of a large group of spectacular ice-enshrouded mountains and rock towers overlooking the Baltoro Glacier. The Ishkamun River is the formal dividing line between the Karakorum and the Hindu Kush, although this is an expedient rather than a justified geophysical boundary.

The Hindu Kush are initially aligned east–west but beyond Tirich Mir, the highest peak at 7,690 m, they trend southwestward with the crest line passing across the international border into Afghanistan. This trend continues to the north of Kabul and, progressively losing height, merges into the complex array of mountains, plateaus, high basins, and deserts of southwest Afghanistan.

This immense mountain region of northern Pakistan is deeply dissected, principally by the Indus and its many tributaries, two of the more prominent being the Hunza and Gilgit rivers. The major access is the Karakorum Highway that follows much of the Indus Gorge before diverting to Gilgit and the Hunza Gorge, reaching the Pakistan–Chinese frontier on the Khunjerab Pass at 4,600 m. Since the opening of the highway in 1978 many branch roads, some surfaced but others requiring four-wheel drive vehicles, have connected most of the hitherto isolated villages to the 'outside' world (Figure 2.8).

The mountains of northern Pakistan and Afghanistan represent a northwestern extension of the Himalaya proper. As mentioned above, the effects of the summer monsoon are progressively reduced from Sikkim and Nepal toward the northwest. Within the mountains of Pakistan, winter precipitation at high altitude predominates. The valley floors experience semi-arid and arid climatic conditions. Gilgit, for instance, at 1,460 m, has very hot dry summers (in 1991 a maximum temperature of 39.5° C was recorded), and its thermal regime, aided by irrigation water from high-altitude melting snow and ice, supports two-crop agriculture. Miehe and Miehe (1998: 102) describe the constraints facing any attempt to assess the natural vegetation zones of the Karakorum on the basis of '[t]he steepness and rockiness of the slopes resulting in small-scale mosaics of water-deficient and water-surplus situations'. Many of the valley floor stations record less than 100 mm/yr, or even below 50 mm/yr, of precipitation. Between 2,000 and 3,000 m precipitation ranges from 400–900 mm/yr and,

Figure 2.8 The Hunza Valley, Karakorum Mountains, northern Pakistan. The old fort of Karimabad is being restored as a UNESCO historic monument. The Karakorum Highway runs across the middle ground on the far side of the Hunza River, out of sight below the town, September 1995.

Figure 2.9 Northernmost Pakistan, Karakorum Mountains below the Khunjerab Pass. View from the Karakorum Highway, showing the extreme aridity of the middle and lower altitudes in this region, September 1995.

where topographically and edaphically feasible, mountain forests are found (Figure 2.9). Above the regional snowline water-equivalent receipts exceed 1,000–2,400 mm/yr. These relatively heavy precipitation amounts account for the extensive glacierization, especially of the higher Karakorum (Flohn 1968).

The Pamir mountains

The Pamir are best described as a complex and compact block of high mountains and plateau extending 500 km east to west and 300 km north to south (37°– 40° North; 70°– 74° East). Some of the highest and most extensive of the mountain ranges of Asia radiate from the Pamir, a geographical phenomenon that has led to the descriptive term 'Pamir Knot' (also known as the 'Roof-of-the-World' throughout the Middle Ages).

The Pamir complex can be divided into five broad physiographic units. At the core is the Academy Range culminating in Peak Communism[1] at 7,495 m and extending for about 175 km from north to south (Figure 2.10). The Trans-Alai Range is a northeastern spur of the main axis that also supports summits close to and above 7,000 m (Peak Lenin, 7,134 m). The Academy Range descends abruptly along its western edge to the Western Foothills. This physiographic unit contains four distinct east–west ridges, including the Peter the

First Range in the north, and the Darvaz, Vanch, and Yazgulem ranges, each separated by deep river gorges that produce much of the headstream waters of the Amu Darya River. The Southeast Pamir is a complex of high mountains and deep east–west gorges, many of which are blocked by earthquake-induced giant landslides (especially the Usoi Dam and Lake Sarez; see Chapters 6 and 10), or have been so blocked in the recent geological past. The Pianj Gorge forms the southern limit of this unit, as well as the international border with Afghanistan, and flows westward into the Amu Darya. To the east the Southeast Pamir tower above the high and dry Pamir Plateau. This is one of the most arid interior regions of Central Asia, receiving less than 100 mm precipitation annually. One of its main features is the extensive saline Kara Kul Lake that has no outlet and is well known as a resting place for myriads of migrating birds. The Chinese Pamir, or Sarikol Range, abut the plateau to the east as a gigantic wall of ice and snow extending to the highest summits of the entire Pamir (Kongur Shan, 7,719 m; Muztagata, 7,546 m).

The moisture regime of the Pamir is dominated by a winter westerly cyclonic pattern that brings heavy snows to the higher altitudes and significant winter and spring rains to the Western Foothills. Garm, at 1,807 m in the Vakhsh Valley north of Peter the First Range, receives about 700 mm annually while at higher levels (2,700–3,200 m) annual precipitation exceeds 1,000 mm.

The permanent snowline of the Pamir is cited by Troll (1972) and Barry (1992) as the highest in the world, averaging 4,800 m in the Western Foothills and 5,500 m in the Academy Range. The high snowfall receipts, accompanied by many days of fog and low cloud that restricts losses by sublimation, account for the large volume of permanent ice and snow on the High Pamir. For instance, Suslov (1961) calculated that the Fedchenko glacier system contained about 200 million cubic metres of ice, making it the largest glacier system in Central Asia (Figure 2.11). The permanently staffed Fedchenko Glacier Station at 4,169 m has frequently recorded a winter accumulation of snowpack in excess of 25 m.

The lower slopes and valleys are semi-arid to arid and human activities are dependent on irrigation based upon rainfall and snow- and glacier-melt from higher elevations. A forest belt lies between 1,500 and 2,800 m on the western slopes and includes maple, walnut, wild plum and apple, juniper, rose, and birch. There are no native pine or spruce. Subalpine meadow, with extensive juniper shrubs, occurs between 2,700 and 3,500 m, giving way to an alpine meadow belt between 3,500 and 4,400 m. The Pamir also support a number of rare mammals, including snow leopard and Marco Polo sheep.

Except for the larger western valleys, the Pamir region is sparsely populated. Khirgiz nomads form the majority of the small population on the Pamir Plateau, and Mountain Tajiks and Turkic groups occupy scattered small villages (*kishlaks*) in the mountain gorges. The most densely populated region is the Pianj Gorge between Khorog, the regional capital, and Rushan, facing Afghanistan across the river. Prior to the civil war (1992–6) the High Pamir experienced an influx of 200–400 alpinists (predominantly Russian) every July. This and the beginnings of trekking and adventure tourism in the late 1980s and early 1990s

Figure 2.10 (opposite top) The high peaks of the Academy Range, Tajikistan Pamir, from above the upper reaches of the Fedchenko Glacier, September 1990.

Figure 2.11 (opposite bottom) The lower reaches of the Fedchenko Glacier, Tajikistan, almost totally covered by moraine debris. The lower slopes of the High Pamir are extremely arid. The features on the slopes just above the glacier are not trees but eroded rock formations, September 1990.

were curtailed by the outbreak of the civil war. However, it provides a suggestion for future development with National Park and World Heritage status possibilities once political stability is achieved. After achieving independence, Tajikistan has experienced not only the economic and environmental destruction of civil war, but separation from the former Soviet Union deprived it of its previous sustained massive subsidy. This has especially affected the people of the Pamir, resulting in widespread return to subsistence agriculture and concomitant pressure on the region's flora and fauna (Breu and Hurni 2003).

3 Status of forests in the Himalayan region

Nepal has lost half of its forest cover within a thirty-year period and by AD 2000 no accessible forests will remain.

(World Bank, 1979)

The status of the Himalayan forests has been a topic of discussion for more than a century. After about 1960 this discourse became more widespread and alarmist until the assumed catastrophic loss of forest cover took the pivotal role in reports suggesting that the entire Himalayan region was facing imminent environmental collapse. These reports were linked together under the title of convenience: *The Theory of Himalayan Environmental Degradation* (Ives 1984, 1987a). Any assessment of the Theory can do no better, therefore, than to take forest condition as a prime point of entry.

Thompson *et al.* (1986) had cast strong aspersions on the early accounts of deforestation. Ives and Messerli (1989) had remonstrated that the scenario of rapid post-1950 forest disappearance, as reported throughout the 1970s and 1980s, was not justified. Hamilton (1983) had previously challenged the conventional assumptions relating deforestation to accelerated soil erosion and downstream flooding. Furthermore, during the Mohonk Conference, Hamilton entered a plea for elimination of the ambiguous and emotive term *deforestation* from the literature unless it be rigorously defined for each different situation in which it was used (Hamilton 1987: 257). Nevertheless, the sparseness and possible unreliable nature of much of the available data was apparent. Despite this degree of uncertainty, it had been recognized during the Mohonk Conference (not unanimously) that most of the widespread conversion from forest to farmland had culminated during the late nineteenth and early twentieth centuries. The actual nature of forest loss, in Nepal at least, was largely the result of government-sanctioned clearing in the Terai and nothing comparable had occurred in the Middle Mountains and the High Himal. These contentions were based upon a small number of individual studies, especially those by members of the staff of the Nepal–Australia Forestry Project (Bajracharya 1983; Mahat *et al.* 1986a, 1986b; Griffin 1987), and through critical analysis of several of the published reports that postulated widespread post-1950 deforestation (Ives and Messerli 1989: 43–87). Tucker (1987), and later Guha (1989), produced similar assessments in terms of pressures on the forests of the Central Indian Himalaya in the nineteenth century. Nevertheless, it had seemed reasonable to recognize that in *certain* areas serious forest loss was occurring. For instance, in the Kumaun Himalaya large-scale commercial logging, legal and illegal, had led to the spontaneous formation of the Chipko Movement (Shiva and Bandyopadhyay 1986). However, in other areas, such as Kabhre Palanchok and Sindhu Palchok districts of Nepal, the Nepal–Australia Forestry

Project (Gilmour *et al.* 1987; Gilmour 1988) had begun to detect actual increases in forest cover, the result of planting by the local farmers on land under their own control.[1] It had also been demonstrated (Byers 1986, 1987a) that some of the statements relating to deforestation in the Khumbu Himal had been extensively distorted. So what can be said of the state of the forests of the early twenty-first century?

Nepal Middle Mountains and High Himal

The staff of the Nepal–Australia Forestry Project have continued their work throughout the 1980s and 1990s and have demonstrated that, without any doubt, significant reforestation together with improvement in pre-existing forest quality is occurring within Kabhre Palanchok and Sindhu Palchok districts (Carter and Gilmour 1989; Gilmour and Nurse 1991; Gilmour and Fisher 1991). Subsequently, Jackson *et al.* (1998) have not only confirmed the earlier work of their colleagues but have also provided additional detail indicating environmental deterioration in specific areas. They demonstrate that between 1978 and 1992 community forestry activities at lower altitudes have had beneficial impacts on the local forests and the overall stability of the land-use system. Shrublands and grasslands have been converted to more productive forest land while some of the least productive *bari* terraces on steeper slopes have been abandoned and allowed to revert to shrubland and forest. Within the formerly degraded forests at lower elevations, qualitative improvements have occurred to the extent that the locally extirpated leopard and other animal species have returned, or increased in numbers. However, it is the areas closest to population centres that have seen the most positive change. Here the efforts to foster community forestry have been most successful, partly the result of the long-term partnership between the local people, the Nepal–Australia Forestry Project, and the Nepalese Department of Forests.

On the upper slopes of the Mahabharat Lekh and on the high hills leading up to the Himal, the situation is much less positive. These areas, and particularly the northern highlands of Sindhu Palchok which are relatively remote, are experiencing environmental deterioration in contrast to the lower elevations. In the higher areas transhumance is still practised and people from various communities come up during the summer to collect forest products such as medicinal plants, roof shingles, and *lokta* (*Daphne* spp) for paper making. And while *bari* terraces are being abandoned in these areas, their reversion to forest, shrubland, or grassland will depend on how the land is used. Over-grazing is a common occurrence and co-ordinated management is extremely difficult to organize.

Jackson *et al.* (1998) argue that their results are significant for the entire area of both districts as the sample sites represent 15 per cent of their total area (400,000 ha supporting a population of about 500,000 people). Despite the relatively short period of survey (14 years), agrarian change has been widespread and pressure on resources has certainly abated at the lower elevations. Nevertheless, sustained population pressure combined with lack of coherent policies and practices continues to result in a steady decline in forest resources on the upper slopes, leading to loss of catchment stability. It is believed that this trend will continue because of the difficulty of establishing collective community responsibility in these more remote areas where land ownership remains ambiguous.

Overall, however, the present situation and future outlook for the two districts is far more positive than those who support the Theory would assume. Jackson *et al.* (1998) find themselves in agreement with Adhikari (1996), who concluded from a study of two villages in Central Nepal, that agrarian restructuring had resulted in a significant containment of population pressure on land resources, especially the forests. Other studies make similar predictions.

The Kabhre Palanchok and Sindhu Palchok district studies could be classed as atypical for Nepal in that they are located in close proximity to Kathmandu and they have benefited from the persistent efforts of the Nepal–Australia Forestry Project. Nevertheless, the studies indicate a far greater degree of resilience than the doom-and-gloomsters would allow. It is demonstrated that in the two districts there are procedures in place that could be applied in other districts further from Kathmandu and would be expected to produce similar results.

The Nepal–Australia project has had the advantage of long-term involvement in a substantial area of the Middle Mountains. This is unusual. However, there have been several intensive studies that include an initial period of investigation and a repeat study some years later. One of the most interesting of these is a study by Jefferson Fox (1993) of the small village of Bhogteni that provides a comparative assessment of forest quality and management in 1980 and in 1990. His results are discussed in some detail as they illuminate several critical points in the deforestation debate. Furthermore, the study includes a large and carefully accumulated database. Fox concludes that, despite a village annual population growth rate of 2.5 per cent, the forests were in much better condition in 1990 than in 1980.

Bhogteni, with a population of less than 1,000, is situated at an altitude of 1,200 m and is an hour's walk north of Gorkha. Forest quality under all three land tenure patterns (private, protected panchayat, and operated panchayat) was remarkably improved over the ten-year period. This observation is based upon data such as number of trees per hectare, volume of wood per hectare, and growth rate per hectare. A change in domestic animal holding patterns resulting in reduced numbers of cows and female buffaloes would have eased forest grazing pressures. However, this was off-set by the increase in human population, although total village fuelwood consumption did not change significantly. The data supporting improved forest quality are impressive: for instance the average annual increment in volume of wood in cubic metres per hectare for all three forest management types increased from 1.93 to 4.76 m^3/ha over the decade.

Fox examines the reasons for this. He proposes no definite answer but draws attention to a range of contributing factors. These include the change in government forest policy from countrywide nationalization to the introduction of panchayat-protected and panchayat-operated regimes in 1976, followed by further strengthening of village control as the system of panchayat democracy was replaced by a more complete form of democratic government from Kathmandu in 1991. Outside assistance was presumably significant; in 1981 the Resource Conservation and Utilization Project (RCUP), funded by USAID, began forest enrichment plantings, and grazing and collection of fodder and fuelwood were prohibited in prescribed areas. In 1982 the road from Kathmandu to Gorkha was completed. This facilitated the import of chemical fertilizers so that agriculture could be intensified without an increase in livestock numbers as source of manure.

All of these factors have been important. Fox preferred not to predict whether or not these changes were sustainable but raised alternative questions instead. These relate to forest policy, the stability of the village forest management committees, and the question of whether intensified agriculture could be maintained with the use of chemical fertilizer. He added that, if the answer to the final question proved to be negative, in order to obtain sufficient quantities of manure farmers would soon have to raise more livestock than the forests could sustain. As an aside, he also contests Thompson and Warburton's (1985a) devastating attack on the value of fuelwood consumption measurements (use of the so-called 67 factor depicting the enormous range of fuelwood estimates). He maintains that, if fuelwood consumption rates can be replicated at ten-year intervals with no significant differences, reliable documentation of consumption is possible; that, while the Himalayan environment may face great uncertainty, fuelwood consumption rates do not provide a good example. However, Thompson and Warburton based their criticism on per capita rates of consumption that had been published, even though they missed those produced by Fox. This appears to be an example of what can be done in contrast to what has been done. Similarly, the grossly unreliable measurements exposed by Thompson and Warburton (1985a) had been used to support the claim for catastrophic deforestation, and it was this claim that could not be justified.

Fox concludes by postulating that the immediate threat to the Middle Mountain lands is not population growth but bad forest management policies. 'Before population can be cited as the cause of forest degradation, forest policies must be implemented that provide incentives (rather then [*sic*] disincentives) for local people to manage forest resources' (Fox 1993: 98). Even so, he adds the now familiar caveat concerning the impossibility of generalizing from one village study to a wider region. This is tempered, however: 'when these [findings] are considered along with Gilmour's (1991) work in the Jhiku Khola watershed, the evidence begins to indicate that, despite continued population growth, more sustainable forest management systems are being established throughout the hills' (Fox 1993: 97).

A group of four papers (Metz 1990; Exo 1990; Zurick 1990; Brower 1990) was published in *Mountain Research and Development* with introductory overviews by Messerschmidt (1990) and Stone (1990). They provide an insightful suite of case studies that examine individual village community land management practices in Middle Mountain and High Himal locations distant from Kathmandu. Brower's observations will be discussed in more detail below in the section dealing specifically with Khumbu Himal. The other three examine successes and failures of the local communities in responding (or not responding) to problems of land (including forest) degradation. Although they do not add significantly to the present effort to ascertain actual loss of forest cover, they all indicate a definite increase in pressure on local resources. Their primary value to the current discussion is the insights they provide about villager efforts to maintain land productivity. In this sense, they support and extend the conclusions of Fox (1993), published later, although all the field studies were undertaken about the same time during the mid-1980s.

The studies were published together for this reason. Each demonstrates that when the villagers are facing a direct threat to their subsistence, such as progressive expansion of gulley erosion or landsliding on privately owned or controlled land, they are fully capable of effective group recovery efforts through household co-operation; they draw upon their considerable understanding of the landscape dynamics upon which they depend. However, when the land under threat is common land or, more particularly, land controlled by the government, no such positive response appears socially feasible, even though successful land rehabilitation would be advantageous in the long term.

Metz's (1990) study encompasses an upper elevation village (Chimkhola) on the flank of Dhaulagiri Himal. Subsistence crops are grown here almost entirely on rain-fed *bari* terraces and the soil fertility is maintained by systematic movement of domestic animals based on an extensive network of *goths* (herding shelters). He indicates that shortage of forest products of all kinds had only recently been felt in Chimkhola. To this the villagers have responded by setting up their own pattern of forest management. Apparent inability to control livestock grazing and browsing, however, seems likely to set the stage for the eventual collapse of community co-operation.

Exo (1990) demonstrates that small NGOs, in providing 'outside' expertise and encouragement, are able to add strength to local community efforts. In contrast, he castigates the large project (RCUP) in Pandong and Takuket panchayats that failed because of its heavy 'top-down' short-term target orientation. He goes on to criticize the (then) prevailing tendency toward centralized control of local and regional resources.

Zurick (1990) explains that, in some village areas, land management efforts were recent and reflected the new and growing recognition of land degradation. In others they represent centuries-old practices that were still being followed. He urges examination of the 'incentive environment' for enlisting villagers to respond to the need for conservation. He also points to the seriousness of land degradation in his village study:

> The management of village slopelands in Phalabang thus provides insights into the nature of local land degradation and the indigenous conservation strategies which exist to reduce or forestall the current high levels of land stress.
>
> (Zurick 1990: 28)

The four studies do not help significantly in determining extent and rates of forest cover loss over time. They emphasize, however, the extensive environmental understanding of the local communities that reinforce similar statements emanating from the early UNU studies in the Kakani area (Johnson *et al.* 1982). They also indicate the serious contrast in management between locally controlled and government-controlled land. This underscores the political and social elements of Himalayan land degradation problems. Since all four researchers spent prolonged periods in their field areas and paid careful attention to collection of reliable data, their insights are especially important.

Smadja (1992) provides yet another perspective on the relationship between increasing intensity of subsistence farming and slope stability in the Middle Mountains of Nepal. She spent a full year studying various processes affecting the different land-use types on the Salme slope as part of a long-term multi-disciplinary project organized by CNRS, Paris. Her field area is located about 50 km northwest of Kathmandu and ranges in altitude from 1,200 to 4,000 m. In places the annual rainfall exceeds 4,000 mm. The area is occupied by Tamang people whose main crops are maize, millet, buckwheat, potatoes, and, more recently for ceremonial purposes, paddy rice at the lower elevations. The main weight of the study concerns the geomorphic effects of the heavy rainfalls on the various land-use types and so will be discussed more fully in Chapter 4. However, she demonstrates that even on areas with the more degraded forest types the heavy rains have little negative impact and that the landscape is remarkably stable, except for a series of major gullies that are geologically controlled. Furthermore, rates of forest conversion appear to have been minimal over the previous several decades. Nevertheless, Smadja and Fort (1998: 398) point out that as the increasing population of the Salme region has resulted in growing dependency on forest products '[w]ith the decrease of forest resources, trees have become more numerous in the [private] fields. In some areas they provide all wood and fodder needs.'

Metz (1997) also raises a vital point in the deforestation debate. The question of vegetation dynamics under natural, or little disturbed, forest conditions had not previously been taken into account. There have been several extensive and highly respected studies of the natural vegetation of the Himalaya (Schweinfurth 1957; Champion and Seth 1968; Stainton 1972; Numata 1983; Ohsawa 1983; Dobremez 1986; Singh and Singh 1987). In the current context, however, Metz has pioneered the incorporation of vegetation dynamics into the assessment of forest cover loss. He investigated seven little-disturbed stands of temperate forest protected by the Bakum Ghyang monastery, some 40 km north-northwest of Kathmandu. The location is in northern Sindhu Palchok district on a homogeneous west-facing slope between 2,400 and 2,900 m. He undertook a detailed standard plant ecological investigation. His three lower elevation stands were dominated by *Quercus oxyodon*; the other four by *Tsuga dumosa*. The first group also supported *Q lamellosa* and Lauraceae species as sub-dominants. The *Tsuga dumosa* stands contained *Q semecarpifolia* and *Rhododendron arboreum*. In the highest stands *Abies spectabilis* was almost as frequent as the *Tsuga dumosa*.

Metz (1997) points out that many scholars of Himalayan vegetation have indicated that forests similar to those in his study were 'climax' (Champion and Seth 1968; Ohsawa 1983; Singh and Singh 1987). This would imply that they are in a steady-state and are maintaining their population structures. Metz demonstrates, however, that in his study all the major canopy-forming species, with the probable exception of *Tsuga dumosa*, were not maintaining themselves, despite being subjected to almost no human harvesting or grazing. He provides five possible explanations. He concedes that browsing may have been more extensive than he had been able to detect. Small bamboos may have blocked regeneration of tree species. Even the regeneration strategies of the forest dominants, several of which have overwhelming mast year seed production, may have affected his test stands. For instance, he cites Singh and Singh (1987) who showed that during mast years in their study area *Q floribunda* produced up to 100 times the number of acorns as they did in 'normal' years. Next he entertains the notion that there may be too few old canopy trees so that there is insufficient mortality to create the gaps in the canopy needed for the establishment of seedlings. He expresses strong preference for the fifth hypothesis which argues that *Quercus* spp are not 'climax' dominants but cohorts of early successional species that had established soon after a large disturbance in the past and that re-establishment of the present forest will have to await a similar event in future – such as extensive fire or very large mass movements.

In conclusion Metz argues for more research, especially 'action research' – trials with teams of workers that include local forest users. This is to ensure that the introduction of any new management techniques result in the regeneration of species that local users and con-servationists desire. Nevertheless, he refers to his own experience and that of several other researchers who collectively caution that degradation of the temperate and subalpine forests is very widespread in the Nepal Himalaya. He maintains that, without improved management, most forests will probably be converted to shrublands within the next 50–100 years. This would cause the destruction of habitat needed to maintain biodiversity and would lead to shortages of subsistence forest products. This is very different from the alarms of Eckholm because it relates to the future, not to the concerns that were being raised during the 1970s and 1980s.

The Khumbu Himal represents a microcosm for the *Theory of Himalayan Environmental Degradation*. This was discussed in some detail by Ives and Messerli (1989: 59–65) based largely on intensive fieldwork by Byers (1987a, 1987b). Some of the early statements of imminent environmental degradation were criticized, not only for being incorrect, but for being politically motivated. Some of the accounts of deforestation in the 1970s were suspected to be part of an effort to induce the government of Nepal to create the Sagarmatha (Mount Everest) National Park. In 1979 the long south-facing mountain slopes in the vicinity of Namche were impressive for their nakedness, for the myriad yak terracettes, and for the widely spaced mature and solitary specimens of *Abies spectabilis* that served to dramatize the absence of forest cover. Much of the more recent research (Byers 1987b, 1996, 1997, 2002 unpub; Brower 1990; Stevens 1993, 2003; Brower and Dennis 1998) has firmly established that the deforestation was effected over the past several hundred years rather than after 1950. It has no significant relationship to recent Sherpa population growth, to the direct demands for fuelwood by trekkers and mountaineers, nor, indirectly, to the extensive construction of lodges and small hotels. Some early alarms had centred on claims that specific mountain slopes had been stripped of their forest cover within the personal experience of a single observer. Most of these have been shown by Byers, by replication of early photographs, to be totally false. The news media, seeking to report impending crisis, tried to indicate that the slopes of Mt Everest itself had been denuded of their forest cover after the time when Tensing and Hillary (1953) reached the summit. In reality, such slopes may have only experienced tree growth as far back in geological time as the early Tertiary Period! (see Chapter 10).

The question now is whether or not the initial rebuttal of the Khumbu 'catastrophe theory' has stood the test of the last 15 years. First, Byers (1996, 2002 unpub.) has greatly extended his research of the 1980s and his repeat photography has been especially revealing. Early photographs taken by such mountaineers as Houston, Schneider, and Müller in the 1950s and early 1960s (as well as his own photographs in 1984 and 1995) have provided the bench-marks for this approach. The conclusions are unequivocal: deforestation since the 1950s at the altitude of the main forest stands has been negligible (Figures 3.1, 3.2, 3.3, and 3.4). Byers has confirmed his earlier suspicion that the actual location of serious environmental degradation is the lower limit of the alpine belt above the settlement of Dingboche in the upper Imja Khola valley where the final stage of the trek to the Everest base camp begins. Here he has demonstrated through the analysis of photo-triplets (1962, 1985, 1995) that almost 50 per cent of the juniper shrubland has been destroyed during the 33 years preceding 1995. The effects of this have been exacerbated by extensive cutting of alpine turf for con-struction material. And because the juniper have been entirely uprooted, the two processes have exposed to erosion a wide area underlain by light friable soils. His erosion study plots of 1984 in this area recorded a rate of soil loss that was 20–40 times greater than that recorded from the erosion plots he maintained at lower elevations below the upper treeline, despite the fact that monsoon precipitation in the Dingboche area is barely half that of the lower plots. Stevens (2003: 266) has recently introduced a new insight into the loss of the subalpine juniper. He reports Hillary had noted that the British reconnaissance team of 1951 had initiated the process of hauling shrub juniper to base camp and the following year the Swiss expedition had used 300 loads (Rowell 1980 in Stevens 2003).

The belated response by the National Park authorities to the incorrect assumption of extensive recent forest losses within the main forest belts was to undertake a programme of tree planting around several of the villages, especially Namche. By 1995 this project had achieved remarkably beneficial results, especially around the Sagarmatha National Park headquarters. Nevertheless, it demonstrates a misplaced response to assumed environmental degradation when measures should have been taken to reduce further damage to the juniper shrubland above Dingboche. The causes of the higher elevation turf and shrubland destruc-tion are probably many: use of juniper for summer fuel and for religious observances by Sherpa herders, and turf for construction (over a long time period but observable within the limits of the replicated photographs, 1962–1995); and increasing use of shrubs for fuel by trekkers and mountaineers and their porters (especially from 1970 onward), despite park regulations that require visitors to bring in their own liquid fuel.

Byers presents strong evidence for human intervention in the Khumbu landscapes long before the conventionally accepted chronology of the arrival of the Sherpas 300–400 years ago. A series of soil profiles, subjected to pollen analysis, radiocarbon dating of contained charcoal, and identification of *Cerealia* grains (implying cereal cultivation) suggest that humans and their domestic animals have been altering the Khumbu environment for perhaps thousands of years prior to the assumed Sherpa arrival. In addition, the pollen analysis also implies that 400–800 years ago a pre-existing warmer and wetter climate was replaced by

Figures 3.1 (upper) and 3.2 (lower) Replicate photographs taken from the north-facing slope of Mount Tamserku, Khumbu Himal, at 4,488 metres. Figure 3.1 was taken in 1961 by Erwin Schneider, Figure 3.2 in 1995 by Alton Byers. The view is toward the west and shows the relatively gently sloping land between Namche (bottom left) and the villages of Kunde (3,840 m) and Khumjung (3,790 m right centre). The Syangboche airstrip is visible centre left. Byers noted thinning of juniper groves and woodland patches, especially on the slopes close to the villages; however, the overall condition of the forest cover has not changed in the 34-year period. The long gulley scar which descends from the top right close to Kunde shows no change, despite repeated reports during the 1970s that this feature was rapidly expanding due to 'deforestation' of the slopes above the village.

Figures 3.3 (upper) and 3.4 (lower) Replicate photographs taken from 4,028 metres on the ridge imme-diately west of Kunde in 1961 by Erwin Schneider and in 1995 by Alton Byers. The dry south-facing slopes above the two villages of Kunde and Khumjung appear unchanged over the 34-year period. Ground observations by Byers revealed that juniper shrubs have certainly been cut or dug up completely, but the claim that these slopes have been 'deforested' since 1950 is not credible. Byers observed large numbers of fir (2,490/ha) and rhododendron (5,935/ha) seedlings on the moist north-facing slopes in the same vicinity. This is contrary to earlier claims that collection of litter and firewood was inhibiting forest regeneration.

the colder and drier conditions of the last several centuries. This, in turn, in conjunction with Sherpa interventions, would be expected to have had a significant impact on the distribution and structure of vegetation types throughout the area.

The foregoing discussion implies the need for an entirely new assessment of the evolution of the Khumbu landscape. Undoubtedly, a thriving and expanding Sherpa population would have effected significant changes over 400 or so years. These would reflect the need for timber and fuel, for the expansion of grazing areas, as well as fields for subsistence crops. From the photographs of Schneider and others it can be stated unequivocally that the major changes in forest cover had already occurred when the first Western mountaineers approached the threshold of Everest in 1950. Brower (1990) and Brower and Dennis (1998) have discussed how, in 1950, a trading embargo with Tibet was enforced by the Chinese closure of the frontier which was followed by an accelerating growth in tourism so that changes in yak grazing patterns were inevitable. The demands for pack animals at a few tourist nodes, such as Namche, not only changed the breeding patterns, but strongly influenced the grazing procedures. Many high-altitude summer pastures were left under-grazed, while areas around some of the villages became heavily over-grazed.

Stevens (1993) has completed an impressive study of the human ecology of the Sherpa people and has analysed the impacts of 'outside' pressures on traditional management practices. There have been many other studies of the Sherpa way of life and its adjustments to the great increase in tourism (see Chapter 7). The adaptability of the Sherpas as well as the significant increase in affluence and general standard of living would have been unexpected even 30 years ago. Detailed examination of this large and important topic, however, falls well outside the scope of the present work. As discussed in Chapter 7, the increased demands for construction timber and fuelwood arising from the tourist boom in Sagarmatha National Park have resulted in illegal logging in the forests south of the park boundary (Stevens 2003; Figure 10.1). And undoubtedly, cutting of trees, both for firewood (Figure 10.2) and lumber, has occurred and is occurring within the park. However, this is on a much smaller scale than had been predicted in the 1960s and 1970s. Stevens (2003) provides a recent discussion of this topic.

The natural, or mountain, hazards that affect life and landscape in the Khumbu are discussed in more detail in Chapter 6. However, as major geomorphic events are often attributed to human impacts on vegetation, and especially forests, the occurrence of catastrophic floods and debris flows resulting from the outbreak of glacier lakes is briefly introduced here. Zimmermann *et al.* (1986) concluded that the very steep slopes of the Khumbu are remark-ably stable, although the recurrent glacial lake outburst floods constitute a dangerous hazard. Their occurrence is most likely the result of a warming climate, glacier thinning and retreat, and the subsequent development of unstable frontal and supra-glacial lakes (Viuchard and Zimmermann 1987; Watanabe *et al.* 1994, 1995; Mool *et al.* 2001a, 2001b).

The Makalu-Barun National Park and Conservation Area was established in 1991. It covers an area of 2,330 km^2, contiguous with and immediately east and southeast of Sagarmatha National Park. It is administered by agreement between the Nepal Department of National Parks and Wildlife Conservation and The Mountain Institute, West Virginia, USA. The conservation area (830 km^2) has an indigenous subsistence population of about 32,000. The park proper, apart from two small enclaves, has no permanent inhabitants but is used extensively by transhumance pastoralists and, much more recently, by an increasing number of trekkers. Byers (1996) undertook a reconnaissance survey of the area's vegetation and human use patterns. This has been followed by further biological survey and detailed satellite and GIS mapping (Carpenter and Zomer 1996; Zomer *et al.* 2001). While the general findings indicate far less human impact than that which has occurred in the Khumbu, a similar pattern of pre-1950 use emerges. There has been long-continued herding at all elevations up to the alpine meadows and there is clear evidence of forest burning over the past several centuries. Nevertheless, while the structure and composition of the forests have changed, a fairly complete canopy remains above about 2,000 m so that little impact on the hydrology and geomorphology is likely to have occurred. During the last 10–20 years, however, there has been burning of juniper and dwarf rhododendron around the Makalu base camp and other camping areas currently used by trekkers and mountaineers. Byers (1996) interviewed local herders, who gave the following reasons for the burning of juniper and rhododendron tracts of up to a hectare in extent: a desire to increase pasture; to create a fuelwood source and income from sale to the Makalu base camp and other camp occupants; and as relief from boredom in the monotonous life of a livestock herder. Below 2,000 m, of course, the relatively large resident population has effected a significant conversion of forest to farmland comparable to other areas of the Middle Mountains, and Carpenter and Zomer (1996) have expressed concern for the increasing pressure on the remaining riparian forest corridors that are so biologically important.

Gautam and Watanabe (2004) have re-evaluated previous determinations of land use/cover change in montane Nepal based on the Land Use Maps of 1978 and the topographical maps of 1992. Their fieldwork was conducted in the Kangchenjunga Conservation Area situated in the extreme northeastern corner of Nepal. The research included reinterpretation of the 1978/79 and 1992 air photographs and comparison of the results with the Land Use Maps of 1978 and the topographical maps of 1992. The original land use map data were digitized using a Geographical Information System to facilitate comparison of the represented cover types with their own air photograph interpretation. After completion of the laboratory work, their air photograph interpretation was verified by field checking and by a questionnaire and interviews with the local people in the four village test sites that they had selected.

The study demonstrated significant discrepancies between the original interpretation and their own results. The most serious and persistent error concerned the assumed rate of deforestation between 1978 and 1992. For instance, the forest area had been assumed to have decreased by 62.5 per cent (23.15 km^2) and the area under agriculture to have increased by 35.7 per cent (1.49 km^2). In contrast, their own air photograph interpretation recorded a forest cover loss of 14.9 per cent (5.45 km^2) and an increase in area under agriculture of only 4.9 per cent (0.21 km^2). These figures are averages for all four test sites. Similar, although smaller, discrepancies characterized attempts to determine changes in grazing, shrub, and barren land use/cover types. Next, interviews with the local people confirmed the results obtained from photograph interpretation and field checking.

Gautam and Watanabe's proposed explanations for the discrepancies are as follows: lack of precision in defining the land use/cover types for the 1978 mapping exercise; inadequate training of the air photograph interpreters; absence of field checking; and limitation of time available for map preparation.

At first glance the results of Gautam and Watanabe (2004) could be dismissed as of little significance – they involve a minute percentage of the total area of montane land in Nepal, and the original maps have been superseded by recent air photography and satellite imagery. However, such a conclusion would miss the very important point of the exercise. The 1978 Land Use Maps have been used repeatedly as a bench mark for estimation of forest losses in Nepal and for the rates of such change. More recent investigations, quoted by Gautam and Watanabe (HMG/FINIDA 1995/96; DFRS 1999a, 1999b; UNEP 2001), have all made use of the 1978 Land Use Maps as a bench mark for calculating rates of land use/cover change. Furthermore, practically all the earlier interpretations of rates of deforestation have been tied to the 1978 Land Use Maps. This has presumably led to a gross over-estimation of the rates of forest loss. In comparison, air photograph interpretation and field work, for example, by Virgo and Subba (1994), Tamrakar (1995), Shrestha and Brown (1995), Jackson *et al.* (1998), and satellite imagery interpretation by Zomer *et al.* (2001), all show much smaller rates of forest loss, or actual increases in forest cover. While it is not feasible to generalize from the numerous studies of comparatively small areas and to provide a figure for countrywide change, the inference is nevertheless highly significant. Gautam and Watanabe speculate that the early determinations of massive deforestation may have been influenced by the widespread acceptance of the Theory of Himalayan Environmental Degradation. Similarly, some of the initial over-estimations of loss may have served to reinforce the Theory. In any event, future attempts to estimate rates of land use/cover change should entail reinterpretation of the 1978 air photographs (LRMP) and a much greater degree of professionalism in the interpretation exercise coupled with field verification.

Eastern Himalaya: Bhutan and northeastern India

Eastward from the Nepalese border lies a region of the Himalaya that exhibits great variability in the degree of human impact on the mountain forests. The natural vegetation of Sikkim has been extensively modified (Karan 1977; Karan and Iijima 1985) and, along with the adjacent Darjeeling region, has undergone the development of extensive cash cropping, including the well-known tea plantations and a great variety of orchards, cardamom, and other marketable crops. The Sikkim–Darjeeling area has been subjected to extensive landsliding during and following torrential monsoon downpours. The 1968 event has already been mentioned, when an estimated 20,000 landslides occurred in a single day (Ives 1970). Starkel (1972a, 1972b) undertook a detailed field investigation of this event and concluded that deforestation had been an important factor, as had the myriad networks of roads (see also Chapter 6).

Reports from Bhutan, however, provide a very different picture despite the very close proximity to Darjeeling and Sikkim. Bhutan, apparently, has experienced insignificant loss of forest cover and the most recent reports (Blaikie and Sadeque 2000: 72) indicate that this small Himalayan kingdom has maintained a 72.5 per cent cover of high quality forest. This assessment may be too optimistic and Hamilton (pers. comm. 22 June 2003) prefers a figure of 60 per cent, not all of which should be classified as in good condition. Regardless, Bhutan is an exception within the Himalayan region; furthermore, the government has been very careful to severely restrict outside influence. On the one hand, this has provided protection against some of the negative aspects of tourism, such as experienced by Nepal; on the other hand it has prevented access to independent and potentially critical scholars (see Chapters 7 and 8). Nevertheless, a recent report by Roder *et al.* (2002), while not directly related to changes in Bhutan's forest cover, is introduced as an example of the conflicts between vested interests in the forests (based on opinion) and scientific fact (or lack of fact). The dispute concerns the supposed environmental impacts of grazing by domestic stock within the mountain forests. The authors assert that foresters, who appear to have the support of government authorities, agitate against grazing. They claim that throughout the Himalaya '[f]orest grazing is frequently blamed for slow regeneration, poor forest conditions and, in extreme cases, for causing ecological disasters' (Roder *et al.* 2002: 368). Although livestock herders disagree, their pleas are seldom effective and '[a]lthough science is often evoked, the arguments on the issue are routinely made without any quantitative proof from either side'.

Since it is estimated that forest grazing contributes between 20 and 24 per cent of the total dry matter required for Bhutan's livestock, this is a serious issue. Furthermore, the transfer of nutrients from the forests to the agricultural land by grazing is vital to maintaining productivity. In the long term this form of nutrient transfer may also accentuate the existing phosphorous deficiency in the soils of the coniferous forests, thereby reducing their regenerative capacity. Roder *et al.* recommend initiation of rigorous research to ensure a dependable database for rational decision-making.

The findings and recommendations of Roder *et al.* (2002), however, prompted a cautionary response from Hamilton (pers. comm. 25 June 2003), although he refers to broad-leaf forests. He maintains that unrestricted grazing in the more humid deciduous forests of eastern Bhutan is a potential threat to future biodiversity. Excellent stands of *Quercus* spp and *Castanopsis* have mainly unpalatable bamboo and undesirable *Symplocus* as understory dominants. Following timber harvesting, either fencing or individual seedling or sapling protection is required to offset loss of biodiversity.

The Himalaya to the east of Bhutan presents a problem for accurate assessment of landscape change, largely because northeasternmost India has been subjected to restricted access since Independence. The area has been periodically destabilized by widespread unrest, tribal warfare, and militant independence movements, especially in Nagaland and Arunachal Pradesh. The situation is even less clear across the border into Myanmar. There have been reports of rapid extension of slash-and-burn agriculture in some areas contrasting with claims of little forest disturbance in others. The entire region remains under the cloud of frontier ambiguity between India and China. Although the dispute is currently dormant, it certainly has not been settled; China has refused to concede that Arunachal Pradesh is an integral part of India. This is a reflection on the manner in which the McMahon Line following the crest of the Himalaya, claimed by India as its established frontier, was drawn up without Chinese consent in the early twentieth century.

Yunnan and Sichuan, southwestern China

Ives and Messerli (1989: 53–7) provided some detail on the condition of the forests in the Hengduan Mountains (western Sichuan and northwestern Yunnan, China). This was discussed in response to the Chinese application of the Theory of Himalayan Environmental Degradation to southwestern China. When China first opened its hinterland to foreign scholars in 1980 they were met with a litany of alarms about post-1950 deforestation and damaging downstream effects. Initial impressions from the Hengduan Mountains indicated that such assumptions were far too extreme (Messerli and Ives 1984; Ives 1985). The extensive areas of the Yunnan Plateau that were conspicuously denuded were purported to be the result of irresponsible post-1950 forest clearance and much of the blame was laid at the feet of Mao Zedong, the Great Leap Forward, the Cultural Revolution, and the Gang-of-Four. It cannot be denied that the period 1950–1980 saw extensive forest depredation throughout China, including Yunnan and Sichuan. However, the initial field investigation indicated that deforestation in Yunnan has had a very long history; in contradistinction, many areas were shown to have had a better forest cover in 1982–5 than 60 years earlier (Ives and Messerli 1989: 53–9).

Extensive fieldwork was undertaken following the initial reconnaissance of 1982, principally in Lijiang County of northwestern Yunnan (see Chapter 7). One of the objectives was to determine the impacts of modernization on the position of the mountain minority peoples, especially the women, and how this in turn was affecting the forest environment of the villages (Swope 1995; Ives and He 1996; Swope *et al.* 1997; He and Yang 1998; Sicroff 1998, Sicroff and Ives 2001).

As in most of the analyses of environmental reports from specific areas of the broader Himalayan region, proponents of both extreme points of view are correct in some instances and in some specific areas. A high degree of uncertainty complicates any comprehensive stock-taking. This apparent dichotomy can be explained by examination of the situation in Lijiang County and surrounding areas of northwestern Yunnan Province in some detail.

Yunnan Province is second only to northeastern China as a source of the country's timber and other forest products. In a timber-hungry country that is modernizing so rapidly, it is hardly surprising that vast areas of forest have been exploited. Reforestation is also wide-spread. More graphically, extensive stretches of the Yunnan Plateau have been stripped of forest cover and denuded of topsoil, in many areas down to the bedrock (Figure 3.5). This was demonstrated during a traverse from Kunming to Lijiang via Dali in 1982 (Ives and Messerli 1989: 60, Figure 3.8). Yet it was concluded that this was the culmination of centuries, if not millennia, of exploitation. In 1985 extensive clear-cutting was occurring in the mountains north and west of Lijiang Town. This was perpetrated both by government commercial logging and by illegal logging by the villagers. A major road provided access from Lijiang Town northward to a Nature Preserve. Reputedly the Preserve was intended to ensure environmental protection for the spectacular Yulong Xue Shan mountain range, including the mixed forests on its more accessible eastern flank. In 1985 the road was frequently choked with trucks, weighed down by logs, at times streaming southward non-stop throughout night and day. There was also extensive illegal logging by minority peoples from a series of small villages situated along the lower slopes of the mountain range. These patterns of forest exploitation were observed throughout much of northwestern Yunnan.

The local people are usually well aware of the environmental and socio-economic losses they will sustain if the mountain forests are seriously depleted. For instance, the Secretary of the local Wen Hai village Communist Party, when shown a photograph taken by Dr Joseph F. Rock in 1925, quickly pointed out the vastly superior forest cover of the present compared with that of 60 years before. He believed that if he were able to submit to the local govern-ment a copy of the 1925 photograph with one taken in 1985, it might off-set the official proposal to build a road to his village. He was convinced that a road would open the village forests to plundering by people living further south.

A series of the Joseph Rock high-quality photographs taken in the 1920s and 1930s were replicated in 1985. This showed that in some areas the 1985 forest condition was much better than that prevailing 50–60 years earlier; in others the reverse was true; in still others the situation was little changed (Ives and Messerli 1989: 54–9; Ives and He 1996). Village elders explained that any interpretation of the forest dynamics must take into account their own efficient management, as well as the accidents of forest fire. A cycle of heavy exploitation in some areas, but with deliberate preservation of scattered mature trees as the future seed sources, was often followed by locally enforced restrictions on the use of forest and meadows and natural revegetation (Figure 3.6). Somewhat similar patterns of change had been observed during a 1982 traverse from Chengdu to Gongga Shan, western Sichuan (Messerli and Ives 1984). The initial research in southwestern China led to the conclusion that the extreme Chinese claims of massive post-1950 forest eradication should be discounted; that a complex situation existed; and that any conclusive appraisal of Yunnan's mountain forests must await more comprehensive investigation.

More intensive field research (1990–5) focused on four groups of villages and involved several hundred interviews of household heads. Questionnaires and open-ended interviews

Figure 3.5 Extensive areas of the Yunnan Plateau have been denuded of forest cover. Wide areas have lost all soil cover down to bedrock. This had been explained as a result of the vicissitudes of fluctuating Chinese Communist actions after 1950. However, village elders insisted that these conditions have persisted at least since their childhood and are probably the result of progressive deforestation through several centuries, April 1985.

were aimed at investigating the life-styles of the extremely poor Naxi and Yi mountain farmers, the status of women, and the pressure on local resources, especially the forests. An attempt was made to determine the local attitudes toward the development of tourism that had been initiated in 1990 by the central governments as the most effective approach to modernization (see Chapter 7).

Some of the more revealing results were obtained from Wen Hai, a Naxi minority village. Divided into lower and upper administrative units, it is situated on a scenic lake at an altitude of 3,110 m, with the summits of the Yulong Xue Shan as impressive backdrop. It lies above the practical upper limit of rice production, yet an insensitive Communist government, from 1950 to 1979/80, had insisted that rural taxes be paid in rice. Thus, this traditionally poor community had become totally impoverished. Their plight was alleviated in 1979/80 when Deng Xiaoping's Household Responsibility System allowed the villagers to choose their own crops and provided temporary relief from direct taxation.

Initial field inspection in 1991 quickly revealed that the healthy condition of the local forests that had been recorded photographically in 1985 had seriously deteriorated through exploitation in the short period of six years. The countrywide problem of forest maintenance has its underpinnings in a century of fluctuating government policy, imperial, national, and Communist. There have been periods of private ownership of forest plots, community/village

Figure 3.6 Extensive re-establishment of a forest of Yunnan Pine (*Pinus yunnanensis*) north of Lijiang is part of the natural long-term cyclic landscape development. The mature trees in the foreground were preserved during the previous round of logging to provide a seed source for generation of seedlings. The eastern face of the Yulong Xue Shan rises above the trees, October 1994.

ownership, commune/collective ownership, and direct central government control, with many intermediate variations. Moreover, some of the changes in forest regulation have been both sudden and short-lived. Thus, there have been periods when the natural response of the villagers was to cut down as many trees as possible before what they considered as their property was taken away from them. In the early 1990s responsibility for forest maintenance devolved upon the villages. In Wen Hai this led to establishment by each administrative unit of a system of household 'self-kept' plots surrounding collective forests (village commons). It had been expected that this disposition would protect the 'collective' centre from illegal cutting. In 1991 it was apparent that the upper village forests were in much better condition than those of the lower village. This was explained by a retired village official: the upper village, perceiving the failure of the original arrangement, abolished the 'self-kept' plots and reverted to a simple collective arrangement. The lower village had persisted with the system that lent itself to abuse. The success of the upper village was widely credited to wise leadership and the demonstration of unselfish example (Swope 1995; Swope *et al.* 1997).

Deterioration in the condition of the forests throughout the Lijiang region continued through the 1990s for a variety of reasons. The small forest service staff was inadequate to enforce forest policy, close family relationships between forest service staff and village residents led to unevenness in application of regulations, remote locations and difficult terrain provided opportunity for continued illegal activities, and so on. However, in certain areas, notably the Yulong Xue Shan Forest Preserve where the growth of tourism was seen to depend upon an attractive environment, stringent application of forest laws certainly had a noticeable effect. But the most serious flaw in the local forest policy was that the villages were responsible for maintenance while administrative control remained in the hands of the central government that had long lost the trust of the villagers. Nevertheless, the 1985 pattern of highly variable forest condition from one location to another persisted until at least 1995.

The photograph (Figure 3.7), taken in 1985, shows the impacts on forests close to another village on the Jinsha Jiang (river) resulting from house construction that had been initiated by the more lenient government policies of 1979/80. Forest clearance, the development of logging skid trails, and their enlargement into rain-fed gullies are apparent. The repeat photograph, taken ten years later (1995) from the same position, shows a remarkable degree of recovery (Figure 3.8).

Eventually, in 1999 the central government placed a ban on all logging in the upper watersheds of the Yangtze (Jinsha Jiang) and Huang He rivers. The reason was the presumed massive deforestation in the mountains that was causing serious downstream flooding and siltation. This was prompted by the recent catastrophic flood events in the lower watersheds and the growing disputes over the justification of the Three Gorges development project. Thus the central government had clearly accepted the Theory as the basis for its draconian response that threatens the well-being of predominantly mountain minority peoples. The government claim that the rapid acceleration in development of tourism throughout Yunnan and other regions of western China would provide alternative sources of livelihood for poor peasants, hitherto dependent on forest resources, is discussed in Chapters 7 and 8. Here the intent is to show that the forest condition in northwestern Yunnan may be deteriorating in some areas, although it is stable or improving in other areas. The situation is highly complex, heavily influenced by rapidly changing and inconsistent government policy and lack of uniform administration of regulations. Above all, the simplistic assumption of massive deforestation in Yunnan and Sichuan since about 1950 is not supportable.

Nevertheless, the most recent policy change is the implementation of the regional logging ban. The impacts of this on the mountain forests will require further assessment in the future. However, early impressions are that this has served to redistribute rather than eliminate forest exploitation by placing the village collective forests under additional pressure (Zackey, pers. comm. October 2002). An additional development is the central government's 2002 application to UNESCO for designation of a vast area of northwestern Yunnan as a World Heritage Site. Termed the Three Parallel Rivers National Park, the recommendations deriving from this process were discussed with the relevant Chinese authorities in June 2003. The Chinese agreed to extend the park's eastern boundary to include the Tiger Leaping Gorge and Yulong Xue Shan and formal designation was approved the following month (Jim Thorsell, senior advisor to the World Heritage Commission, pers. comm. 11 August 2003). This designation as a World Heritage Site will certainly have a considerable impact on the entire province of Yunnan. The logging ban and the World Heritage designation will be discussed in greater detail in Chapters 7 and 8.

A recent publication on the impacts of assumed deforestation in the upper Yangtze watershed presents a somewhat conflicting view (Lu *et al.* 2003). They undertook detailed analysis of the monthly and extreme daily water discharge and sediment transport for 12 stations covering the entire upper Yangtze watershed. These authors write that there are firm grounds for a causal relationship between extensive deforestation and changes in the patterns of water discharge and sediment transport over the long period of record (1957–87). This conclusion is qualified by the difficulty of discriminating between a variety

Figure 3.7 Lijiang County, Yunnan Province. The lower mountain slopes above the small village have been almost denuded of forest cover and monsoon rains are beginning to enlarge the logging skid trails into the village. This local deforestation began after 1979 when Deng Xiaoping's relaxation of central government control permitted the construction of new houses. The view is toward the west across the Jinsha Jiang, a few kilometres above the Tiger Leaping Gorge, May 1985.

of possible causes, such as changing climate, reservoir construction, and water consumption. Furthermore, they provide no evidence upon which their assessment, that 'massive deforestation also occurred in these tributaries', is based (Lu *et al.* 2003: 63). Nor is there any reference to the much longer (centuries) history of deforestation that large parts of the watershed have experienced (Messerli and Ives 1984; Ives 1985; Ives and He 1997; Swope *et al.* 1997).

Mountains of northern Thailand

Environmental and cultural studies in the mountains of northern Thailand are discussed here because they demonstrate a pattern similar to those described in the Nepal Middle Mountains, Khumbu, and northwestern Yunnan. UNU's project on 'Highland-Lowland Interactive Systems' in 1978 resulted in formal collaboration with Chiang Mai University, facilitating access to its field station at Huai Thung Choa and the surrounding region in the mountains northwest of Chiang Mai. This provided an introduction to the villages of ethnic minority peoples and the problems they were facing (Figure 3.9). Thai authorities perceived these problems to be the result of the rapid population growth of local minority peoples, that is, the Thai version of the Theory – that mountain minority population growth, swidden

Figure 3.8 Telephoto view onto the same village shown in Figure 3.7 taken ten years later. After the 1979 construction phase the lower mountain slope that had been so severely denuded of forest cover has shown impressive natural reforestation. While still visible, the logging skid trails that were being converted into active gullies ten years previously were also beginning to heal. This example of environmental recovery serves to guard against the popular tendency to assume that deforestation is not only widespread but irreversible, November 1995.

agriculture, reduced forest fallow, and deforestation (in this case with opium production) were leading to downstream flooding and siltation. Interdisciplinary study that followed resulted in a series of publications to be added to those based on the research of several other groups (Ives *et al.* 1980; McKinnon 1983; Chapman and Sabhasri 1983; Hurni and Nuntapong 1983; Ives 1983).

The various research projects ranged from sociological village level studies of different ethnic minority peoples, such as Karen, Lisu, and Hmong, to an intensive investigation of soil erosion on different traditional and potential cash crop plots (see Figure 4.2; Hurni and Nuntapong 1983). From physical and biological investigations, it was demonstrated that introduction of a variety of cash and subsistence crops to replace the opium poppy (Figure 3.10) and conversion from swidden to intensive terrace agriculture were feasible. However, socio-economic obstacles, especially the heavy requirement of labour for terrace construction and the massive disruption of traditional culture that this would entail, precluded this approach. The study indicated first hand the problem of imposing 'scientific' solutions on traditional cultures. It also illustrated the willingness of a mainstream dominant culture, the

Figure 3.9 A Lisu village of thatch and bamboo, Chiang Mai Province, northern Thailand, cut into a forest clearing and surrounded by many small swiddens. In 1978 the houses were traditional in construction and function. By 1983 most of the houses had been replaced by 'improved' structures of cinder blocks and tin roofs – hot and damp in summer and cold and damp in winter, and decidedly not aesthetic, April 1978.

Thai, to use ethnic minorities as convenient scapegoats; thus the following statement was published with confidence:

> (1) existing Thai development policies may not be adequately based upon the behavioural patterns and perceptions of the local people; (2) the perceptions of the highlanders by the Thai and outsiders are not necessarily accurate, and in particular the ethnic northern Thai are responsible for much more deforestation than are the highlanders; (3) the role of *Imperata* grassland is not perceived clearly and the problems are neither technical nor scientific, but social and economic; (4) agro-forestry, if appropriately applied to existing subsistence systems, has considerable potential for resolving the traditional conflict between forester and farmer. The overwhelming conclusion, however, is that the problems of the hill country of northern Thailand, and of many similar areas, are neither scientific nor technical. Rather they are *people problems*. Throughout the developing mountain world solutions imposed by outsiders will frequently fail; success depends upon the degree to which the local people are enabled to take the initiative.

(Ives 1983: 311)

Figure 3.10 Early attempts to induce the hill people of northern Thailand to convert from opium production to profitable cash crops. In this large-scale experimental nursery and commercial production site (Huai Thung Choa, Chiang Mai Province), both cut flowers and seeds for garden plants are cultivated. The slope beyond has been largely deforested and invaded by *Imperata cylindrica* and the last remaining opium poppy swidden in the immediate area can be seen at the top right. Dependency on refrigerated truck transport for the cut flowers proved an obstacle and there was competition with opium as a light-weight, high-value cash crop, October 1979.

There was no doubt that progressive conversion from secondary forest to *Imperata* grassland was occurring in some areas. This was most likely the result of severe reduction in the length of the periods of forest fallow following cropping due to increased pressure on available land from accelerating population growth. However, enforced reforestation by the Thai Forestry Service using fast-growing pines was further reducing available land and inciting the local people to deliberately destroy plantations by setting fires. Nevertheless, significant sheet erosion and rill wash were also apparent on several very steep maize–opium swiddens of the Hmong and Lisu. Here slopes exceeded 25 degrees and minimal soil conservation measures were in effect. To Western eyes a massive amount of erosion had occurred during the previous monsoon season as indicated by up to 8 cm of soil being washed away from the root stocks of the maize plants (Figure 3.11). Hillsides that had been invaded by *Imperata* grass swards, on the other hand, appeared highly resistant to soil loss (Figure 3.12).

Subsequent research in the same general area by Savage (1994); Forsyth (1994, 1996, 1998); Ganjanapan (1996); Rerkasem (1996); Roder (1997); Renaud *et al.* (1998); Schmidt-Vogt (1998), and others, has provided a much more comprehensive understanding of the situation in northern Thailand, with implications for Laos and Vietnam.

Figure 3.11 Close-up view of the root system of a maize plant in a Lisu swidden, Huai Thung Choa, northern Thailand. The swidden was being used by Lisu farmers for a summer monsoon maize–winter poppy rotation. The steep, unprotected slopes and light soils were highly susceptible to erosion. More than 6 cm of soil was lost from this swidden during the preceding summer monsoon rains, October 1979.

Schmidt-Vogt (1998) carried out a combination of ecological and cultural studies in three small villages in northern Chiang Rai district (1990–1992). Two villages, Ban Tun at 1,100 m, and Ban Huai Sai at 950 m, are occupied by Lawa and Karen peoples respectively. They practise forms of secondary swidden agriculture that involves short cultivation periods followed by long forest fallow. The third is the Akha village of Ban Aze at 980 m; the Akha traditionally practised primary forest swiddening which involves a more intensive and longer period of cultivation and a very long fallow. Schmidt-Vogt's intent was to test the orthodox view that swidden agriculture in general leads to environmental degradation, loss of forest cover, soil erosion, and downstream siltation and flooding. He argues that such an assumption is a political construct that frequently results in severe authoritarian bias against forest users, especially swiddeners. Similarly, he insists, quoting Kunstadter *et al.* (1978) and others before them, that there are many forms of swidden and they are often related to the ethnicity of the practitioners. He postulates that secondary forest growth is not a simple or uniform phenomenon and it produces a wide variety of vegetation types dependent upon the form of swiddening. It follows that many secondary forest formations cannot be classed as degraded forests without thorough investigation. His survey involved vegetation studies of all successional stages in each of the three village areas. For the later successional stages

Figure 3.12 Chiang Mai Province, northern Thailand. Population growth and extensive reforestation with pine plantations by the Thai National Forest Service reduces the length of swidden forest fallow and encourages invasion of *Imperata cylindrica* grassland. Although a much less productive form of land use (thatch, farm implements, and limited grazing for domestic animals), the coarse grassland is resistant to soil erosion, contrary to assumption of the Thai authorities, October 1979.

he recorded species composition and stand structure (height, diameter, and crown cover of trees). These studies were backed by interviews in Thai, or assisted by an interpreter, both in the field and in the villages. He was able to demonstrate that 10–15-year-old secondary forests contained as many, or more, species than did the nearby primary forests or mature secondary growth forests, except for the Ban Aze swiddens. This is likely a reflection of the different form of swidden and the fact that, in contrast to the Lawa and Karen, the Akha do not leave 'relict emergents' as seed sources during and following burning for new fields. The stand structures for all three villages were dense, although those of Ban Aze were of shorter stature. This led Schmidt-Vogt to endorse the earlier conclusions of Zinke *et al.* (1978) and Kunstadter (1978) that certain forms of swiddening produced a great diversity of useful species as part of the secondary regrowth during fallow. He also concludes that the contribution of swiddening to denudation and downslope sedimentation has been greatly over-stated. He believes that the perception of the causes of increasing water shortage in the lowlands is misplaced and is probably the result of increased use of water *in the lowlands.* His observations indicate that transport of sediment derives largely from the erosion of the river banks and not from the swidden fields.

The results of Schmidt-Vogt (1998) are in agreement with those of Forsyth (1994, 1996, 1998). Forsyth noted that in his field area, also in Chiang Rai district, much of the downstream sediment transfer originates in gullies, typical of hilly country underlain by granites, which had developed before the introduction of agriculture. In addition, the recent extension of a relatively dense road network throughout the mountains of northern Thailand has provided a major source of sediment. Nevertheless, soil fertility continues to decline, although 137-Caesium determinations of soil losses (Forsyth 1994) indicate that they remain within acceptable levels.

Forsyth conducted a land classification of approximately 12 km^2 around the village of Pha Dua. This is a Yao minority settlement that was founded in 1947 by ten households numbering about 110 people. By 1995 these numbers had risen to 118 households and about 900 people. The Yao are considered primary shifting cultivators (corresponding to Schmidt-Vogt's 'primary forest swiddeners'), originally from Laos and central China. After establishing their new village in 1947 they became permanent settlers and so, by the early 1990s, were beginning to feel the effects of land shortage and declining soil fertility.

Forsyth produced a slope map, dividing the area into four slope categories overlain with an index of 'predicted erosion'. He had access to air photographs taken in 1954, 1969, 1977, 1983, and 1987 and was able to obtain information on the historical frequency of cropping from interviews with the villagers. He established four categories of predicted erosion by multiplying ranked slope-steepness category with ranked historic frequency of land use. They ranged from 'least eroded' to 'most eroded'. Only 2 per cent of his research area fell into the 'most eroded' category compared to 26 per cent as 'least eroded'. This demonstrated that the farmers were avoiding the steeper slopes (which accounted for most of the area designated as 'most eroded') in favour of more intensive use of the flatter land. He asserts that the Yao were well aware that the steeper slopes were susceptible to soil erosion and this was why they avoided them. However, as the village population expanded the flatter slopes were subjected to increasingly more intensive use and crop yields progressively declined. From this it becomes apparent that land shortage and nutrient losses constitute the primary problem of the Yao, in contradiction to the more common view that population growth is leading to cultivation on steeper slopes, soil erosion, and the downstream damage that is assumed to derive from this.

In general, Forsyth concludes that lowland assumptions concerning the environmental impacts of upland farming due to rapid population growth may 'reflect lowland fears about resource degradation, and long-term political conflicts between upland and lowland communities' (Forsyth 1996: 388). He characterizes his research as a hybrid study seeking to combine local indigenous knowledge with standard scientific techniques. This facilitated a test of, and additional challenge to, the Theory of Himalayan Environmental Degradation as applied to northern Thailand. His results 'indicated the adaptability of local communities and the folly of some western assumptions' (Forsyth 1996: 389).

Renaud *et al.* (1998) note that many government and non-government agencies continue with little success to introduce soil conservation techniques to areas undergoing perceived environmental degradation. They argue that this lack of success results from persistence in introducing 'technical packages' to farmers who reject them for one or more of a number of reasons: agronomic inadequacy; poor financial returns; lack of capital; labour constraints; lack of land tenure security; and poor interactions with extension services. The authors criticize many of the agencies for persisting with their technical approaches despite decades of failure, or very limited success. A large part of the problem is the apparent inability or unwillingness of the central government to adopt a socially positive attitude toward the hill people.

Nevertheless, Renaud *et al.* (1998) accept the claim that extensive deforestation and forest degradation has occurred in northern Thailand over the last 50 years.

Ganjanapan (1996, 1998) comes to a similar conclusion, stating that '. . . the natural environment, especially forests . . . has been depleted at an alarming rate' between 1985 and 1995 (Ganjanapan 1998: 72). He then launches an attack on what, from the details he provides, can only be described as a perfidious governmental attitude to both its ethnic minorities and its approach to environmental conservation. He indicates that 'the state of the environment is not an end in itself but a discourse in the struggle for the control of environmental resources' (Ganjanapan 1998: 72). He criticizes the government's increasingly stringent measures to control the country's forests. In 1989 legislation was introduced to evict all forest dwellers who would not practise 'permanent and sustainable' agriculture, permitting the use of military force to relocate villagers from newly established conservation forests. In 1992 the total area of conservation forests was increased to embrace 25 per cent of the national territory. This was further expanded to 27.5 per cent the following year – the first time land declared as conservation forest exceeded the area actually occupied by forests. The establishment of new national parks and conservation forests with extensive reforestation was accelerated. Ganjanapan protests that, while these policies were imposed on the ethnic minorities with increasing severity (forced evictions, fines, and jail terms), business interests were favoured. Community forests, including sacred groves, that had long been protected and well managed by the villagers were subjected to forced evacuation and then turned over to commercial interests for logging. There is no legal recognition of ethnic rights over forest land despite generations of use. Ganjanapan maintains that this form of repression will never succeed in preserving the remaining forests of the northern Thai mountains. It will serve only to accentuate local conflict and place added pressure on those community forests that are still being managed by the villagers.

There remains the question of the downstream impacts of this actual and perceived land-cover change in the mountains of northern Thailand. Alford (1992a), working independently, analysed a large pre-existing hydrological database. He was able to demonstrate that there was no evidence to support the claims of significant downstream impacts as a result of land-use and/or land-cover changes in the mountains. Alford's work will be discussed in more detail in Chapters 4 and 5.

India: Central and Western Himalaya

Westward along the Himalaya from Nepal the climate becomes progressively drier as the impact of the summer monsoon is reduced. Nevertheless, the Indian Central Himalaya, especially Kumaun and Garhwal, can be compared with Nepal. The conflicts over access to forest resources have been especially virulent in Kumaun and Garhwal and have been partially responsible for the breakaway of the mountain segment of Uttar Pradesh and the formation of the new state of Uttaranchal in 1999 (Mawdsley 1999). It is here that the internationally renowned Chipko Movement was born and proved highly influential in persuading Indira Gandhi's government to declare a ban on the cutting of green timber.

Since the Mohonk Conference of 1986 a great deal of environmental research has been undertaken in the Kumaun and Garhwal Himalaya, and also in Himachal Pradesh, immediately to the west. An overview of the relevant aspects is presented here. Much of it is the work of the G.B. Pant Institute for Himalayan Environment and Development and the Central Himalayan Association for Mountain Research and Development. However, other hill universities, government institutes, private individuals and international research groups have also participated (see especially Berkes *et al.* 1998; Gardner 2003). As will be seen, there are many conflicting views, although some of the disagreements are probably due to the different research methods employed and the wide environmental contrasts between the areas under study.

In general there is widespread support for the major conclusions emanating from the Mohonk Conference. However, the more recent work indicates the need for some modifications to several of the conclusions drawn by Ives and Messerli (1989).

Many researchers in the Central Indian Himalaya emphasize the negative impacts of the extensive road construction that followed India's 1962 border war with China although this does not deny the equally important positive impacts, in terms of enhanced social facilities and improved communications for the mountain peoples (Rawat and Sharma 1997). In particular, attention has focused on the direct connection between road construction, much of it inferior, and the increase in landsliding. While this was discussed by Ives and Messerli (1989: 119–21), a much fuller assessment is now available. This also has important repercussions on the discussion of sediment production and its downstream transfer, and the topic will be taken up again in Chapter 4. Suffice to say here that the dramatic increase in accessibility afforded by the 10,000 km of new roads (Tejwani 1987), and many more since 1987, has affected most aspects of mountain life. Despite the government's ban on logging, illegal trafficking in timber has been facilitated. Out-migration of mountain people, whether daily, seasonal, or permanent, has accelerated. Tourism has developed apace. And a pervasive shift from subsistence farming to cash cropping and off-farm wage-earning has occurred. All of these developments have had profound effects on the use of the natural resources and especially on the forests.

Haigh (1982a, 1984) provided some early accounts of the impact of the road construction on landslide incidence. Bartarya and Valdiya (1989) and Valdiya and Bartarya (1991) produced a detailed assessment of landscape change in the Gaula watershed of the Kumaun Lesser Himalaya in the vicinity of Nainital. They mapped 550 active landslides ascertaining their causes: slope angle, bedrock structure, extent and type of vegetation, and road construction. They affirm that severe deforestation has occurred, with forest cover being reduced from 69.6 per cent to 56.8 per cent of total area over a period of 22 years. They show that this is one factor in the increasing incidence of landsliding and soil erosion and the extensive drying up of springs and reduced stream flow during the dry season. The geomorphic implications will be taken up in Chapter 4. Nevertheless, Valdiya and Barcharya emphasize the difficulty of determining the relative importance of the various anthropogenic causative factors and they show a very significant relationship between landslide incidence and bedrock type and structure, with a preponderance of landslides occurring on areas underlain by the major zone of tectonic instability, with faulted and crushed rocks. The research results are further complicated by the reduced receipt of precipitation during the ten-year course of their studies.

Mukerji (1993) examined the changing environmental condition of a 30-km section of the Siwalik Hills close to Chandigarh. From a study of Mughal documents he shows that the hills were densely forested during the Muslim period (sixteenth to eighteenth centuries). Even in the early part of the nineteenth century travellers reported extensive forest cover in the upper valleys. It appears that the political stability brought on by British administration after the middle of the nineteenth century led to rapid development, especially a strong demand for timber in the lowlands to the south. This resulted in progressive deforestation exacerbated by over-grazing so that by the early 1960s practically all vegetation had been removed from the upper slopes and ridges and only thorn shrubs and grasses remained at lower levels and along stream courses. Since then a major undertaking, funded by the World

Bank and state agencies, has brought about a significant reversal. Water harvesting with construction of small reservoirs has been the core of the programme. The area under cultivation by 1985 had increased to 16–18 per cent of total area and over 5,000 ha had been reforested. Mukerji described this process as 'the greening of the Siwalik Himalaya'.

Sharma and Minhas (1993) describe severe pressure on forests in Kinnaur, Himachal Pradesh, leading to accelerated soil erosion. However, they also provide historical data to demonstrate that the forest losses occurred over a long period. They refer to Forest Department records and cite a Mr Barnolis, British colonial officer, who reported that, between 1859 and 1863, 30,000 mature deodars (*Cedrus deodara*) were cut and that, between 1882 and 1892 a further 100,000 trees, principally deodars and blue pine (*Pinus wallichiana*), were felled so that the lower slopes were 'almost denuded'. They give today's forest cover as 7 per cent of the total watershed area, about half of which is classified as 'protected forest'. The existing (1990) moratorium on tree felling has helped to curtail further forest loss. Nevertheless, since it did not prohibit forest access by the local people for pasture, fodder, fuel, and timber, including crate making to package local horticultural products, some deterioration has continued.

Rawat and Rawat (1994a, 1994b) also demonstrate extensive loss of forest cover in the Central Indian Himalaya and infer accelerated erosion and modification of the hydrological regime. M.R. Scott (pers. comm, 24 April 2003) has undertaken an assessment of the erodibility of a sector of the Siwalik Hills. Attention is also drawn to the extensive loss of forest cover well before 1950. The resulting sediment production and downstream transfer is considered further in Chapter 4.

In contrast to several of the previous reports, Berkes *et al.* (1998) and Duffield *et al.* (1998) refer to the stability in the basic distributions of land cover types over the last hundred years in the Upper Beas River watershed, Kulu, Himachal Pradesh. Even the dramatic increase in orchards and horticulture, for which this area has become famous, has not significantly encroached upon the forest lands. Nevertheless, villager interviews revealed concern over the decline of the most desirable tree species in terms of timber and fuel supplies. The authors credit the comparative environmental balance of the Kulu Valley to the extremely favourable conditions for the local people that were incorporated as part of the 1865 Forest Law, together with subsequent community management of the extensive common lands. Nevertheless, burgeoning tourism, increasing avalanche and landslide activity, and extensive use of pesticides imply a progressive shift toward environmental instability in the future (see also Gardner 2003; Chapters 6, 7, and 8).

Studies by Hoon (1996) and Chakravarty-Kaul (1998) have demonstrated the occurrence of significant environmental problems at high altitude in the Kumaun Himalaya and in Himachal Pradesh respectively. Hoon's field area is the northeastern corner of the Kumaun Himalaya close to the Tibetan and Nepalese frontiers. She made a comprehensive study of the life-cycle of Bhotiya people that includes both transhumance and nomadic forms of herding. She indicates that restricted access to their traditional winter grazing areas lies at the core of a series of problems that now threatens their way of life. The transhumance herders for generations have practised winter grazing at middle altitudes between 500 and 2,000 metres in the forest lands in the vicinity of Almora and Pithoragarh, while the nomads have used the lower altitudes of the Terai plains. Each group takes its flocks slowly up to summer pastures; the transhumants have cultivated summer crops close to their high-altitude villages, both for direct use and for trade across the international border into Tibet.

In both Hoon's and Chakravarty-Kaul's study areas the access to winter forest grazing is being curtailed. In part this is due to the expansion of cultivated areas into the forests by the settled agriculturists and the development of modern infrastructure, such as reservoirs and roads. Furthermore, government action to protect the remaining forests is resulting in reforestation with chir pine and fir (both inimical to grazing) and extension of protected forests. Efforts by the government to reverse the rapid disappearance of the forests and so check soil erosion has turned the Bhotiyas into 'invisible victims of government decisions both at the international and national levels' (Hoon 1996: 96–7). She maintains that the Indian government is not even aware of the constraints that have been imposed on the Bhotiya and they, in turn, are not equipped to respond. At the high altitude extreme of the herders' range, serious over-grazing is occurring. This is largely the result of the influx of the herds of the Gaddis who are being forced out of their traditional lands in Himachal Pradesh onto the Kumaun alpine meadows. Border closure with Tibet after 1962 also placed pressures on the Bhotiya by eliminating the important trading component of their seasonal life-cycle. The international agreement that led to border reopening in 1992 was beneficial but imposition of a tax of 40 per cent on all items that cross the border, added to the Chinese 10 per cent tax, represents a severe hardship. Hoon records the precipitous decline in the number of Bhotiyas who are able to continue their traditional way of life. She regrets the prospect that this environmentally sound form of mountain living is dying out as the Bhotiyas are forced into assimilation.

Chakravarty-Kaul (1998) investigated the recent pressures facing the traditional way of life of the transhumance Gaddis in Himachal Pradesh. Again, developments in the plains, including reservoirs, irrigated agriculture, urban expansion, and intensification of cash cropping have reduced access to winter pasture. At the same time, the herders' payment for winter grazing in the form of providing organic fertilizer during the process of grazing is being rendered obsolete as the permanent agriculturists apply more and more chemical fertilizers. Thus the early summer movement of the herds up through the forest belts must begin progressively earlier because of restricted winter pasture, yet movement onto the alpine pastures is controlled by the season. This enforced delay in the upward transfer of the herds adds to the grazing impacts on the intermediate forests. Finally, summer grazing is prolonged as long as possible which in turn is leading to over-grazing of the alpine meadows. Chakravarty-Kaul defines the problem in large part as a result of poorly thought-out government policy and actual prejudice against the transhumance herders. This, she explains, is leading to the break-down in the age-long synergism between herder and the permanently settled cultivators (Uhlig 1995) to the detriment of most members of both groups as well as to the alpine pastures, the winter grazing areas at low altitude, and the forests along the transfer routes. Additional information on the transhumance Bakrwals of Jammu and Kashmir is provided by Casimir and Rao (1985).

In terms of deforestation the case studies introduced by Hoon and Chakravarty-Kaul are somewhat tangential to the main line of the present discourse. Nevertheless, they indicate that deforestation is occurring and outline some of the forces that are driving it. They also show that government assumptions and lack of understanding of the ways of life of the herders are leading to policies that are unsound, both environmentally and culturally.

Northern Pakistan Himalaya and Karakorum

The Kaghan Valley is situated in the western Himalaya in Pakistan. Schickhoff (1995) has described it as transitional between the heavily monsoon-influenced Himalaya and the western mountains that receive their precipitation principally from winter westerly circulation at high altitude. It has been selected for the current discussion because of its transitional position and because of the significance of the detailed work carried out on the status of its forest cover (among other aspects) by Schickhoff (1993, 1995, 1998).

The valley extends approximately 100 km and drains into the Jhelum, a tributary of the Indus. Annual precipitation recorded at Balakot, 991 m, on the lower valley floor is 1,545 mm, 56 per cent of which falls between June and September. In 1990 forests occupied 16 per cent of the total area of the watershed; village and agricultural land occupied another 50 per cent, glaciers, alpine meadows, and high rock peaks the remainder. Schickhoff determined that at the beginning of the nineteenth century the valley was very thinly populated and that forest cover was about twice that of 1990. The valley had been taken under British control in 1847 and the accompanying political stability resulted in a rapid growth in population. This, in turn, produced a wave of deforestation over the next 20 years. The destruction of the forests was so rapid that, with the establishment of the Forest Department in 1864 and the passage of the Indian Forest Act the following year, stringent measures were put in place which slowed down this process. The local people were still allowed forest access for acquisition of timber and fuel for their own use and intensive pastoral activities at high altitude caused a progressive lowering of the timberline. Nevertheless, Schickhoff was able to conclude that the dense forests on the north-facing slopes, protected by Forest Department decree after 1873, have remained virtually intact to recent times. He estimates that there has been a forest cover loss of only 4 per cent in the temperate conifer stands. In essence, the landscape of the Kaghan Valley has changed little over the last 100 years despite considerable population growth. Between 1901 and 1981, for instance, the valley's population increased from 37,000 to 110,000. Forest preservation has been possible because of a progressive shift to cash cropping, off-farm wage earning, and use of chemical fertilizers that has off-set dependency on forest products.

Schickhoff goes on to consider other western Himalayan valleys. He quotes early nineteenth-century travellers who report extensive forests in the upper valleys, slow population growth until about 1850, and then rapid development following establishment of British rule and growing demands for timber in the lowlands. This in turn led to rapid exploitation of the forests. Tucker (1983) also describes the rapid extension of agriculture and trade fostered by the British and the general integration of the mountain regions into the market economy of the mountain foreland throughout much of the western Himalaya. After suppression of the Indian Mutiny (1857/58) the road and rail network expanded rapidly. The prized forest species was the *Cedrus deodara* and the seeming unending demand for railway sleepers had a massive impact on many of the forest areas. Even so, photographs taken in the Kaghan Valley at the turn of the twentieth century by Watson, a colonial officer, showed a similar landscape to that of today. Tucker (1983: 103) refers to the early 1940s as 'a second era of massive cutting of India's forests', the result of heavy demands during the Second World War.

In contrast to the Kaghan Valley, Schickhoff (1995) concludes that massive forest exploitation in the Siran Valley amounted to a reduction of 45 per cent of forest area between 1979 and 1988 alone. This occurred despite a similar history of forest management and was likely due to lower and more gentle topography and greater ease of access.

Nüsser and Clemens (1996) and Clemens and Nüsser (1997) provide detailed information on the Rupal Valley on the south flank of Nanga Parbat. They describe strongly degraded south-facing slopes that the German Himalayan Expedition recorded as having a 20–25 per cent cover of juniper forests in 1934. By 1990 there was almost no forest left. However, the north-facing slopes have retained a favourable cover of temperate coniferous forests. They believe that the Rupal Valley, despite rapid population increase, can still meet the local demand for timber and fuel although they estimate that the rate of forest use in the early 1990s was exceeding natural replacement. In particular, as the 'Line of Control' between India and Pakistan is approached, the large military units there are having a strong negative impact despite a logging ban. In general, they conclude that '[n]either the abundance of forest resources nor the property laws . . . have served to motivate the villagers of the Rupal area to evolve suitable strategies for forest management' (Clemens and Nüsser 1997: 258).

Proceeding further into Pakistan's Northern Areas, an extensive study of forest resources by Schickhoff (1997, 1998, 2004) is significant. He investigated forest stand structure and species composition of 62 representative test areas spread throughout the region. The exploitable forest stands are usually found between 2,500 and 3,000 m, mostly on north-facing slopes. Except for the private and commercially operated forests of Chilas and Darel/Tangir, all are 'Government Protected Forests'. The widespread destruction of these forests has been facilitated by the recent road construction together with extensive illegal logging, both for local use and for black market sale. Furthermore, contractors cutting in the private forests frequently exceed the legal yield and there appears to be a complete disregard for sustainable management. Schickhoff describes the situation as catastrophic and predicts a near-future loss of all accessible forest cover.

There are many supporting references to this assault on the forests of northern Pakistan (Grötzbach 1990; Kreutzmann, 1994, 1998). The under-staffing of the Forest Department is part of the reason, and corruption is rife. Allan (1987) also refers to the serious forest losses resulting from the Pakistan government's unwise settlement decisions when swamped by 3.5 million refugees from Afghanistan during the Soviet invasion. He cites Chitral, Bajaur, Dir, and Manshera as especially heavily impacted areas. Apparently deforestation has been exacerbated as Gujars and other Pakistani used the confusion caused by the millions of refugees as an opportunity for illegal logging. A somewhat similar situation had occurred following the Partition of 1947 in the context of resettlement of large numbers of Muslim refugees fleeing from India. Haigh (1991) also reports extensive deforestation in Swat and Dir.

Without doubt, loss of forest cover in northern Pakistan has been, and continues to be, catastrophic. However, it would be extremely difficult to predict a proportionate increase in soil erosion and downslope sediment transfer in a region of such pronounced aridity. The total area originally under forest cover can hardly match the area of naturally bare slopes that undergo extensive mass movement during the occasional rainstorms (Hewitt, 1993). However, this does not imply that the loss of forest resources is unimportant, both in terms of the local people's need for forest products as well as for their aesthetic value.

Tajikistan and the Pamir Mountains

Tajikistan is a predominantly mountain state. A large proportion of the mere 2.3 per cent of the country that is forested occurs in the east and north, between 1,000 and 3,000 m, especially on north-facing slopes; if the extensive orchards are included the total estimated coverage is increased to 4–5 per cent. The focus of the present discussion is the eastern half of the country – the Pamir Mountains. The lowlands of the southwest and the mountain valley floors are semi-arid. With increasing elevation the western slopes of the Pamir receive greater annual amounts of precipitation so that at mid-elevations rainfall totals approximate 1,500 mm. This is enough to support extensive open mountain forests. At the highest levels (up to 7,000 m) precipitation is mainly in the form of snow and nourishes a great accumulation of permanent snow, ice caps, and glaciers – some of the longest valley glaciers outside the Polar regions. This is the source of Tajikistan's most important natural resource – water. The Pamir are deeply dissected and the numerous small villages (*kishlaks*) scattered at intervals along the floors of the mountain valleys and gorges depend on glacier- and snow-melt for irrigation. Thus they are usually located on alluvial cones created by the glacier- and snow-melt streams while the lower valley slopes between the villages are either barren rock and scree or else sparsely vegetated. In this sense, they offer landscapes comparable to those of the Hindu Kush in Afghanistan and the Karakorum in northernmost Pakistan.

Little information is accessible on the extent of forest cover in Tajikistan. The best available indicates that total forest cover has been increasing by 2,000 hectares per year between 1990 and 2000 and that represents a 0.5 per cent annual rate (FAO: www.fao.org/forestry), although the civil war of the early 1990s has also taken a heavy toll. Recent publications (Badenkov 1992, 1998; Cunha 1994, 1995; Merzliakova 1998) all refer to the depopulation of the mountains that resulted from Soviet policy. The main period of forced resettlement was in the 1940s and 1950s when the Red Army moved many of the mountain Tajiks to the cotton fields of the southwest as a source of cheap labour, and burned their *kishlaks* in the process. Several episodes of forced migration have occurred, the most tragic following the major Khait earthquake of 1949 (8.0 plus on the Richter scale). This earthquake and the major rockfall/landslide that it caused eliminated more than half the *kishlaks* in the northern Surhob valley, tributary of the Vakhsh, and caused more than 28,000 fatalities (Yablakov 2001). Survivors were forcibly transferred to the cotton fields. Badenkov (1997) writes in general terms about 'large-scale deforestation, over-grazing, and cultivation on steep mountain slopes' that have led to accelerated erosion and landslides. Nevertheless, the long period of depopulation appeared to have left the Pamir forests as some of the best preserved in Central Asia (Badenkov 1992). With the introduction of Gorbachev's policy of *perestroika* in 1987 a return migration to the Pamir began and many of the ruined *kishlaks* were rebuilt. However, this brief period of repopulation was brought to a near standstill in the early 1990s as civil war erupted in the newly independent Republic of Tajikistan. The last decade, however, has witnessed renewed pressure on the small areas of forest and shrubs. This is due in large part to the cessation of massive Soviet subsidies in 1990 when Tajikistan became independent. A collapsing hydro-electric system and stoppage of coal deliveries to the mountain kishlaks has reduced the mountain people to dependency on forests and shrubs for fuel as their economy is reduced to a subsistence level (Breu and Hurni 2003).

Recent overflights across the Pamir and northeastern Hindu Kush and extensive land excursions by jeep suggest there is limited, yet important, forest cover, much of it open mountain woodland. Without more accurate survey it would appear that, whether or not deforestation has occurred, the area affected would be insignificant in terms of any landslide activity and sediment transfer. The large number of natural landslides, rockfalls, and debris flows, induced by heavy rainstorms and/or the frequent earthquakes, as well as the periodic outbreak of glacier lakes, would overwhelm in significance any anthropogenically induced erosion.

Afghanistan

It is tempting to extend the same line of reasoning to Afghanistan, as well as the northernmost reaches of Pakistan. Afghanistan is a special case. Lack of secure access can hardly disguise the severe depletion of forest resources that the warfare and unrest of the last several decades have brought about. During the most recent episode, the graphic TV coverage of aerial bombardment of Taliban and al-Qaeda forces frequently illustrated widespread destruction of mountain forests. Nevertheless, regardless of the forest damage that has undoubtedly occurred over the last 50 or 100 years, in terms of per cent cover of the country's total area, there has been miniscule impact of deforestation on erosion and sediment transfer downstream.

Summary

This chapter may well confuse even the careful reader. To a degree, that was one of the intentions of writing it. It should indicate that some areas of this vast and complex mountain region have been ruthlessly stripped of most forest cover while other areas still have well maintained forests, or have been reforested to various degrees. Yet other areas were denuded 100 or 200 years ago, or even in the more distant past. Some forests, undoubtedly, have been subjected to cycles of clearance and revegetation, either by natural regrowth or by plantation. Still other areas, that were reported to have been denuded since 1950, had experienced no significant forest clearance, the reports having been deliberately distorted. The more arid northwestern sector of the region under review had such a modest forest cover in proportion to total land area at the turn of the nineteenth century that neither good forest management nor total eradication of that which existed in the mid-twentieth century would have had a significant impact on accelerated erosion and downstream sediment transfer on a regional scale. This does not detract from the seriousness of the loss of forests *per se* in areas such as northern Pakistan.

Nevertheless, the question of forest quality, in the sense of over-use by subsistence farmers, such as excessive lopping for fodder and fuel and exposure of the forest floor by litter collection and over-grazing, has not been discussed. The deterioration in forest quality is widespread throughout the entire region. Although this is an extremely important aspect of environmental stress it has not been analysed systematically because of the general absence of reliable data (Dipak Gyawali pers. comm. 22 November 2003).

These general considerations strongly reaffirm the original rejection of the claims that widespread deforestation was occurring, one of the two central driving points of the Theory of Himalayan Environmental Degradation (Chapter 1, pp. 4–5). Its corollary is that the subsistence farmers cannot in any way be cited as the dominant cause of deforestation. Some farmers certainly have been responsible for forest removal as their numbers have increased throughout the second half of the twentieth century, more in some areas than in others. The basic cause, however, where deforestation has occurred, appears to have been a combination of poor government policy and management, marginalization of the mountain minorities, corruption, and commercial logging without replanting. In many areas, such as parts of Yunnan and the western Indian Himalaya, deforestation occurred so long prior to 1950 that they are not directly relevant to the present discourse, except in so far as to serve as contributions to the extremist position of environmental degradation. In other areas, for instance Sindhu Palchok and Kabhre Palanchok districts in the Nepal Middle Mountains, it is the local farmers who have initiated a process of tree planting on private land. In the Kulu Valley, for instance, very long-term local management of the commons has ensured the preservation of luxuriant forests to this day.

To this must be added a series of secondary considerations, many of which were introduced during the Mohonk Conference by Lawrence Hamilton (1987). Deforestation is not inherently *bad*. Much depends on the mode of tree removal and the land use that replaces the forests. The condition of the forest floor is more important than the crown density in terms of soil erosion. Large water drops from a high canopy that attain terminal velocity can detach soil particles more readily than rainfall from an open sky. It is worth repeating, if not fully accepting, Hamilton's plea for an embargo on the use of the term *deforestation*, which has been so seriously misused that it has become suffused with emotion and, therefore, meaningless without qualification.

In this chapter the emphasis placed on the wide variations in assessment of the rates and timing of changes in forest cover has omitted reference to the connection between the condition of the mountain forests and the sustainability of mountain agriculture into the future. This relates directly to the well-being of the poor rural farming communities and will be examined as part of Chapter 9. Nevertheless, the increasing risk to rural livelihoods, in terms of food insecurity and coping strategy responses, remains central to any discussion on Himalayan environmental deterioration and political stability (Bohle and Adhikari 1998; Chapters 8 and 9). The next chapter discusses the geomorphology of terraced and cultivated mountain slopes that account for much of the actual historic (long-term) conversion from the original forest cover.

4 Geomorphology of agricultural landscapes

> It would not be fair to blame the people of these valleys on the Himalayan fringe for the frequent
> landslides that occur here. In turning the steep slopes into fruitful fields they have neither been
> lazy nor neglectful.
>
> (Tilman 1952)

The long-standing debate over the relative importance of *natural* processes of erosion and those induced by subsistence agriculture in the Himalaya has been indecisive because of the lack of reliable, representative, and consistent (replicated) data. It has also been clouded by emotion, by the personalities of institutions and individuals, and a wide-ranging Western-based assumption that a combination of poverty, assumed ignorance and uncontrolled population growth must inexorably lead to soil erosion and serious downstream consequences. As with the *deforestation* debate (Chapter 3), the assumed acceleration of soil erosion and landsliding was a major focus of enquiry at the Mohonk Conference.

Thompson *et al.* (1986) provided a break-through when they identified extensive post-1950 forest depletion as a mountain myth. Nevertheless, the weight of precise field information used to question the myth came predominantly from the work of the Nepal–Australia Forestry Project (Mahat *et al.* 1986a, 1986b, 1987a, 1987b) and the theoretical contributions of Hamilton (1983, 1987). To this was added the results from the UNU teamwork (Johnson *et al.* 1982; Kienholz *et al.* 1983, 1984) and the overall criticism of reports that supported the notion of massive post-1950 forest destruction (Ives 1984; Ives and Ives 1987; Ives and Messerli 1989). However, even if the assumption of extensive deforestation since 1950 can be dismissed, it does not necessarily follow that accelerated soil erosion on steep terraced slopes also can be automatically discounted. The previous chapter has added substantially to rationalization of the *deforestation* debate as it stood in 1989. This chapter will attempt to do the same for the debate concerning *accelerated soil erosion and landsliding*. Both issues are fraught with uncertainties of representativeness and scale in place and time. By 1989 it had been demonstrated that a significant amount of sound geomorphological data and theory could be mustered to counteract the contemporary prevailing and unsupported assumptions. Nevertheless, it was felt that access to reliable and relevant data was still insufficient to provide an adequate basis for unequivocal contradiction.

Anyone walking through Nepal's Middle Mountains, Kumaun's Lesser Himalaya, or the Darjeeling Hills would see at a glance that a significant proportion of the land was under crops, regardless of whether the total area was expanding, contracting, or remaining stable. The great majority of the hill walkers (trekkers) would be afoot only in the pre-monsoon or post-monsoon seasons when the numerous landslides (debris flows) would stand out like the proverbial sore thumb. These casual observations, nevertheless, remained to distract from the actual situation and required a factually based response.

By the late 1980s the research by Starkel (1972a, 1972b), Brunsden *et al.* (1981), Ramsay (1985), Carson (1985), and others, and that of the UNU research team from Kakani and Khumbu was available. Also to hand were the impressions derived from widespread field observation. These many sources of new information, together with application of the established principles of geomorphology, provided an adequate base on which to initiate a challenge to the assumption that the presumed deforestation led to increased landsliding and soil erosion and accelerated runoff during the summer monsoons. Nevertheless, it was recognized that more exacting research would be needed to confirm that there was no direct relationship. Subsequent publication of a large body of progressively more sophisticated and more detailed geomorphological and soils research over the last 15 years has added significantly to the discussion. There are the extensive studies in the Jhikhu Khola watershed organized by Hans Schreier, by Kegang Wu and John Thornes in the Likhu Khola watershed, and the Royal Geographical Society teamwork in the same watershed led by Rita Gardner and John Gerrard. In the far western Himalaya and Karakorum important results have been obtained by several teams of researchers under the leadership, respectively, of Kenneth Hewitt, Jack Shroder, and Irmtraud Stellrecht. These represent the major team efforts, to which must be added the work of K.S. Valdiya and the Central Himalayan Research Association, the G.B. Pant Institute, and researchers from several other Indian hill universities. These, and many more that will be cited in relevant parts of the text, will form the basis of the present chapter.

Forms of agriculture in the Himalayan region

The Himalayan region has experienced a considerable range of traditional agricultural techniques that have evolved over generations of subsistence mountain living. Most forms constitute integral parts of mixed farming systems; others were synergistic with transhumance and nomadic herding. Although this chapter is not intended as a treatise on Himalayan farming systems, the main forms are introduced briefly as they are the very basis for the geomorphologic investigation of slope stability and instability.

Slash-and-burn, jhumming or swidden agriculture

This subtitle subsumes a great range of techniques. Two of the best sources for a detailed understanding of swidden agriculture are Kunstadter *et al.* (1978) and McKinnon (1983). Swiddening generally implies the periodic cutting and burning of small clearings, ideally in primary forest, and their planting as fields with a wide variety of edible, or otherwise useful, crops, possibly more than 20 different species in a single field (Figure 2.6). A field may be cropped from between two and ten years, depending upon ethnic group and method of swiddening practised, and then abandoned to fallow as other fields are opened up in the forest. Under traditional conditions, that is low human population in relation to available forested land, forest fallow would be maintained for 20 years or more. But as populations have increased land shortages have occurred forcing progressive reduction in the period of fallow. As the fallow period was shortened to less than about ten years, secondary reforestation was inhibited and *Imperata* grassland replaced the forest cover and led to loss of soil nutrients. Slash-and-burn agriculture (the very term lends itself to emotive and negative responses) has been a very widespread traditional form of subsistence agriculture. Until the 1940s and 1950s it was widespread from the Indian Central Himalaya, throughout Nepal, and eastward through Bhutan into Myanmar, Thailand, and Southeast Asia, in general, including the Chittagong Hill Tracts in Bangladesh.

Although there are almost as many forms of swiddening as there are ethnic groups, over the last 30–40 years there has been widespread abandonment of swiddening and conversion to sedentary and intensive agriculture, sometimes coerced by governmental authority. Swiddening has long been regarded by central authorities, aid agencies, and conservationists as environmental anathema (Figure 4.1). Only the swiddeners' impacts on downhill movement of soil need be discussed here, and this rather selectively. Forsyth (1994) has argued persuasively that the Yao, among whom he worked in northern Thailand, avoided steep slopes as land shortages increased; however, the Hmong and Lisu in Chiang Mai province certainly cut and burned and planted their crops on very steep slopes. A Lisu poppy field, 60 m in length and with a 30° gradient lost 4–8 cm of soil in a single season. This converts to 150–300 t.ha.yr. Such losses would be reduced to half by construction of the traditional oblique drainage ditches that would reduce slope length and drain off much of the excess rain water. Nevertheless, such a swidden, used for four or five years, would require an 80-year fallow to be sustainable, which indicates how such farming methods become untenable as population pressures continue to increase and fallow periods have to be reduced.

Figure 4.1 Gerardo Budowski walks across the still hot ground after a recent burning for a new Lisu swidden, Chiang Mai Province, northern Thailand. The mature trees provide a seed source for reforestation after the swidden has been cropped for 3–4 years, April 1978.

In the Huai Thung Choa area of northern Thailand, Hurni supervised the careful instrumentation of slopes under both traditional and experimental cash crops (Hurni and Nuntapong 1983). Five of the six study plots, 20 m in length with a 20° gradient, were planted with traditional upland rice/poppy, four with various agro-forestry combinations, and the sixth was left fallow. The local Lisu farmers tended the plots. Soil losses were 89, 22, 13, 10, 10, and 58 t.ha.yr. respectively (Figure 4.2). On this basis, Hurni (1982) concluded that, while these soil losses were significant, they appeared to lie within manageable limits in terms of prospective conversion to sedentary and intensified agriculture. This finding was based on the assumption that rapid weathering in a monsoonal tropical climate would keep pace in terms of production of new soil, given reasonable soil conservation practices. As mentioned in Chapter 3, the rates of soil loss were not the obstacle to agricultural conversion; the limitations were social and economic (Hurni and Nuntapong 1983). While some forms of swiddening persist throughout northern Thailand, Laos, Myanmar, and the eastern Himalaya, in general, slash-and-burn agriculture is fast disappearing as the population continues to grow and pressures on available land continue to rise. Coercion by central government authorities has also been an important factor in the suppression of swiddening, although it persists in the eastern Himalaya and in Southeast Asia, often in modified form.

Figure 4.2 The main soil erosion test plots constructed by the UNU–Chiang Mai University research project under the supervision of Hans Hurni, University of Berne, Switzerland (see text for explanation), July 1981.

Mixed terrace farming

It is generally agreed that extensive swiddening occurred across large areas of the Nepalese Middle Mountains during the first half of the twentieth century and this form of agriculture persists in many of the more marginal areas. However, Western visitors to Nepal over the last 40 years have eulogized the spectacular flights of terraces that can been seen widely across the mountain slopes below about 2,000 metres. Many are cut into extremely steep slopes, some in excess of 30°. The slopes also seem to lend themselves to excessive landsliding (clearly apparent to the naked eye on a casual basis during the pre- and post-monsoon seasons) and soil erosion, basically unseen, but assumed to be pervasive, if not destructive.

There are two types of terrace that have very different agricultural and geomorphic characteristics (Figure 4.3): irrigated terraces (*khet*) and rain-fed terraces (*bari*). The khet give the lower and middle slopes a distinctive form of architectural beauty (see Figure 1.3). They are cut horizontally to contain standing water and sweep along the contours and were originally developed primarily for rice (paddy) cultivation. Thus, at certain times of the year they are flooded; as the rice shoots grow and mature they present a range of startling shades of luxuriant greens. A second crop of rice may be grown, and in the dry season, wherever sufficient water is available, winter wheat is planted. Careful inspection reveals that the terraces are built to encompass extensive controlled drainage systems since any overflow from one terrace level must be trained through the next below, and successful paddy culture depends on almost continuous availability of water. Furthermore, any significant spill would risk breaching the low outer retaining wall (bund) to release a slump or landslide and divert essential irrigation water away from terraces lower in the system. Their use is labour-intensive and they are the most productive element of the entire farming system. If and when breaks occur, leading to slumping during a monsoon downpour, all available labour will be marshalled to limit and repair the damage. In the UNU Middle Mountain study area of Kakani, Nepal, khet terraces occupy valley floors and lower mountain slopes up to a general altitude of about 1,750–1,800 metres. In a few places they occur as high as 2,050 metres. The terrace risers may be as high as two metres and terrace width may be up to 100–200 metres, or much less, depending on the overall slope of the land.

Above the khet is located a very different form of terrace – the rain-fed *bari* (Figure 4.3 and Figure 4.4). These slope out from the mountain side and have no bund. Their risers are on average about two metres high, but may be as high as four metres or even more where the mountain sides are very steep. In the Kakani area these slopes are generally between 20° and 40°. Closer to the ridge crests the gradients are gentler (10–20°). The outward slope of the bari terraces may be as little as 5° but can exceed 30°. In contrast to the khet, they are engineered so that excess rain flows off them. They are often angled gently across the contour and have shallow ditches at the base of the risers so that some of the excess water is channelled laterally along the length of the terrace into mountain streams and gullies. At their lower level of about 1,750 metres they inter-finger with the khet, although small groups of bari can be found at much lower levels where the slopes are too steep to permit construction of khet. The bari terraces are usually used for growing millet, buckwheat, and maize and are likely to be fallow and brown in winter. They are less productive than the khet and are much more susceptible to erosion – sheet wash, rill erosion, slumping, and landsliding: hence the contrast in geomorphic characteristics between the bari and the khet. In the Kakani area, bari occupy 23 per cent and khet 26 per cent of the total land area. Other land-use types are pasture (10 per cent), shrubland (17 per cent), deciduous forest (21 per cent), coniferous forest (2 per cent), and other uses – roads, villages (3 per cent).[1]

Figure 4.3 Kakani, near Kathmandu, Nepal. The contrast between the rain-fed *bari* terraces (sloping outward with no bunds, and fallow dry for the winter season) and the irrigated *khet* terraces (horizontal and with bunds 20–30 cm high) is evident. The *khet* support a crop of winter wheat, dependent on irrigation from a small mountain stream, and are primarily used for summer paddy. The Trisuli Road cuts across the upper section of the photograph. This is a surfaced road that becomes an ephemeral steam bed during heavy monsoon rains. At sharp bends in the road flood waters spill over onto the otherwise carefully manipulated terraces and cause debris flows, centre right, April 1978.

Superimposed upon this physical differentiation, however, is a cultural distinction that has shaped the landscape from the beginnings of human occupation. Most of the khet land, at least in the Kakani area, is owned or controlled by Chhetri, Newar, Brahmin, and Balami (a Newar sub-group). The higher altitude bari are mainly farmed by Tamang people who are, therefore, much more marginalized in terms of living standards. Service castes are distributed throughout the Kakani area (Johnson *et al.* 1982; Gurung 1988).

On the slopes below the village of Kakani a maximum gradient for bari was measured as 38° on a mountain gradient of 52° (Kienholz *et al.* 1983), which indicates the marginal and geomorphologically precarious nature of Tamang farming. During the winter the bari terraces are trimmed back with special tools and the eroded risers are sharpened. This serves to repair damage from slumps, rill erosion, and the summer burrowing activities of rodents; it also eliminates weeds that would otherwise compete with the crops for nutrients. The earth removed from the risers is spread across the surface of the next lower terrace. In this manner and over a period of years an entire bari terrace system is gradually moved horizontally into the hillside as the landscape is moulded by human manipulation (Figure 4.4 and Figure 4.5).

Figure 4.4 The *bari* terraces are built to slope outward from the hillside so that excess rainwater can run off. They are susceptible to soil erosion, including landslides, and are repaired during the winter dry season and are fertilized organically. This view shows how the terrace risers are cut back and sharpened by hand tools, April 1978.

Similarly, the khet terraces are carefully groomed; their bunds are repaired and strengthened and organic manure is spread across their surfaces. Recently, especially where there is road access, as in the Kakani area, chemical fertilizers are being used.

Beginning in 1979 and extending over four years, the UNU team undertook the first research on slope dynamics in association with cultural geography and land use in Nepal's Middle Mountains. This significantly advanced our understanding of the stability/instability of these spectacular terraced slopes that, under the Theory of Himalayan Environmental Degradation, had been claimed to be the primary source of soil erosion, landsliding, and downstream flooding and siltation. The research team developed a firm sense of empathy with the subsistent farmers as their skills and environmental understanding came to be appreciated. It became apparent that they recognized both the dangers and advantages presented by landslides. For instance, the local people created their own landslides by diverting mountain streams to facilitate the cutting of new terraces in the soft run-out deposits that this type of action produced.

The much more detailed research in the Jhikhu Khola and Likhu Khola catchments in the early 1990s adds convincing weight to the claim that the fifth point in the eight-point scenario of the Theory is insupportable. This more recent work, therefore, will be examined after presentation of an overview of some of the results from research in the 1980s. The earlier work is revisited here because it provided the benchmark for so much of the subsequent research.

Figure 4.5 A young Kakani farmer with the traditional implement for *bari* terrace repair, April 1978.

Mountain slope geomorphology

As a basis for this discussion it is necessary to have a broad understanding of the rates of mountain building and regional denudation within the context of geologic time. With widespread acceptance of the theory of plate tectonics, the dynamics of continued mountain building was clarified. Several estimates of present-day rates of Himalayan uplift have been published. Zeitler *et al.* (1982) indicate a rate of uplift for the Greater Himalaya of about 1 mm/yr; Low (1968) estimates 1–4 mm/yr since the close of the Lower Pleistocene; Iwata *et al.* (1984) about 1 mm/yr for the Nepalese Himalaya; and Zeitler *et al.* (1982) about 9 mm/yr for the Nanga Parbat region (see also Shroder 1998). Geophysical work by the Chinese Academy of Sciences (Liu and Sun 1981) on the Tibetan Plateau and the northern slope of the Himalaya has estimated a rate of 4–5 mm/yr over the past 10,000 years that is continuing today. Measurements in the vicinity of Garm, Tajikistan, indicate that the Peter the First Range, on the western slope of the Pamir Mountains, is rising about 15 mm/yr. The crude estimates of regional denudation rates, to be introduced below, barely match those of the uplift estimates. Current uplift, therefore, equals or even exceeds denudation in some areas, implying that the Himalaya–Ganges system over the past 10,000 years and longer has continued to be an extremely dynamic section of the earth's crust.

The uplift of the Himalaya and the Tibetan Plateau, as the Indian plate thrusts beneath Central Asia, has caused enormous masses of eroded material to be deposited in the foredeep to the south which, over the past several million years, has produced the great plains of the Indus, Ganges, and Brahmaputra; drill holes have penetrated more than 5,000 m of alluvial sediments beneath the Ganges Plain (Sharma 1983). On a more recent time frame, the Sapta Kosi River has shifted its channel across its extensive alluvial fan, which forms much of Bihar State, through a distance of more than 100 km during the past 250 years. Thus, there is abundant evidence of massive erosion and regional denudation and equally massive sediment transfer and deposition that has occurred over the past million or more years. Present-day evidence and geophysical hypothesis would indicate that the height of the Himalaya is equal to, if not higher than, that of a million years ago, so that the relief energy between the crestline and the Ganges Plain has remained undiminished over recent geological as well as historical time. Thus, without very convincing evidence to the contrary, it would seem reasonable to argue that the contribution of human interventions over the past three or four decades, or even centuries, has been insignificant when balanced against these natural processes. Before this point is examined further it will be helpful to bear in mind a few broad concepts and to reiterate some of the dilemmas facing geomorphic and hydrological research.

1   The Himalaya–Brahmaputra–Ganges–Indus system is one of the world's most dynamic mountain-building and sediment-transfer systems, processes that have continued unabated over recent geological time and will likely continue into the future.
2   These processes – the endogenous, tectonic/ isostatic – and the exogenous – climatic/ weathering/ hydrological – have created an unstable landscape of the utmost complexity.
3   Given the massive scale of relief, from more than 5,000 m below sea level to nearly 9,000 m above, the enormous variations in climate, vegetation, and topography, and the variability of major geomorphic events in time and space, the present database is inadequate for determination of actual rates of activity of the various processes affecting the land surface at the local or micro-scale. Therein lies the difficulty for precise determination of the impacts of human intervention, including deforestation, land-use changes, and manipulation of water flow, and their differentiation from the natural processes as a proportion of the total rate of change.

Regardless of this obstacle to geomorphic evaluation, some important contributions can be made provided they are set in the context of recent geological time. It is also important to question the widespread tendency during the last half century to assume that problems of erosion affecting either the plains or the mountain slopes themselves are entirely, or largely, the result of human intervention, in terms of misuse of the Himalayan environment. This does not imply that loss of topsoil is not serious and limiting in terms of sustainable mountain agriculture.

Some critical definitions are needed. First, soil erosion, a widely used term, should be restricted to the secular loss of soil, especially the A-horizon, or what has been substituted by farm practices, in which most organic matter is concentrated. Soil erosion occurs in entirely natural environments as well as in areas transformed by human intervention; for the latter the term 'accelerated erosion' is preferable; it implies a combination of natural processes and human-induced processes. Soil erosion should be distinguished from mass movement, which is the down-slope movement of the mass of fractured and weathered bedrock, the weathering mantle, on which the topsoil forms as the end-product of that weathering process. Mass movement includes such almost imperceptible, continuously operating processes as soil creep, and at higher altitudes, frost creep and solifluction, by which the weathering mantle moves downhill under the influence of gravity at rates of a few millimetres per year. The rates of downhill mass movement are greatly influenced by soil moisture content and, hence, rainfall and snow-melt. Mass movement also includes more dramatic, intermittent processes such as landslides, mudflows, rock falls, and rockslides with short, long, or indeterminate recurrence intervals.

Soil erosion, whether natural or accelerated, and mass movement, in practice, grade into each other, but to facilitate a clearer understanding of the relationships between the types of terrace agriculture and natural slope processes, they should be retained as conceptually separate processes in landscape change. It should also be recognized that the weathering mantle and the topsoil are continually being formed as the bedrock is reduced by a combination of mechanical, chemical, and biochemical processes. In certain instances in nature the topsoil and weathering mantle may be shed from a slope (for instance, during a cycle of landsliding) and the partially weathered bedrock exposed. This type of rapid mass movement will usually be followed by a long period during which the weathered mantle and its vegetation cover will be regenerated. On steep mountain slopes climax, or zonal, soils may never develop because the slopes are too unstable to allow a mature soil cover to evolve. In these circumstances agricultural terraces actually reduce slope instability and the soils developed on them are at least partially, and in some cases largely, the product of human labour. In other words, the immature azonal mountain soils receive much of their organic matter from the addition of crop residues and domestic animal fertilizer. Soil loss is characteristic of all natural slopes as well as those modified by human activities; it becomes a problem for subsistence farmers only when the decline in soil productivity due to topsoil losses cannot be compensated for by addition of nutrients from organic fertilizer and the continued accumulation of inorganic matter as rock weathering proceeds.

Effective erosion, if the soil or weathered material is to be moved out of the immediate field area, be it a small watershed or hillslope segment, requires the assistance of a transporting agent. The most effective agent of transport is running water. Thus information on the relationship between mass movement on slopes and water transport is important. The river itself is an agent of erosion as well as an agent of transport and, by cutting its channel and undermining its banks, the river is the principal force in maintaining local relief energy in an orogenically active mountain range. The river is also responsible for deposition of its transported load, and hence for the formation of the plains. Glaciers and wind are effective transporting agents but will receive little attention here because of their minimal spatial significance in the intensely used Himalayan belts – the Middle Mountains and the Siwaliks.

Denudation is a term used to describe the overall, theoretical, lowering of the landscape resulting from the erosive and transporting activities of all operating processes. In practical terms this is usually calculated as mm/yr in surface lowering averaged over entire regions, usually watersheds (drainage basins), despite the fact that actual surface lowering will be extremely variable in space, as it depends upon many factors, including the underlying bedrock lithologies and structures, and slope angle, as well as the local incidence, for instance, of heavy rainstorms. Regional denudation estimates are derived in two ways: they are obtained from numerous determinations of the rates of the erosion processes characteristic of a watershed multiplied by the total area under study, or they are extrapolated from measurements of sediment being moved out of the watershed through the main stream channel averaged over the total surface area of the watershed. Determination of total sediment discharge involves the long-term monitoring of rivers where they exit a particular watershed. In either approach the sediment delivery ratio must be taken into account. This is the ratio of sediment yield in a river – the actual volume of material transported out of the watershed – to the gross sediment production upstream, much of which goes into temporary storage within the watershed. Mass-movement data cannot be directly translated into rates of denudation because much of the material moved remains within the watershed in storage, for instance in the form of lake sediments or as accumulations (talus, glacier moraines, land-slide deposits, footslope colluvium) lower on the slopes or on the valley floors. The weathered material that remains in a particular watershed, of course, provides necessary substance for agriculture.

Material transported by a river is classified as the suspended load, the load carried as dissolved salts, and bedload. The sediment yield that is actually measured is often only the suspended sediment. Even this is difficult to measure accurately for rivers that experience enormous variations in volume over the course of the year and from year to year. Such high-level fluctuation is especially characteristic of rivers in monsoonal climates where low flow in late winter and spring may be several orders of magnitude below peak rainy-season discharge. The problem of measurement is exacerbated when we consider that Himalayan river channels are frequently dammed by landslides; the ensuing ephemeral lake, when it breaks the dam, will produce a peak discharge and be capable of carrying much larger sediment loads, sometimes an order of magnitude, or more, than that of normal summer monsoon peaks.

Even these periods of high sediment yield may be totally eclipsed on some rivers by peak surges resulting from the outbreak of ice-dammed and moraine-dammed lakes. The critical importance of such catastrophic floods to Himalayan water resource development has been discussed by several workers (Hewitt 1982; Xu 1985; Galay 1986; Ives 1986; Vuichard and Zimmermann 1986, 1987: see Chapters 6 and 10).

Measurements of bedload have not been recorded on any Himalayan river, even under conditions of 'normal' summer flow. An estimate is usually made for this component of the total sediment transfer in calculating the design life of reservoirs. It is now thought that bedload has been grossly under-estimated systematically throughout the Indian and Nepal Himalayan foreland. Despite this, hundreds of millions of dollars have been expended on dam and reservoir construction with the design life over-estimated two-, three- and fourfold. Finally, little is known about the amount of the dissolved salts in rivers in the Himalayan region. This lack of knowledge is of critical importance as dissolved salts in running water may amount to a significant proportion of total sediment discharge, especially in warm climates.

A final component of the mass movement/sediment transfer/soil erosion discussion is the influence of surface vegetation cover. While the dispute over degree and rate of forest removal is a primary consideration, as emphasized in Chapter 3, in the present context most significant are the types of crops and changing degree of surface cover according to agricultural type and cycle, including dates of planting, transplanting, harvesting, and the time interval between harvesting and the emergence of a subsequent crop. These site variations that have major influences on raindrop impact and surface runoff add complications to any attempt to construct a complete slope sediment budget. Plantation agriculture, especially tea plantations, presents a critical set of soil erosion circumstances, the more so when they are abandoned. Gerrard (pers. comm. 2 June 2003) proposes that the interval of change from one type of system to another is when the more serious soil loss problems seem to occur.

It is now appropriate to consider the data that were available on sediment yield, erosion, mass movement, and denudation in the late 1980s.

Estimates of denudation rates in the Himalaya

Any attempt to understand the significance of rates of denudation that have been determined for the Himalaya must be based on a general perspective of the region's physiographic divisions, bedrock geology, and tectonic structure. To keep this section within reasonable limits the discussion is restricted to the Himalaya *sensu stricto*.

The entire central part of the mountain system can be divided into nine strike-oriented physiographic units that trend approximately from west-northwest to east-southeast (see Figure 2.3). The four northern units constitute the southern portion of the Tibetan Plateau, the Tibetan Marginal Range (6,000–7,000 m), the Greater, or High, Himalaya (with numerous summits exceeding 7,500–8,500 m – maximum 8,848 m), and the Inner Himalaya (sometimes referred to as Trans-Himalaya), a system of high plateaus and valleys lying between the two great mountain ranges. These four units contain the 'high mountain' belt and, except for very small areas in the more deeply cut valleys, extend above the montane forest belts and timberline at approximately 4,000 m. The high mountain landscape (High Himal) is extremely complex, heavily sculptured by glacial erosion during the Late Cainozoic and supports a considerable cover of snow and ice today. Lithology ranges from sedimentary and metamorphic rocks to granite intrusives: the whole has been subject to extremely complex faulting, folding, and overthrusting. The higher, north-facing slopes tend to be semi-arid to extremely arid because of high altitude and rainshadow effects. The south-facing slopes intercept the summer monsoon although precipitation decreases with increasing altitudes above about 3,000 m, and the very large precipitation totals are generally confined to the outer ranges and plains. Until recently Cherrapunji, south of the Himalaya in Assam, claimed the world's record annual rainfall total (11,615 mm). However, long-term records from high-mountain stations, or any precipitation data at all, are rather scarce. Several more recently established stations are showing totals in excess of 5,000 mm/yr; nevertheless, these are all located on south-facing slopes exposed to the monsoon.

The southern flanks of the Greater Himalaya merge via a transition belt with the so-called Middle Mountains which have been the traditional centre of high population densities in the Himalaya for centuries. The transition belt, the Middle Mountains, and the northern flank of the next, outer unit, the Mahabharat Lekh, are collectively referred to as the *pahar* in Nepal (*pahad* in Mahat *et al.* 1986a). The upper northern sections of this tripartite division remain largely under montane forest (2,900–4,000 m), below which is the belt of intensive agriculture. Lithology is extremely varied, and includes sedimentary rocks, metamorphics, and granites. Extensive areas are underlain by phyllites and schists and these are deeply weathered, rendering them highly susceptible to mass movement and to disturbance by human activities, especially road construction.

The Mahabharat Lekh, together with the Siwaliks (or Churia Hills), constitute the Lesser Himalaya, or outer ranges, bounded against the Ganges Plain (Terai) by the Main Frontal Thrust (a major tectonic translocation). Highest summits of the Mahabharat Lekh approach 3,000 m; the Siwaliks are much lower. Between and within them occur the famous 'dun' valleys, tectonic depressions that support some of the richest agricultural land, extending eastward from Dehra Dun in the west and including the Rapti Valley at Chitwan to eastern Nepal. The lower, outer duns are locally referred to as the Inner Terai. The Siwaliks are composed of the youngest, Tertiary strata and contain some of the most easily erodible bedrock (including unconsolidated sands and gravels) of the entire Himalayan Region. Where these ranges have been extensively deforested, as in parts of Uttaranchal and Himachal Pradesh, there has been extensive movement of sediments and development of badland topography.

The Siwaliks abut the Ganges Plain abruptly at an altitude ranging from about 150 m in the east to 500 m in the vicinity of Dehra Dun. The alluvial sediments of the Ganges Plain can be divided into the high Terai (Nepali, *barbar*, or porous place), composed of massive coarse-grained alluvial fans and torrent fans, and the low Terai, underlain by finer sediments of the Ganges flood plain proper. As mentioned above, these sediments are several thousand metres thick in places and represent the vast accumulations of Himalayan weathering products over the past several million years.

Central to an understanding of the efficiency of erosional processes in the Himalaya, as well as the significance of the available estimates of denudation rates, is the concept of slope-channel coupling (Brunsden and Thornes 1979). Bearing in mind the great complexity of the landscape, the multitude of micro-climates, and the problems of representativeness of any study site in both time and space, a review can now be attempted of the denudation rates that have been derived for the Himalaya region.

Analysis was undertaken of selected denudation rates for the Himalaya as a whole, or for various watersheds and areas of differing size. Despite the wide range (from 0.5 to 20.0 mm/yr), and the inherent ambiguities discussed above, these figures are high when compared with rates from other parts of the world. The Alps, for instance, experience an overall rate of 1.00 mm/yr; amounts above about 6.0–7.0 mm/yr are among the highest ever recorded or estimated worldwide. At the outset, therefore, they are suggestive of a very dynamic environment, especially when coupled with the high rate of tectonic uplift and the intensity of seismic activity.

The studies of Brunsden *et al.* (1981), Starkel (1972a, 1972b), and Ramsay (1985) are some of the very few pre-1989 systematic attempts to derive denudation rates from actual field measurements. Brunsden *et al.* provided especially valuable insights for a section of the Lesser Himalaya in eastern Nepal, including observations from two field surveys for a road alignment between Dharan on the Terai and Dhankuta at 2,200 m. Theirs was the first comprehensive description of active hillslope and fluvial processes. They concluded that the relative relief of their study area had been increasing throughout the past million years or so because stream downcutting exceeded lowering of the local interfluves. Valley side slopes, therefore, had progressively lengthened and their parallel retreat was being effected by stream undercutting and landsliding. Thus, according to Brunsden *et al.* (1981), debris mobilized on the steep slopes flowed directly into the rivers, giving a very high sediment-delivery ratio, and was subsequently moved out onto the Ganges Plain by fluvial transport.

The two-period fieldwork of Brunsden and his colleagues was serendipitously timed to bracket a summer monsoon period with high rainfall amounts and a flood peak with an estimated ten-year recurrence interval. Thus they were able to study the impact of such a condition on the steep slopes of their field area. They concluded that the Lesser Himalaya of eastern Nepal had one of the highest rates of denudation in the world. The present land-forms were regarded as characteristic forms in equilibrium with current processes and with the controlling tectonic, climatic, and base-level conditions. Storage and transport processes and the linkages between them were identified. In particular they reasoned that one of the consequences of the very efficient slope-to-stream channel sediment-transfer system was that delivery of weathering mantle to the valley floor was not a continuous process. Debris tended to move in waves during storm events, causing pulses of heavily silted water to move downstream. This conclusion indicated that the system of erosion in the Lesser Himalaya of eastern Nepal contrasted with that prevailing in Barsch and Caine's (1984) alpine (mid-latitude) field areas in terms of high sediment-delivery ratio. Brunsden's field area is characterized by a high sediment-delivery ratio and the latter by a low ratio.

However, quantification of the various elements of the Brunsden *et al.* (1981) system, and especially the development of even a preliminary sediment budget for a single small watershed, was not then attainable.

Ramsay (1985) undertook the most exacting and effective study available prior to 1989 of a specific watershed in the Himalaya. His assessment of erosion processes in the Phewa Valley, near Pokhara, Nepal, is discussed in the following section. Within the context of his attempt to convert erosion process data, together with an extensive review of the geomorphological literature on denudation rates available at that time, however, his conclusion is an appropriate summing-up:

> The author . . . would like to emphasize that his own figures for failure age, volume, and frequency (and hence the estimate of surface lowering by landsliding [2.5 mm/ yr]), should never be used without the prefix 'based on a small sample'. They are probably as valid, or as invalid, as the estimates made by Caine and Mool (1982), and Starkel (1972a and b).
>
> (Ramsay 1985: 130)

Mass movement

As emphasized above, the ubiquity of landslides and other catastrophic slope failures throughout the Himalayan system had been not only widely reported but viewed with alarm because it was assumed to be the direct consequence of human land-use and land-cover changes over recent decades. This section will review some of the work on mass-movement processes and attempt a preliminary conclusion concerning their causes – natural (geological) or accelerated (induced by human activities).

Kienholz *et al.* (1983 1984) produced detailed maps of land use, geomorphic processes, and mountain hazard assessment on a scale of 1:10,000 for a small section of the Middle Mountains (Kakani–Kathmandu). Caine and Mool (1981, 1982) examined the channel geometry and stream-flow estimates and the landslide activity of the same area, as part of the United Nations University Mountain Hazards Mapping Project.

Caine and Mool (1981) concluded that the two mountain streams that they studied in detail were similar in their dynamics and form to streams in other parts of the world. From this they inferred that hydrological concepts and models derived from long records and detailed studies in mid-latitude regions could be applied to problem resolution in tropical and sub-tropical mountains, where few direct observations were available. They also demonstrated that the direct impact of human intervention on the two fluvial channels was limited to a constraint on channel width in the uppermost eight kilometres. This resulted from the construction of agricultural terraces, especially masonry walls, that caused the stream to adjust by cutting a deeper channel. As the stream size increased beyond about eight kilometres from the source, local landscape modification by the subsistence farming community was limited by the available technology and was not adequate to influence the lower reaches.

Caine and Mool (1982) also concluded that landslides occupied about 1.0 per cent of the land surface of the map area. The total volume of landslide debris was calculated as more than 2.2×106 m³ and the mean age of the landsides was about 6.5 years, that is, number of years since initial failure. They proposed a rate of lowering of the entire surface (denudation rate when data from other erosive processes were added) of about 12 mm/yr (recalculated by Ramsay (1985) as 11.3 mm/yr).

In the Phewa Valley, Ramsay (1985) calculated that landslide density was 1.6 per km² with 95 per cent of the landslides being small, shallow failures. The total area affected was 0.7 km², or 0.5 per cent of the area. If the area of the landslide deposits was added to that of the areas of mass movement failure (as in the case of the Kakani fieldwork), this would represent a total of 3.25 km², or 2.7 per cent of the catchment area. Ramsay challenged Caine and Mool's estimate of the rate of landslide expansion per annum and noted that many landslide scars in the Phewa Valley appeared to be healing. This view was supported by Ramsay's observation that the age of the small, shallow failures tended to cluster at slightly less than eight years, suggesting that they would likely heal within a decade. Caine and Mool's estimated mean age of 6.5 years was close to Ramsay's estimate. When this is taken in conjunction with Ives's (1987b) repeat photography (1978–1987), it would appear that the contribution of landsliding to overall denudation, regardless of the actual cause of the landsliding, had been over-estimated. Nevertheless, it had not been possible to take into consideration the positive impacts of the farmers that would further strengthen this conclusion (Kienholz *et al.* 1984).

Ramsay's larger areas of failure, which he termed mass-movement catchments after Brunsden *et al.* (1981), contrasted markedly with the larger number of small failures. Their average age was about 24 years and they appeared to be enlarging rapidly due to feedback mechanisms, such as the extension of bare ground and over-grazed areas that produced a much higher proportion of immediate runoff during heavy rainstorms. These were certainly influenced to some degree by human intervention in the form of mismanagement of land for agricultural purposes.

An annual denudation rate of 2.5 mm/yr was calculated from landslide activity alone (Ramsay 1985). This did not include the effects of soil creep, sheetwash, rifling, gullying, and solution. However, neither did it take into account the sediment-delivery ratio and, as Caine and Mool (1982) and Kienholz *et al.* (1984) noted, a large proportion of the material transported by mass movements remained within the small watersheds and was not transferred to the larger rivers. Thus it appeared that there were either differences in interpretation by Caine and Mool (1982), Ramsay (1985), and Brunsden *et al.* (1981), or else differences in actual field conditions between their three study areas. Several other studies had been undertaken with a view to determining the regional importance of landsliding, and these were reviewed by Ramsay (1985) and Carson (1985).

Laban (1979) carried out an airborne reconnaissance of most of Nepal south of the High Himal. Using a light aircraft he made observations on the number of slope failures per linear kilometre of flight as viewed from one side of the aircraft. The data were categorized according to ecological province. Despite the limitations inherent in his method, he was able to determine that road and trail construction was associated with 5 per cent of all landslides observed; that some of the most densely populated and extensively terraced areas contained the lowest frequency of landslides; and that specific regions and lithologies displayed the highest frequencies, for example, the Siwaliks and Mahabharat Lekh, and areas underlain by the deeply weathered phyllites. He concluded that geological structure and lithology were far more important (accounting for more than 75 per cent of all landslides in Nepal) in terms of landslide occurrence than were human land-use changes. A similar conclusion was reached following an extensive field and remote-sensing survey in a section of the Garhwal Himalaya (B.C. Joshi 1987).

Wagner (1981, 1983) used a statistical analysis of the general characteristics of a hundred landslides, mainly along roads in the Middle Mountains. He developed a site-specific landslide hazard assessment and mapping approach based upon use of 'equatorial Schmidt projections', a geometric tool for studying the intersection of geological planes. Wagner concluded that the overriding cause of landslides and rock slides was geological (natural) and not human. These conclusions, derived from the work of Laban (1979), Wagner (1981, 1983), B.C.

Joshi (1987), and the synthesis of Carson (1985), need to be carefully qualified. It could be misleading to refer simply to the stipulation that landslides were caused primarily by the impact of rainstorms on certain lithologies (i.e. that the cause was overwhelmingly natural); it is the relationship between land use and lithology that needs to be resolved.

Brunsden *et al.* (1981) and Caine and Mool (1982) attributed human activities as an important cause of gullying. However, they also emphasized the importance of the material, particularly the brittle behaviour of the weathered, *in situ* bedrock. Caine and Mool (1982) did not find a direct relationship between heavy rainstorms and landslide occurrence, noting that a high water-table (within less than 2 m of the surface) was a primary controlling factor. Thus heavy rainfalls early in the summer monsoon season did not produce many landslides. As ground-water recharge progressed and the water-table approached the surface, land-slide intensity accelerated. An opposing view is that landslides are most common early in the summer monsoon season because that is the time of minimum surface vegetation cover. The discussion serves to emphasize not necessarily conflicting viewpoints, but possible differences between specific areas, a common obstacle to any balanced overview. For the Kakani area, at least, the conclusions of Caine and Mool (1981) have been verified by several seasons of observation and are supported by many informal expressions of opinion by local informants.

Carson (1985) summed up by noting that many Himalayan landscapes appeared to experience a period of relative slope stability when mass-wasting was minimal. A brief period followed, characterized by excessive instability, when large numbers of landslides occurred almost simultaneously on otherwise long undisturbed slopes. Frequently the triggering mechanism was an unusually severe rainstorm with a long recurrence interval, such as the Darjeeling 1968 disaster (Ives 1970; Starkel 1972a, 1972b), or a somewhat smaller event, such as the ten-year rainstorms of the Dharan-Dhankuta area of eastern Nepal (Brunsden *et al.* 1981). It was also assumed that major landslide cycles may have been triggered by large-scale seismic shocks, with or without rainfall. Carson (1985: 11) observed that villagers often told stories about extreme precipitation events, occasionally coincident with earth-quakes, that triggered large-scale slope failures. He cited the area southwest of Banepa in the Nepal Middle Mountains where the villagers maintained that all the visible landslides had occurred during two heavy rainfall events, one in 1934 and the other in 1971. They maintained that, in spite of more recent heavy rains, no major landsliding had occurred since the 1971 summer monsoon. Another case was the heavy rainstorm of September 1981 when mountain slopes near Lele, Lalitpur district, Nepal, failed *en masse* in spite of a thick natural regrowth of shrub. Photographs taken in October 1988 showed that within seven years natural revegetation had virtually concealed the landslide scars.

Certainly, the above commentary provides the impression that many of the large-scale landslide occurrences not only required a major triggering event, but may have been followed by long periods of inactivity. The long period of relative stability was presumably required for accumulation of a new weathered slope mantle as a necessary condition for the next major failure. Seismicity was undoubtedly important, but lack of adequate earthquake records precluded any quantification of its influence. There were also several statements to the effect that presence or absence of a forest cover would have had little influence on the scale of a major landslide cycle (Carson 1985). Moreover, it was conjectured that forested slopes merely postponed the occurrence of a major cycle and, by so facilitating the production of a deeper debris mantle, ensured larger-scale events with longer recurrence intervals. In contrast, deforested areas given over to rain-fed agriculture, and especially grazing and over-grazing, were destabilized more frequently but by small, shallow failures. Thus the integrated effect over a longer time-scale may have been comparable. Carson (1985: 35–6) concluded that '[m]ass wasting processes are not usually directly related to man's activities. Consequently, intervention by man to reduce mass wasting can be very expensive with less clear-cut results.' He also stated that '[f]looding and sedimentation problems in India and Bangladesh are a result of the geomorphic character of the rivers and man's attempts to control the rivers. Deforestation likely plays a minor, if any, role in the major monsoon flood events on the lower Ganges.'

Erosion rates in the High Himalaya

There has been little precise information on rates of erosion and mass movement processes in the High Himalaya. The prevailing assumption was that the erosion rates, and hence the denudation rates, must be very high because the high mountains displayed some of the greatest relief on earth, including very high-angle glaciated slopes, and they were also influenced by tectonic instability.

In part to off-set this paucity of information, and in part to test the assumptions that high rates of erosion and mass wasting could be substantiated, Byers established 35 soil erosion study plots in Khumbu Himal and made precise observations at each of them at weekly intervals throughout the entire 1984 summer monsoon (Byers 1986, 1987b, 1987c). Total precipitation recorded for 1984 was slightly above the 28-year mean of 1,048 mm for Namche, so that the data obtained were reasonably representative. Byers employed a standard geomorphic approach to the problem, using unsophisticated instrumentation that could be maintained easily in an isolated high-mountain area. The 35 study plots of 5 × 5 m extended through a 1,000 m altitudinal interval (3,440–4,412 m) above and below upper timberline. They were arranged in a stratified, replicated design to permit comparison between (1) north-facing forest (*Abies spectabilis/Betula utilis/Rhododendron* spp.), (2) south-facing scrub grassland (*Cotoneaster microphyllus/R. lepidotum/*grass forb), and (3) variations in elevation. The fixed instrumentation consisted of a rain gauge, plastic sediment trough, erosion pins, and two sets of painted marker pebbles. Ground cover was determined for each plot by direct measurement at the time of instrumentation, and seasonal vegetation changes were assessed on adjacent 400 m² quadrats using a point-intercept method (800 points). Weekly observations included air and ground temperature, soil capillary pressure, notes on plot changes, and disturbance caused by local people, and weather.

Preliminary analysis of the data led to the following findings. Summer monsoon precipitation was modest (Khumjung: 3,790 m, 725 mm compared with a 15-year average of 773 mm) and decreased with increasing altitude. However, local rain-shadow effects were marked, so that the simple relationship of diminishing total precipitation with altitude was frequently masked (compare the Khumjung total with that of 1,002 mm for Tengboche that was located in the open main valley of the Dudh Kosi–Imja Khola at 3,867 m). More significantly from a geomorphic point of view, rainfall intensities were low (mean value of 29 measured 30-minute maximum rainfall intensities for Khumjung was only 3.03 mm/hr; and between 1968 and 1984 24-hour interval rainfall amounts exceeded 25 mm on only 13 occasions with an absolute maximum of 54 mm). Cumulative precipitation increased sharply during the summer monsoon but, at most of the study plots, sediment yield was slight. The one exception was the plot at Dingpoche (4,412 m, in an over-grazed pasture) that demonstrated a high sediment yield. This was representative of an alpine meadow–juniper shrub area with a sandy soil subjected to heavy human disturbance, including up-rooting of juniper shrubs (see Chapter 3, p. 49).

The low sediment yields were partially explained by the vegetation cover data. Many areas had 50 per cent or less ground cover in the pre-monsoon season, which may be why trekking-season visitors had made alarmist reports of environmental degradation and anticipated severe erosion during the summer monsoon that they did not stay to witness. In contrast, Byers observed that pre-monsoon showers and light rains early in the summer monsoon season produced a rapid spread in ground vegetation cover and hence considerable protection against rain-drop impact.

Although Byers monitored only surface erosion and did not consider such processes as debris flows, talus creep, and the higher-altitude glacial and fluvial systems, when the findings were coupled with the hazards mapping results, it was concluded that one of the world's highest mountain regions was far less active geomorphically than had been suspected and than had been generally assumed by proponents of the Theory.

A further topic that was considered in Khumbu Himal by the UNU research team was the catastrophic outburst of glacial lakes (Icelandic: jökulhlaup – glacial lake outburst floods: GLOFS). A moraine-dammed lake drained catastrophically on 4 August 1985 in a tributary valley of the Bhote Kosi (Galay 1986; Ives 1986; Vuichard and Zimmermann 1986, 1987). This provided a remarkable opportunity for before-and-after assessments and prompted a preliminary enquiry into the recurrence interval of such events. The process and timing of lake formation and drainage is considered in Chapter 6. Here emphasis is placed on the downstream transfer of the sediment produced. At least three jökulhlaup, and possibly five, have occurred in the Khumbu within living memory.

Vuichard and Zimmermann (1987) developed a sediment budget for the Khumbu outburst of 4 August 1985. They concluded that most of the material moved was redeposited within about 25 km in the stream channel and only about 10–15 per cent (the finer fraction) was transported out of the area. Their data indicated that 900,000 m³ material was removed from the moraine dam but most of this was redeposited within the first two kilometres below the breach. Much more material was picked up from the stream channel and valley sides further downstream. The peak discharge was calculated at 1,600 m³/sec some three kilometres below the source and it attenuated downstream.

It is difficult to develop long-term sediment-transfer averages from the scanty information available. Nevertheless, the magnitudes of these events are sufficiently high and their recurrence interval, on an areal basis, quite small (5–15 years) so that they may prove to have been a significant source of sediment for deposition at lower altitudes. Also, even if only a fraction of the material is far-travelled during the actual event, rivers subject to periodic jökulhlaup may well prove to have been major source areas for subsequent sediment flux during peak monsoon rainstorms as the coarse material dumped in their channels would be carried further downstream. The spectacular dynamics of the Sapta Kosi, for instance, and the development of its vast alluvial fan that occupies much of Bihar State, may have been influenced to a considerable degree by the occurrence of jökulhlaup in several of its headstreams (Dudh Kosi, Sun Kosi, and Arun).

Impact of road construction on sediment production

For comparative purposes, the sediment production from road construction, mainly in the Indian Central Himalaya, is reintroduced. Research by Haigh (1982a, 1982b), Valdiya (1985, 1987), and Tejwani (1987), and others in the late 1970s and 1980s produced some significant information on numbers of landslides and volumes of sediment yield.

Tejwani (1987) assessed data from Bansal and Mathur (1976) to calculate that ten small-to-medium landslides were caused by slope instability from each linear kilometre of road construction. Prior to the 1962 border war with China, the Indian Himalaya were accessible primarily by foot trails. The only available roads were those that led to famous hill stations, such as Mussoorie, Simla, Nainital, and Darjeeling, established during the British Raj. An extensive programme of road construction was prompted by the shock of the Chinese military presence on the Himalayan frontier. Construction was undertaken in great haste after 1962 and military expedience outweighed any concern for careful planning and sound engineering.

According to Tejwani (1987), more than 10,000 km of highways were added to the existing road network in the Indian Himalaya. He estimated that poor alignments and ill-considered design produced soil losses of 1.99 million tonnes per year. This he equated to slope movement of 1.99 tonnes of sediment per linear metre of road per annum (recalculation indicated a misplacement of the decimal point, so this figure should read 0.199 tonnes/m/ yr). Valdiya (1985, 1987) indicated that 44,000 km of new roads had been constructed, and calculated that during the construction phase an average kilometre of road had required removal of 40,000–80,000 m^3 debris. Following completion of road construction, extensive slope instability produced enormous volumes of debris that were dumped on the road bed and further downslope during heavy monsoon rainstorms in the form of debris flows, rockfalls, rockslides, and mudflows. Valdiya provided the following estimates of annual debris production per linear kilometre of road bed for three specific highways: Western Himalaya, Jammu–Srinagar: 724 m^3; Central Himalaya, Tanakpur–Tawaghat: 411 m^3; and for the Eastern Himalaya in Arunachal Pradesh: 719 m^3. From these figures he derived an average annual debris yield of 550 m^3 per km and a total debris production of 24×10^5 m^3 for the 44,000 km road network.

In practical terms, the slope instability caused by the new roads produced frequent highway blockages and enormous maintenance costs, especially during summer monsoon periods. And since the standard method of road clearance was to dump the debris over the road side and down the slope below, this in turn further extended the area of instability. It also caused destruction of downslope vegetation cover as well as the agricultural terraces of local subsistence farmers, who were usually not compensated for their losses.

Narayana and Rambabu (1983) also determined that unsatisfactory highway alignment and poor design had produced an enormous amount of debris each year. Valdiya (1985: 24) stated that '[t]he damage to the ecological balance [of the Himalaya] is mostly man-made or is caused by human negligence'. He accounted road construction, over-grazing, and reckless deforestation as the primary causes. He concurred with Narayana and Rambabu who had stipulated that road construction in seismically and tectonically unstable bedrock was 'the most important factor'. Haigh (1982a, 1982b, 1984) had also investigated the incidence of landslides along the Mussoorie–Tehri road in what has since become the separate state of Uttaranchal. In 1977 and 1978 he measured 470 debris outfalls and postulated that the frequency of movement was related to depth of road cut, steepness of slope, degree of forest cover, geological structure, and lithology.

Each of the above researchers assumed that landslides resulting from road construction were responsible for massive increases of suspended load in the local headstreams of the Ganges. Furthermore, a significant proportion of this was assumed to be carried as far as the Bay of Bengal, so contributing to extension of the delta and the development of New Moore Island (24×11 km) offshore (Valdiya 1985: 20–4).

Regardless of the weight of argumentation and the relative precision of the estimates of debris dumped onto specific sectors of the road bed each year, no accurate assessment of the proportion of the debris that entered the local rivers is available. Many of the landslides ran part-way down the slope below the road bed and did not reach even the first-order stream channels. It follows, therefore, that while road construction and road maintenance problems in many parts of the Himalaya, and the attendant destruction of local slopes, were environmentally and economically catastrophic, no data are available to support the claim that any of the debris so produced reached the plains, let alone the Bay of Bengal. Nevertheless, the landslide scars associated with the Himalayan roads are very evident even to the casual observer, and it is reasonable to conclude that sediment yield would have been proportional to the density of the road network. The fact that Laban (1979) concluded that only 5 per cent of Nepalese landslides were attributable to highway construction is presumably a reflection of the much lower density of the mountain highway network in Nepal at that time.

The data on road construction and induced slope instability are significant. The broader interpretations of the effects of roadworks on flooding and sedimentation on the Ganges Plain and on the seaward portion of the delta in the Bay of Bengal had been supported by many workers (Haigh 1982a, 1982b, 1984; Tejwani 1984a, 1984b, 1987; Valdiya 1985; Narayana 1987). These were questioned by Ives and Messerli (1989: 119–22). Nevertheless, it could not be concluded that the effects of man-made erosion (accelerated erosion) were negligible on a regional scale. The seriousness of reckless road construction from a physical point of view was not doubted. Much avoidable damage had been, and was being, perpetrated. The costs of keeping the roads open were significant in themselves, as were the economic losses from repeated and widespread road closures. The development of between 10,000 and 50,000 km of road had produced a major socio-economic impact on the Himalaya which included greater accessibility of hitherto remote forests for commercial logging, and ease of movement of people both from the mountains to the cities of the neighbouring plains and from the plains to the mountains. The process of road extension has continued to the present throughout the Himalayan region.

The situation before 1990: a summary

Ives and Messerli (1989), as indicated above, had contested the early claims that deforestation and farming on steep and unstable slopes were leading to massive soil erosion and landsliding. It was further postulated that there was no significant body of evidence to indicate that sediment transfer from such slopes, resulting specifically from peasant farming activities, was entering the main streams for transfer onto the plains and so into the Bay of Bengal. Similarly, there was no rigorous support for the popularly held conviction that increased peak river discharge during the summer monsoons, resulting from increased surface runoff because of deforestation, was significantly affecting the patterns of flooding on the plains. If any degree of connectivity could be established between sediment movement on mountain slopes and the main rivers, especially the Ganges, then the impacts of road construction and other large-scale interventions may well prove more influential than indigenous farming. Nevertheless, the issue was by no means settled. The assumptions on which the Theory was based have continued to receive support. In any event, more rigorous research was required. A great deal of such research has been undertaken over the last 10–15 years. It will now be assessed and compared with the research results introduced above.

Recent research on the Middle Mountains of Nepal

Wu and Thornes (1995) studied the hydrological effects of terrace management in an area immediately north of the UNU research site of Kakani. They raised three questions:

1   What is the sub-surface water movement on the two types of terraced mountain slopes (khet and bari)?
2   Under what conditions will slopes fail?
3   What is the relationship between slope stability and terrace hydrological behaviour in terms of water management for agriculture?

They conducted field surveys during the 1992 summer monsoon in the Likhu Khola watershed about 25 km from Kathmandu. The Likhu Khola flows into the Trisuli River, a tributary of the Ganges. The watershed is underlain by gneisses and mica schists that are deeply weathered and extends from about 700 to 2,500 m above sea level. The main species of the natural vegetation are *Shorea robusta* (Sal), *Pinus roxburghii, Schima wallichii,* and *Quercus* spp. The authors regard the Likhu Khola as representative of the Nepal Middle Mountains, in terms of both natural vegetation and cultural characteristics, so that their findings are believed to be relevant to a much larger area.

Four field sites were chosen, two on khet terraces, two on bari. All were on north-facing slopes near the intersection of three different tectonic zones. The average slope angle into which the bari terraces were cut was 26°; that of the khet terraces was 24°. They measured maximum slopes up to 60°; nearby grassland had an average slope of 36°.

The average rainfall in the area ranged from 1,500 mm to 3,000 mm. The Kakani climatological station is situated on the ridge crest immediately to the south at 2,703 m and had recorded an average annual rainfall of 2,800 mm, with over 90 per cent falling between June and September, with July (681 mm) and August (703 mm) being the wettest months. During the 1992 field season July, with 1,194 mm as recorded at Kakani, proved the wettest month. Individual rainfall events were of high intensity and frequently exceeded the soil infiltration capacity. The authors indicate that the hillslopes of their field area were prone to mass failure. General conditions, except for differences in the underlying rock types, are similar to the UNU Kakani research area (Ives and Messerli 1981; Caine and Mool 1982; Kienholz *et al.* 1983).

Wu and Thornes (1995) found that it would take four days to fill two hectares of khet terraces with water from the serving irrigation canal and up to 12–15 days if the supply of water was limited. Since the heights of the terrace bunds are about 20–30 cm and the terraces are filled only to about 10 cm, this implies that two hectares of khet terraces, acting as a reservoir, would detain water, and hence surface overflow, for hours, if not days. The total 'reservoir effect' would equal a rainstorm of 100–200 mm (see Figure 1.3). They point out that as khet land occupies 26 per cent of all agricultural land of the Middle Mountains of Nepal, the checking effect on water flow, and hence the capacity to prevent erosion, is highly significant. Furthermore, this soil conservation effect of the terraces also has an important impact on stream discharge and so reduces the risk of downstream flooding. Despite this, small slope failures are frequent, especially on the khet lands, although they are usually promptly repaired as the entire irrigated slope depends on the controlled flow of water through all the terraces to secure a good rice crop.

The bari terraces have an entirely different hydrological function. Surface erosion is a problem during the pre-monsoon rains when vegetation cover is negligible, and occasionally in August after the maize has been harvested and a new crop has yet to be planted. Rill erosion and debris flows are much more common on the bari terraces than on the khet. Piping can also be a problem (Caine and Mool 1982; Kienholz *et al.* 1983), and when the roof of a pipe collapses initiation of gullies is an immediate threat. Wu and Thornes (1995) corroborated the conclusions of Johnson *et al.* (1982) that when large debris flows occur among bari terraces during periods of intense rain, the ensuing damage often must be left unattended until the following winter season when labour becomes available. They also quote unpublished research by Brown (1985) who had determined soil erosion rates on a variety of land-use types in eastern Nepal in tons/hectare/year:

bare, denuded land	13.350
old reforested landslide	8.146
grassland	2.547
bari	2.885
khet	0.088

Wu and Thornes reached preliminary rather than unequivocal conclusions: that the construction, farming, and general management of terraces on hillslopes in the Likhu Khola watershed does not change the hydrological behaviour of the slope. The effects of this indigenous land management pattern can be regarded as positive rather than negative in that individual terrace failures generally do not contribute any sediment to the downslope, nor augment streamflow. Nevertheless, Wu and Thornes (1995) indicate that larger failures involving several, or even dozens, of terraces were much more complicated and were beyond the scope of their survey. In essence, there is close agreement with the results of the earlier UNU work in the Kakani research area immediately to the south.

Concurrently with Wu and Thornes (1995) the Royal Geographical Society launched a major study of the geomorphology and land-use management of the Likhu Khola watershed (Gardner and Jenkins 1995; Blaikie 1995; Gerrard and Gardner 1999, 2000a, 2000b, 2000c; Gerrard 2002; Gardner and Gerrard 2001, 2002, 2003). Fieldwork was carried out in 1991, 1992 and 1993. Intensive studies, including measurements on numerous soil erosion plots, were made of the relationships between rainfall, runoff, soil erosion, and landsliding on khet and bari terraces and on uncultivated land; the nature and management implications of landsliding on khet terraces and the overall influence of landsliding on the shaping of the landscape were investigated.

Erosion study plots were located on 11 bari terraces (Figure 4.3 and Figure 4.4), generally paired on adjacent terraces, and 15 plots were established on a variety of non-cultivated land. Most of the erosion plots were located close to climate stations. A total of 86 fields were surveyed and over 400 landslides were examined and mapped. The team monitored 912 rainfall events on the non-cultivated plots and 530 events on the bari terraces during the 1992 and 1993 monsoon seasons. The erosion study plots were constructed to conform to the natural topography of the slope, not on the standard rectangular plot model. This divergence from standard plots might account for some differences between these results and those derived from other studies (Gerrard pers. comm. 2 June 2003). The determination of possible soil losses from the irrigated khet terraces was obtained through mapping and individual site analysis of all failures that occurred during the three-year investigation (Gerrard and Gardner 2000a).

The vast majority of the landslides were very small slumps on the khet terrace risers; the bari terraces experienced most of the larger landslide events, although these were comparatively few in number. Three degrees of connectivity between the slope movements and the river system were defined: failures classed as exhibiting low, medium, and high connectivity with the river system. Low connectivity failures were small mass movements that never extended beyond the terrace systems on which they occurred; they accounted for more than 90 per cent of total failures and can be dismissed from any consideration of Likhu Khola denudation. Medium connectivity failures may result in material being carried into the river system but not during the immediate period of failure; this category of failure requires prompt repair. High connectivity failures are those that do leave the terrace system, either by flowing into a ravine or by being transported directly into the river system. However, only 4.4 per cent of all failures fall into this category. Gerrard and Gardner estimate that if all the material from the high connectivity failures, half that of the medium failures, and none of the low connectivity failures leave the terrace system and exit the watershed, the total soil loss would amount to 0.48 t.ha.yr. This is a very small loss and well within manageable limits. The much larger events, especially from the abandoned land and the degraded Sal forest land, occur on land cover categories that account for a small percentage (5.8 per cent) of the total area of the watershed. Again, total soil loss per annum would be sustainable under the exist-ing agricultural regime. From a management point of view, while virtually all the failures that occurred on the khet terraces were repaired promptly, as more and more of the bari were being converted to khet, the resulting increase in labour for khet repairs may become a limiting factor in the future. This is a more serious aspect of Likhu Khola land use than that of actual soil erosion with the potential for negative effects downstream. Nevertheless, Blaikie (1995) found that 56 per cent of the local farmers were producing enough paddy rice to export significant quantities, although 25 per cent of these would sell the valuable rice and purchase for domestic consumption cheaper cereals, such as maize, millet, and buckwheat.

Gerrard and Gardner calculated average annual losses from slope failures from the large data bank accumulated in tons/hectare/year, as follows:

khet terraces	0.48
bari terraces	3.65
grassland	1.86
forest	0.80
scrub and abandoned land	23.95

From these calculations and the map of land use they were able to estimate a combined denudation rate for the four sub-catchments within the Likhu Khola watershed of 5.55 t.ha.yr. This refers specifically to slope failures. On the non-cultivated land the runoff coefficients varied from 1–2 per cent under grassland and mixed broadleaved forests to 57–64 per cent on the bare sites. Soil losses ranged from 0.1 t.ha.yr for grassland and undisturbed forest plots, to 3–10 t.ha.yr for Sal forest in various stages of degradation, to over 15 t.ha.yr for bare sites. Soil losses from bari terraces ranged from 2.7–12.9 t.ha.yr. The highest soil losses were associated with red, fine grained soils.

Changes in land use over the previous 30 years were determined. Cultivated land had increased at the expense of grassland and there had been extensive conversion of bari to khet terraces. Forest cover had increased slightly in extent but had declined in quality.

In contrast to the preliminary findings of Gilmour (1986) and Gilmour *et al.* (1987, and as reported in Ives and Messerli 1989: 84–5), Gardner and Gerrard (2002) determined that runoff was generated on all land-use types by rainfall events of relatively low magnitudes and intensities, but did not necessarily lead to significant soil loss, the relationship between runoff and soil loss being highly complex. Their overall conclusions were 'that landsliding is a significant contributor to overall denudation but that large-scale landsliding is not a major problem in the study area' (Gerrard and Gardner 1999: 253–4). They also concluded that the few large failures were 'probably unrelated to recent human activity, although the long-term change in land use from forest to grassland in the upper south-facing catchments, where most of the larger debris slides occur, may be relevant' (Gerrard and Gardner 1999: 254).

In terms of the geological time-scale of landscape evolution, Gerrard and Gardner estimated that well over half of the area had experienced landsliding at some stage in the past. However, the majority of the larger failures were extremely old and part of the natural evolution of the area. From this they caution that:

> . . . estimates of total soil losses are significantly less than the hypothetical figures quoted in the past (e.g. Carson, 1985), suggesting that there may be less need for concern about soil erosion on agricultural terraces in the Himalaya than hitherto believed.
>
> (Gardner and Gerrard 2003)

They also propose that there may be a need for re-evaluation of the degree of land degradation that is occurring in such areas.

Despite some discrepancies among different workers, and these may be due to varying field conditions and monsoon characteristics, the work of the Royal Geographical Society extends and corroborates many of the conclusions deriving from the earlier work of the UNU team (Ives and Messerli 1989).

As part of a multiple watershed study for determination of water and sediment transfer from the High Himal to the Ganges plains, Schreier initiated the Jhikhu Khola watershed project in 1988. It has been continued through to the present, although the Maoist Insurgency restricted access to the field data loggers during much of 2002 (Hans Schreier pers. comm. 20 June 2002).

Jhikhu Khola is a tributary of the Sunkosi River that flows into the Ganges. Situated in the Nepal Middle Mountains approximately 40 km from Kathmandu, its watershed area is 11,000 ha. It is densely populated and intensively farmed with much of the area under double, and even triple crop rotations. Questions had been raised about the sustainability of such intensive use, characteristic of many similar areas in the Middle Mountains. It had been assumed that such watersheds were incurring serious nutrient depletion and substantial soil erosion losses. For example, the Jhikhu Khola was experiencing increasingly intensive extraction of biomass from the forest floor to support local agriculture. This had prompted concern for the long-term maintenance of soil fertility within the forest despite extensive reforestation in the 1980s.

Thus a long-term interdisciplinary project was initiated with application of remote sensing and GIS and extensive automatically recording field instrumentation. It is the soil erosion studies that will be examined in the present context. An overarching question was raised by the planning team – how long would it take for sediments to travel from the upper watershed some 500 km downstream for eventual deposition on the delta of the Ganges? This question remains unanswered but, on the basis of many years of research in the Jhikhu Khola watershed, Schreier and Wymann von Dach (1996) speculated that the answer would be on the order of thousands of years.

Three levels of nested research sites were established. The highest and smallest was an erosion study plot of 70 m^2 involving two bari terraces in the upper watershed. A stream gauging station was set up below. Four kilometres further downstream a 520 ha site with stream gauge and sediment trap was established. Last, the outflow into the Sunkosi River was instrumented so that the entire watershed became the 11,000 ha site. Runoff, sediment quantity, and phosphorous content were monitored. The project also involved a network of 50 24-hour rain gauges and four tipping bucket gauges. The results presented here are drawn from a small part of the overall study encompassing the monitoring of over 100 rainfall events during three monsoon seasons.

The Jhikhu Khola experiences a monsoon climate with 70 per cent of the annual precipitation occurring between June and September. The data from the dense rain gauge network were used to demonstrate the episodic nature and great spatial variability of the rainfall, even within the confines of a small watershed. Another highly relevant point was that the pre-monsoon rains produced individual storm intensities and total amounts comparable to monsoon-period storms. This partly explains some of the otherwise surprising patterns of sediment transfer. First, the sediment rating curves showed a distinct difference between pre-monsoon and monsoon events in that sediment losses were much higher during the pre-monsoon period. In fact, 60–80 per cent of annual soil loss was pre-monsoonal. This was because, at that time of the agricultural cycle, terraces were bare of vegetation and therefore were unprotected. Furthermore, the farmers were leading off as much as possible of the water and sediment that was discharging from the bari terraces and transferring it onto the irrigated khet. This represented a formidable water and sediment control system and included 72 small check dams. When a very severe monsoon storm struck, over 70 per cent of the check dams were destroyed. Thus, when this occurred there were large soil losses from the two upper sites, but even then no sediment was transferred entirely out of the watershed and into the Sunkosi River. In effect, the research team was able to trace sediment movement from the upper to the lower watershed but not out of the research watershed, so intensive and intricate was the manipulation by the farmers.

From the hundreds of soil samples collected, it became apparent that both sediment and nutrients were being transferred from the bari terraces onto the khet. Wymann (1991) calculated an inverse relationship within the khet terraces between nutrient level and altitude. The most fertile terraces were those at the bottom of the slopes because they had received maximum sediment and nutrient transfer from above.

During the early years of the Jhikhu Khola project a catastrophic event occurred less than 30 km southwest of the study area. The Kulekhani watershed had been dammed for hydro-electric power production. It was contributing 45 per cent of Nepal's total electricity in 1993. Between 19 and 21 July of that year the Kulekhani watershed experienced a very severe rainstorm. Intensities reached 70 mm/hr and 540 mm in 24 hours. Literally, hundreds of landslides released. Dhital *et al.* (1993) measured a landslide density of $47/km^2$, the largest number occurring on grassland and forest areas.

The Kulekhani hydro-electric project had been designed for an expected useful life of 60–70 years, based on the assumption that the annual sediment yield from the watershed above the dam would be 11.t.ha. In practice, measurements between 1984 and 1993 had been 20–45 t.ha.yr. During the 19–21 July rainstorm alone the sediment yield amounted to 410–500 t.ha. In three days the design life had been reduced from 60–70 years to 10 years, yet the rain gauge network in Jhikhu Khola registered no perturbation. Schreier and Wymann von Dach (1996) also drew a comparison between their Jhikhu Khola experience and Messerli's theoretical discussion of scale (in Ives *et al.* 1987). Messerli stipulated that it was necessary to view the Himalayan problem in the context of scale and proposed a system of micro- ($< 50 km^2$), meso- ($50–20,000 km^2$), and macro-watersheds ($>20,000 km^2$) as a basis for reckoning the relationships between human and natural effects. He suggested that the human effects would be distinguishable only at the micro-scale. The Jhikhu Khola watershed data indicated that even in this micro-scale system it was difficult, if not impossible, to distinguish between human and natural processes.

Discussion

The common denominator of the foregoing account of mountain slope agricultural geomorphology is *extreme variability* – variability in time and space, in bedrock type and structure, altitude, slope angle, and in aspect, vegetation cover, inter-annual and intra-annual precipitation patterns, farming practices and ethnic groups, and degree of access to a road network. To this must be added the major environmental contrasts across the length of the Himalaya–Karakorum–Hindu Kush–Pamir mountain systems and the many differences across the international borders of the region. Thus it is hardly surprising that there are apparent disagreements among the research findings of the individual research workers. Even so, such disagreements are of a secondary nature; the conclusions that have been drawn from the substantial post-1989 research do not differ significantly in the overall findings from those presented in more preliminary form by Ives and Messerli (1989). There is, however, a greatly enlarged database involving detailed research from several additional field sites, so that the conclusions can now be stated in much firmer terms.

It can be argued, therefore, with much more confidence than in 1989, that the extension of terraced agriculture on steep slopes that had originally been under dense forest was not a primary cause of the hitherto assumed acceleration of soil erosion and landsliding. Nor were the rural farmers contributing significantly to increased surface runoff and hence downstream flooding and siltation. The Middle Mountain region, from which most of the environmental degradation was presumed to have originated, is a landscape that has been extensively shaped by fluvial erosion and landsliding over geological time. That the landslide-moulded slopes with very steep gradients have been so extensively terraced and maintained is a remarkable feat of human ingenuity, intensive labour, and persistence. The great increase in our understanding of indigenous slope management leads to the inevitable conclusion that the farmers are actually slowing the rate of erosion, and hence, denudation, rather than augmenting it. Several studies have shown that there is very low connectivity between slope material transfer and the river system. A high proportion of the material moving downslope is redeposited on the lower slopes. Nevertheless, it must be borne in mind that this is contested by Brunsden *et al.* (1981) based on their work in eastern Nepal, and by Starkel and Basu (2000) based on their work in the Darjeeling Himalaya. Explanation of this apparent contradiction, at least in part, may be related to differences in field techniques, in environmental differences between the areas surveyed, and even in the inter-annual and intra-annual precipitation characteristics. However, one of the critical contrasts in the study of Brunsden *et al.* (1981) with those undertaken closer to Kathmandu is that the major river in their field area cuts against the grain of the Himalayan structure, causing rapid incision and considerable valley slope instability. Much of the Middle Mountain drainage is aligned with the structure. This is certainly the case of Likhu Khola, Jhikhu Khola, and to a certain extent Kolpu Khola. These valley sides, although steep, are much more open and the influence of river under-cutting is much less which emphasizes the complexity of the Himalayan region and the dangers of generalization (John Gerrard pers. comm. 2 June 2003).

There is a further consideration. Although Blaikie (1995) and the Likhu Kola research team, as well as the Jhikhu Kola team, refer to off-farm sale of crops, there is growing evidence that an overall change is occurring in the very basis of slope farming that, if continued, will result in a significant reduction in the area of subsistence mountain farming across much of the Himalaya. For specific areas in Nepal, Malla and Griffin (1999) have illustrated remarkable changes in land use and farm economy in the last two decades of the twentieth century (see Chapter 9). These include fundamental changes in type of domestic animals retained

– reduction in the numbers of cows and increase in the numbers of buffalo and goats to provide increased milk and meat production for cash sales – accompanied by increased stall feeding and reduced dependency on the forests for fodder. Additionally, trees are increasingly being grown for the cash value of their timber, upper bari terraces are being converted from crops to trees, and the entire local assessment of the forests is changing. As part of this process, there is a significant increase in availability of off-farm labour opportunities that is influencing the abandonment of low-value crops, partly because of the labour shortage. Also, chemical fertilizers are being introduced. These changes are directly related to road access and road construction, and to the growth of new small urban centres that continues to accelerate.

It is apparent that the mountain farmers are showing a great deal more flexibility than they have been credited with in the past. From this point of view also, the proponents of the Theory who envisaged a collapse of the Himalayan farming systems have not allowed for the remarkable adaptability of the farmers.

These changes in farm economy and land use do not necessarily affect the weight of evidence invalidating the earlier contention that expansion of cultivation on steep mountain slopes was leading to increased surface runoff and downstream flooding and siltation. They are introduced, nevertheless, to demonstrate that economic and agricultural change is an important part of the Himalayan scene. This will be discussed in more detail in Chapter 9. Chapter 3 and the present chapter have effectively discredited two of the major driving points of the Theory of Himalayan Environmental Degradation. It remains to examine what is happening on the plains. Losses from flooding in the lower watersheds of the Ganges, Brahmaputra, and Meghna rivers have increased significantly throughout the second half of the twentieth century and there have been a series of catastrophic floods within the last 30 years, for instance, in 1987, 1988, and 1998. The causes will be the central topic of Chapter 5.

5 Flooding in Bangladesh

Causes and perceptions of causes

God sendeth down water from Heaven, and causes the earth to revive.

(Qur'an, XVI, C.625)

Introduction

Flooding in Bangladesh has been featured prominently in the Western news media since long before the country gained its independence from Pakistan in 1971. The worldwide perception of the Indian monsoon is inextricably linked to two of the great rivers of the subcontinent, the Ganges and Brahmaputra. Cherrapunji in the Indian Meghalaya highlands, with an annual average precipitation in excess of 11,000 mm, is perceived to lie close to the centre of the flood threat despite being located in the upper reaches of the Meghna watershed. Most of this huge amount of water falls between June and September, a characteristic of the entire flood plain region of the Ganges, Brahmaputra, and Meghna rivers. Would not a person from western Europe or North America contemplating an early August first visit to Cherrapunji, envisage in the mind's eye a column of water nearly six times the height of an adult? Notwithstanding this association between inconceivably heavy rainfall and flooding, it has long been assumed that the widespread destruction of the Himalayan forests has served as the driving force for the progressively larger and more frequent catastrophic floods that have been assumed to be a feature of the second half of the twentieth century. This assumption, of course, has been a prominent point of discussion throughout this book.

There has been a general unconditional acceptance of the truism that forest-covered terrain absorbs water during heavy rainfalls and releases it slowly over a much longer period, thus modifying the rate of river discharge. It would be axiomatic, therefore, that massive clear-cutting of trees in the upper watersheds of a major river system will result in accelerated runoff during heavy rainstorms and more extensive and frequent flooding downstream. Since most of the tributaries of the Ganges and Brahmaputra originate in the Himalaya, then surely the well known orographic accentuation of monsoon rainfall anchors the linkage between Himalayan deforestation and increased flooding in Gangetic India, Assam, and Bangladesh. Accordingly, and to cite only one of numerous examples, the *Basler Zeitung* in its 15 September 1998 issue published the following statement:

> The severe floods in eastern India and Bangladesh are not the result of a natural disaster, but of the ruthless exploitation of the forests which has been practised over many centuries in the Himalaya.

The journalist appears not to have considered the heavy rainfall in Bangladesh itself. The location of Cherrapunji, with its long-held world record of maximum annual precipitation, is more than 150 km south of the Himalaya in the Meghalaya Hills, India, at 1,313 m, and although there are several rival claimants to this distinction the point is well based. It is significant that the watersheds of the two great rivers are shared by Bangladesh, India, Bhutan, Nepal, and Tibet (China). The upper riparian states of India, Bhutan, Nepal, and Tibet contain the mountainous upper watershed sections where climatological stations are sparse and river gauging stations totally inadequate for analytical investigation. The lower riparian states of Bangladesh and India (which is both upper and lower riparian) have a dense network of climatological stations and river gauges. However, both countries have kept their accumulated databases, or parts of them, top secret. The Government of India classifies much of its hydrological data at the same security level as its military strategy. The opportunities for systematic research, therefore, are somewhat constrained. Furthermore, during part of the period of severe flooding in 1988, Bangladeshi authorities withheld daily precipitation totals, usually accessible to the global climatological network (Rogers *et al.* 1989: 74). This is a primary indication of some of the political overtones of flooding in northeastern India and Bangladesh. It is generally conceded, for instance, that international disaster relief is roughly proportional to the perceived magnitude of the disaster. In the same context, it has also been reasoned that political convenience tends to identify upstream neighbours as the source or cause of a catastrophic flood. Hence the early concern of Ives and Messerli (1989) for the Nepalese subsistence farmers who served unjustifiably as a convenient scapegoat for downstream flooding.

Moreover, the flood plain of the Ganges, and especially several of its major Himalayan tributaries, have been subjected to large-scale river works: dams for hydro-electricity and irrigation; for large diversions out of the watershed for irrigation and urban uses; and for diversion using the Farakka Barrage on the Ganges, situated just upstream from the Bangladesh frontier, for flushing out the Hugli, distributary of the Ganges, in an attempt to reduce siltation of the port of Calcutta. In addition, many kilometres of embankment have been constructed along all three main rivers and along several of their tributaries. The political implications of any interpretation of the causes of monsoon flooding within the region, therefore, are significant. In the world-at-large, humanitarian and environmental concerns tend to predominate. Hence, when Lawrence Hamilton made the remark at the Mohonk Conference that 'it floods in Bangladesh when it rains in Bangladesh' (Hamilton 1987) the officials of the lower riparian states did not applaud.

Ives and Messerli (1989) stated that post-1950 deforestation in the Himalaya, even if proven, would not contribute significantly to flooding and siltation in Gangetic India and Bangladesh. However, this statement was not adequately supported by analysis of river flow and precipitation data, although the more exacting research of Rogers *et al.* (1989) provided additional support. Thus, it was decided to extend the UNU project on *Highland–Lowland Interactive Systems* downstream from the Nepal Himalaya as far as the Bay of Bengal. Bangladesh was chosen as the major area for investigation because of the greater ease of access to essential data. The research was carried out under the direction of Bruno Messerli, co-ordinated by Thomas Hofer, and involved five years of teamwork in the field. The early, preliminary publications (Hofer and Messerli 1997; Hofer 1998) already make it clear that Hamilton's intuitive statement about 'rain on the plain' was essentially correct. The following account is derived largely from the final report of the UNU/SDC funded research in Bangladesh (Hofer and Messerli 2002). The publications of Rogers *et al.* (1989) and Alford (1992b) have also been consulted extensively.

Bangladesh: a flood plain and a delta

Bangladesh is cited in United Nations agency assessments as one of the poorest countries in the world. It is characterized by extremely high population densities, low literacy rates, high infant mortality, an overwhelming dependency on subsistence agriculture, and all the other attributes of poverty and under-development. The total area of about 150,000 km² has to support over 120 million people and the birth rate is very high. Sixty per cent of the entire country lies less than six metres above sea level; roughly half lies within the deltas of the Brahmaputra and Ganges, and 66 per cent is flood plain (Wohl 2000: 9). Ahmad (1989) noted that, during a 'normal' monsoon, water entering Bangladesh via the Ganges, Brahmaputra, and other rivers is sufficient to form a lake over 10 metres deep that would equal in area the entire country. Yet this claim appears to take no account of the rain that actually falls on Bangladesh.

Other features of Bangladesh (Figure 5.1) include the immense delta of the combined Ganges, Brahmaputra, and Meghna rivers passing southward into the Sunderbans, an intricate network of river channels, ponds, and lakes, and some of the world's most extensive tracts of mangrove swamp. Much of the delta is subject to tidal surges and devastating cyclones. In addition, there is the monsoon that deposits an average of approximately 1,500 mm of rain each year over the entire country. During a so-called 'normal' monsoon some 20 per cent of Bangladesh is inundated (average for 1954–1998 is 21.12 per cent: Thomas Hofer pers. comm. 9 July 2003). Immediately north of the northeastern frontier the Meghalaya Hills rise to over 1,000 metres; they extend about 250 km from east to west and have a maximum width of close to 120 km. The only highland of note within Bangladesh occurs in the extreme southeast (the Chittagong Hills, with heights reaching 600–1,000 metres) and because of the location can be excluded from consideration in this chapter.

Worldwide, floods are generally considered as disasters and this is one of the perceptions that has coloured attitudes toward Bangladesh, leading to its depiction as a 'development fiasco'. This is highly unfortunate for many reasons. One reason is best explained by stating the need to define the term *flood*. During a less than average monsoon, with the area under water much reduced, Bangladesh faces a *real* disaster. Less than average rainfall results in insufficient water for the fast-growing flood-adapted rice varieties and jute. Another consequence is the deprivation over vast areas of the life-giving fertile sediments that the rivers deliver in most years, with the exception, of course, when the sediments are coarser, such as sands, which can be quite infertile. A further danger is a winter shortage of water accompanied by widespread dry-season crop failures.

In a brief account of the recent geological history of the territory today occupied by Bangladesh, Hofer and Messerli (2002) demonstrate how the flood plain and delta have evolved over the last million years or so. During the maximum of the last Ice Age, about 18,000–25,000 years ago, with world sea level as much as 120 metres below its present level, the rivers were entrenching their beds. Because of their steeper gradients they carried coarser sediments than today and so were depositing coarse sands and gravels. As world sea level rose with the melting of the last ice sheets, the river gradients were reduced and masses of clays and silts were deposited. Of course, most of the sediments came from the Himalaya, which were undoubtedly forested. By about 5,000 years ago, the height of the so-called Holocene warm period, the forest cover would be complete in the Central and Eastern Himalaya to an altitude of 3,500–4,000 metres, the height of the present-day upper treeline. Much of the plains area would also have been forested. Hofer and Messerli confidently imply that the long period of geological time, effectively a period with no human intervention, or with only minimum impact during the last few thousand years, would also have experienced

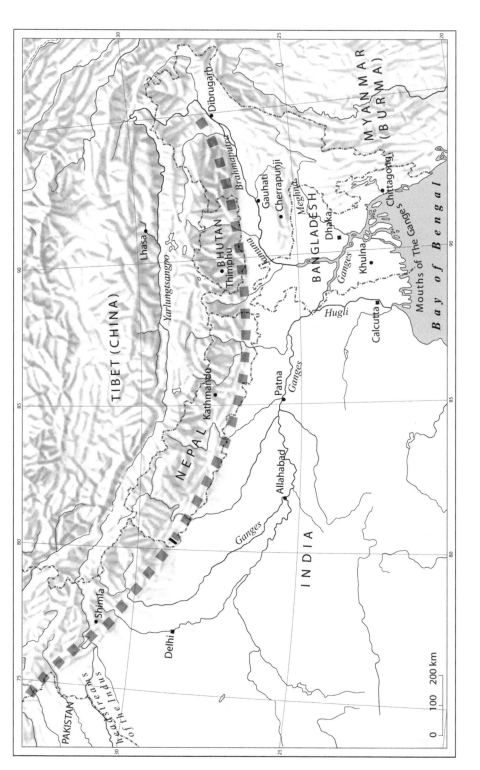

Figure 5.1 The drainage basins of the three major rivers: Ganges (Ganga), Brahmaputra, and Meghna. Map base modified after Hofer and Messerli (2002). The line of small squares denotes the Himalayan Front.

heavy periodic flooding and massive siltation comparable with that of today. This strongly supports similar earlier statements by Ives and Messerli (1989) and Rogers *et al.* (1989). The precise nature of these processes cannot be assessed nor is this necessary for the current discussion. Nevertheless, it must be emphasized that the fundamental make-up of Bangladesh is a product of erosion in the mountains, sediment transfer by the great rivers, and periodic flooding and deposition of sediments on the plains and delta. The sediments deposited by these processes in places are several thousand metres thick (Rogers *et al.* 1989: 7). They have filled in what otherwise would have been a deep extension of the Bay of Bengal.

Curry and Moore (1971), quoted in Rogers *et al.* (1989) and based on a 1968 seismic investigation, estimated that the undersea fan of sediment derived from the Himalaya and deposited in the Bay of Bengal by the Ganges and Brahmaputra rivers was 1,000 km in width, possibly 12,000 m thick at its maximum[1] and 3,000 km from north to south – that is, extending well south of Sri Lanka. Such deposition would require that 7,000 mm of material (i.e. denudation) were eroded from the entire area of the Himalaya and transported into the Bay every thousand years.

Geological time, however, may appear somewhat remote, for instance, to the diplomat inconvenienced by the temporary inundation of Dhaka International Airport in 1988. Therefore, the question must be asked: have the periodic extreme monsoon floods become more extensive and more frequent during the last 50 years, and if so, is this the result of human disturbance of the regional environment, especially deforestation of the Himalaya?

Flooding during the eighteenth and nineteenth centuries

There are few sources of information on flooding during the eighteenth and nineteenth centuries. The Imperial Gazetteer of India for 1908 catalogues the major floods of the preceding 200 years, although it gives no systematic explanations. Nevertheless, the records show two distinct periods of high flood frequency, 1770–1790 and 1860–1890. Each period is preceded and followed by longer periods of low frequency. More information is available for the Ganges than for the Brahmaputra. While this may be due to differing hydrologic characteristics of the two rivers, it may also be the result of the remoteness of the Brahmaputra in relation to British India, centred on Delhi and Calcutta. Hofer and Messerli favour the interpretation that the information is a reflection of variations in monsoon rainfall characteristics.

Additional information is available from the district gazetteers, particularly for Bihar and Bengal. Here 1767, 1770, and 1787 are designated as years of severe flooding, and 1787 is especially significant. This is the year of major changes in river courses: the Tista changed its course to become a tributary of the Brahmaputra rather than the Ganges; the Brahmaputra, until 1787 confluent with the Meghna, reoccupied an old channel to join the Ganges. Exceptional flooding combined with heavy rains also characterized the year 1871. Similarly, in 1885 North Bengal experienced exceptionally heavy rains accompanied by high floods along the Ganges. It is worth noting that, if an event comparable to that of 1787 were to occur today, because of the much larger population and the extensive developments on the flood plain, much more severe losses would be experienced than those attributed to the exceptional floods of 1987 and 1988.

Although the information available from the eighteenth and nineteenth centuries is descriptive and intermittent, it raises the suggestion that the magnitudes of monsoon floods may have been comparable to those of the last two decades. It also allows introduction of the relationship between actual flood losses and density of population and infrastructure. This relationship becomes much more pertinent to the central questions of this chapter during the last 50 years and is discussed in more detail below.

Flood records of the twentieth century

Rogers *et al.* (1989) had originally drawn attention to the importance of checking the historical record of flooding in India and Bangladesh. Their preliminary exploration of the available data drew them to the 1770s, 1780s, and 1790s when high-magnitude floods were frequent. They pointed to the fact that back-to-back major floods, such as those of 1987 and 1988, appear quite often in the record (Rogers *et al.* 1989: 34–5). They also stated that it was reasonable to believe that the 1988 floods were the worst ever experienced up until then, if ranked according to economic losses. However, they quoted Shukla's (1987) analysis of Calcutta's rainfall records back to 1829, which demonstrated that the nineteenth century must have experienced floods of equal physical magnitude. Shukla (1987) also concluded that flood levels for selected recent years showed that the 1988 flood was not of significantly greater magnitude than two of the four severe floods of the previous 35 years.

After 1890 much more information on flooding became available. Hofer and Messerli (2002) examined all available data for the years 1890–1998 which they divided into two periods: 1890–1950 and 1951–1998. The bulk of the information from the earlier period, while much more abundant than previously because of the increase in infrastructure and services, was still largely descriptive and qualitative. However, several of the long-term series of rainfall records were initiated in the early part of the period. After 1950 many more quantitative data were accumulated. In addition, information on the actual areas affected by the floods, river discharge, and sediment load entered the record. On the negative side, India began to classify and restrict access to much of the data.

During the period 1890–1998, 11 major floods were recorded. The data demonstrate that there were periods of frequent high-magnitude flooding alternating with longer periods with no excessive inundation. For example, several large floods occurred during the period 1906–1922, and again between 1954 and 1998. The intervening period of 1923–1954 was exceptional for the virtual absence of above 'normal' flooding. If the record for the period 1923–1998 is reviewed on its own, then it would be reasonable to conclude that frequency and magnitude of flooding has increased throughout the twentieth century. On the other hand, if there were data available only for the period 1890–1950, then it would be tempting to postulate a reverse trend of fewer and lower magnitude floods over the 60-year period. In other words, examination of flood records to this point does not support the contention that flooding has become progressively more severe nor that high-magnitude floods have become more frequent throughout the twentieth century. Moreover, it also illustrates how data can be selected to prove exact opposites – that is, the data can be manipulated to demonstrate whatever the writer wants the reader to accept. This illustrates Thompson and Warburton's (1985a) 'uncertainty on a Himalayan scale'.

Another problem identified by Hofer and Messerli (2002) relates to the reliability of the data. For example, the 1954–1998 flood data for Bangladesh has been compiled by the Bangladesh Water Development Board. There is a lack of strict consistency in the methods of record taking and in the criteria applied to defining flood magnitude in different areas of the country. For specific regions the reported flood dimensions might have been slightly exaggerated, with the intent to influence the scale of flood relief from national and international agencies. Nevertheless, such inconsistencies are minor and do not reduce the broad relevance of the available data which are significant and warrant full analysis.

One of the most notable findings from the analysis by Hofer and Messerli (2002) is that, after 1975, the records show a marked increase in inter-annual variability. Thus, while the highest magnitude flooding that can be firmly identified occurred during the 1975–1998 period (note especially the high-magnitude floods of 1987, 1988, 1998), some of the least extensive floods occurred during the same period. Correspondingly, the 'average flood' was much less in evidence between 1975 and 1998.

The general characteristics of monsoon flooding in Bangladesh have attracted considerable discussion and search for causes. While it is beyond the scope of this chapter to attempt any exhaustive review of this literature, some examples are informative in terms of their ambiguity and contradictory nature. For instance, Hughes *et al.* (1994: 21) state: 'Ironically, embankments serve to prevent "normal floods" whilst failing to prevent "abnormal floods", thus reducing the beneficial functions of normal flooding.' Brammer (pers. comm. in Hofer and Messerli 2002) strongly opposes this interpretation. He states that the design of the lateral river embankments has produced a form of flushing sluices that can be controlled to let water onto or off the flood plain as required by the local people. And embankments do not prevent flooding from direct heavy downpours. Much has been written about the failure of embankments, implying unwise construction. Yet it is only the Brahmaputra right embankment that has suffered repeated breaching while embankments along the Ganges and the Tista (with one exception) have held, even during the record high floods of 1987 and 1988. In addition, Rogers *et al.* (1989) refer to a study by Stewart (1989) who investigated two samples each of 300 households, one inside the Meghna–Donagoda embankment, the other outside, for the period October–November 1988. He found that the average material damage incurred inside the embankment was more severe than that experienced outside (on the river side). It appeared that the households 'benefiting' from flood protection suffered higher flood damage than those beyond the protected area. This is a frequent feature of Bangladesh flooding. The embankments are often several kilometres distant from the river. After a flood has peaked, the area on the river side of the embankment will usually drain rapidly while the flood waters on the landward, protected, side will remain ponded.

Yet another debate over the cause of the high-magnitude flooding has centred on the loss of the beels (seasonal and perennial lakes) and swamps that have served as natural storage for excessive flood waters. Over time, not only in Bangladesh but also higher up the Brahmaputra in Assam, many of these areas have been claimed for agriculture, or constricted by road and embankment construction. Thus storage during periods of high magnitude flooding has been progressively reduced and this is presumed to have induced larger-scale flooding downstream. This is controversial. Hofer and Messerli (2002) also refer to the construction of polders that have further reduced flood storage capacity.

Rogers *et al.* (1989) recommended against heavy engineering solutions. In particular, they caution against the extremely high costs that would be involved, not only for the initial instance of construction, but also for maintenance, as the trained river will also tend to deposit sediments in certain sections and thereby raise its bed, so the embankments will have to be constantly raised. If and when a breach actually occurs, then the danger from the ensuing flooding will be much greater.

Hofer and Messerli (2002) summarize the discussion to this point with the following statements:

1   High-magnitude flooding on the Ganges and on the Brahmaputra has not occurred simultaneously in any of the years investigated.

2   In the three Indian states of Uttar Pradesh, Bihar, and West Bengal (Ganges system), except for 1971 and 1978, floods have not simultaneously reached the same magnitude in any one year.

3   When serious flooding occurs in Uttar Pradesh and West Bengal (Ganges system), Assam (Brahmaputra system) usually has no above-normal flood.

4   The Bangladesh flood records correlate positively with those of Assam and negatively with those of Uttar Pradesh and West Bengal.

5   The records fail to show a single year when flood magnitude increases downstream from Uttar Pradesh along the Ganges and into Bangladesh. Floods that developed high on the Ganges plain, even in the Himalayan foothills, do not seem to travel downstream with increasing magnitude. For instance, in 1982 Uttar Pradesh experienced major flooding, yet the flood magnitude diminished progressively downstream. At the same time, Bangladesh experienced one of its least extensive flooding events for the entire 1954–1994 period.

The records and their analyses, together with many more details provided by Hofer and Messerli, fail to fit any model that implies deforestation of the Himalaya induces excessive runoff from the mountains to propagate as downstream flooding on the plains. The complex regional and inter-annual patterns of high-magnitude flooding prompt the speculation that variations in regional and inter-annual monsoon rainfall patterns may be the real driving force.

Twentieth-century rainfall patterns

Any discussion of rainfall patterns across the Ganges, Brahmaputra, and Meghna drainage basins must take into consideration the general characteristics of the summer monsoon circulation. Webster (1987), for instance, comments that while the monsoon circulation can be seen as macro-scale phenomena that are driven by ocean-atmosphere interactions, the actual rainfall does not derive from a single large precipitation system. Rather, it is produced by a series of discrete rainfall events that operate at smaller scales. Synoptic and meso-scale features that are embedded in the general monsoon circulation determine variations in inter-annual rainfall totals and in the production of the specific heavy rainfall events capable of causing floods. The timing and location of such events in relation to the flood stage of the different rivers are critical factors. This partly explains the difficulties facing any systematic analysis.

Hofer and Messerli (2002) selected for detailed analysis eight climatological stations with long-term records. They were chosen to provide regional geographic representation and more or less complete 100-year records. These are: Delhi, Allahabad, Kathmandu, Darbhanga, Dibrugarh, Gauhati, Sylhet, and Bogra; that is, from Delhi in the west to Dibrugarh on the upper Brahmaputra in the northeast, and from Kathmandu in the Himalaya to Bogra on the main Bangladesh flood plain.

The records show a generally increasing trend in total monsoon rainfall receipts from about 1920 onward throughout the century. The Brahmaputra–Meghna system demonstrates this most strongly, while for the Ganges system the trend is much less distinct. From this alone, an increase in the intensity of flooding in Bangladesh throughout the same period should be expected. However, there have been significant fluctuations. For instance, as indicated above, the conditions at the end of the twentieth century are similar to those at the beginning; and the incidence of flooding is similar for both periods.

Hofer and Messerli also raise the possibility that the increase in precipitation amounts during the last two or three decades of the century may reflect the current change in climate that is believed to be occurring worldwide. Theoretically at least, an increase in the temperature gradient between the ocean and the subcontinent should produce higher monsoon totals in terms of lengthening of the monsoon season (early start and late withdrawal) and/or an increase in diurnal intensities.

One of the more diagnostic comparisons between magnitude of flooding and monsoon rainfall amount relates to the outstanding flood years of 1987 and 1988. Hofer and Messerli (2002) state that during the peak of the 1987 flooding, the rainfall in the north and northeast of Bangladesh was 8–10 times higher than the 1950–1981 average for the corresponding period. Unfortunately, comparable data were not available for the Indian states of Uttar Pradesh and Bihar, although no significant flooding was reported for either of those years.

Rogers *et al.* (1989: 12) concluded from statistical analysis of monthly rainfall data from 11 stations in the lower Ganges–Brahmaputra basin dating back 150 years that no statistically significant trend could be detected, neither in total monsoon rainfall, positive or negative, nor in rainfall cycles or periodicities. They also indicated that the 1988 Calcutta total monsoon precipitation of 1,322 mm fell well within the 150-year range.

A vital result of the data analysis is the demonstration of a close coincidence between the increase in inter-annual variability of precipitation with increase in the variability of the actual area affected by flooding. The strength of this crucial relationship between rainfall amounts and patterns of flooding is most pronounced in the Brahmaputra–Meghna system. Therefore, a reasonable conclusion (perhaps, working hypothesis, at this point) is that flooding intensity during the last quarter century is a direct response to fluctuations in monsoon rainfall. Furthermore, there is no evidence to support the claim that the floods have been caused primarily by human interventions in the Himalaya. The effects of human interventions on the flood plains, however, are discussed below.

Twentieth-century discharge trends

Hofer and Messerli (2002) selected seven river gauging stations for analysis of discharge based on the need to obtain records for as long a period as possible and to provide regionally differentiated results. The Himalayan foothills and the northeastern area of Bangladesh were selected along with key points on the three main rivers. The period 1955–1995 was used for analysis of all the stations except for the Ganges for which Hardinge Bridge records from 1934–1995 were available. In 1975 construction began on the Farakka Barrage on the Ganges and this must be taken into account. The other stations were: the Karnali and the Narayani rivers of Nepal and the Tista of West Bengal; the Kushiyara River at Sheola, to represent the hills of northeastern Bangladesh and their drainage into the Meghna; the Bahadurabad station on the Brahmaputra in Assam; and the Bhairab Bazar station on the Meghna in Bangladesh. All station records were analysed for total monsoon and August discharge and for peak daily discharge.

The analysis of the Karnali, Narayani, and Tista gauge records showed that only the maximum daily flow figures on the Narayani indicated a rising trend that was statistically slightly significant. The two remaining variables for the Narayani and all three variables for the Tista showed non-significant rising trends. The Karnali trends were all negative, although they were not statistically significant.

These three data sets demonstrated that there was no significant increase in river discharge from the Himalayan foothills for the period 1955–1995. This compares with a similar conclusion drawn from analysis of the discharge records of the Indus tributaries: Sutlej, Chenab, Beas, and Jhelum (Hofer 1993).

Alford (1992b) argued that examination of the impacts of land-use practices on long-term variation in river flow is best considered through analysis of fluctuations in the mean annual stream flow records. He undertook a study of the annual trends of gauging station data within the Karnali, Narayani, and Sapta Kosi basins in Nepal, all tributary to the Ganges. The hydrological database available provided up to 30 years of continuous record. Alford found no apparent increase or decrease in mean annual flow. Inter-annual variations were often as much as 50 per cent of the mean flow for the period of record. As an example, the decade of the 1970s was a period of generally higher-than-average flow, but this did not persist in all the sub-basins examined. Some sub-basins that did experience a modest increase in flow volume into the mid-1980s, such as the Dudh Kosi and Arun, tributaries of the Sapta Kosi, ended the decade with values below the long-term average. In this sense he demonstrated that any changes in land-use practices could not be identified on the scale of individual gauged sub-basins of the Nepal Himalaya.

Analysis of the low flows of the Narayani and Sun Kosi rivers from 1964 to 1985 showed virtually no year-to-year variation (Alford 1992b). In contrast, peak flows showed very high year-to-year variation; in addition, they were frequently negatively correlated. Alford pointed out that the extreme mountain topography precludes the possibility of relating regional variations in precipitation or changing land-use and land-cover type to variations in stream flow.

On a larger scale, Alford's analysis indicated that the Nepal Himalaya runoff and Ganges river floods and droughts were negatively correlated. Maximum runoff from the Nepalese mountains in general coincided with minimum Ganges discharge as recorded at the Farakka Barrage. His hydrological analysis represents an extensive undertaking so that only a few points more relevant to the specifics of the current discussion have been introduced.

In a recent compendium on inland flood hazards (Wohl 2000) there is repeated emphasis on highland–lowland hydrological relationships although little precise analysis. However, there is a statement to the effect that rainfall patterns in the upper basins of the Ganges and Brahmaputra seem to have little impact on severe flooding in Bangladesh. At least any impact that may occur will be negligible compared with the inter-relationships between locally heavy monsoon rainfall and climatologic and hydrologic factors already in place in the lower basins of the river systems (Cenderelli 2000: 73–103).

Hofer and Messerli's (2002) research showed that the Kushiyara River, draining the hilly land of northeastern Bangladesh, experienced a statistically significant positive trend for the entire monsoon and August discharges, and for the daily peak figures. This increase indicates that progressively more water over time has been added to the Meghna to which the Kushiyara is tributary.

The three big rivers all showed slightly positive trends for the entire monsoon and maximum daily discharge data with intensity increasing from west to east. However, the Ganges trend is negligible, while that of the Meghna is conspicuous, although neither is statistically significant. The Brahmaputra analysis falls midway between the other two. For all three stations there is an increase in the inter-annual variability resulting from both increases in the maximum values and decreases in the minimum values.

To conclude this analysis, attention is drawn to Hofer and Messerli's stipulation that the overall non-significant trends correspond with the positive trend in monsoon precipitation totals, its regional differentiation, and increasing inter-annual variation. Furthermore, in contradiction to the general assumption, there was no demonstrably significant increase in the monsoon discharge of the Ganges for the 60 years preceding 1995. Moreover, there appear to have been no monsoon season effects on the Ganges discharge from the construction of the Farakka Barrage, in contradistinction to repeated claims by Bangladeshi authorities that the barrage has facilitated practically 50,000 cu secs of flood waters being diverted into Bangladesh. The effects of the barrage are rather to deprive Bangladesh of dry season water by diversion away from Bangladesh.

Perceptions of increasing flood damage in Bangladesh

The Theory of Himalayan Environmental Degradation has as one of its components the assumption that increases in periodic flooding in Gangetic India and Bangladesh have resulted in accelerated losses in human lives and property throughout the twentieth century. This is a completely different claim to that of increased frequency and magnitude of flooding over the same period. Consequently, it should be rigorously isolated from discussion of the actual physical characteristics of flooding and examined for its own sake. Thus the patterns of changing human use of the flood plains over time warrant close consideration. This will entail a large number of variables including: population trends; intensification of agriculture; expansion of urban centres and all associated infrastructure, including roads, railways, airports, and large-scale industrial and utility structures; and the business and banking systems. Next there is the wide assortment of riverine structures: canals; embankments; pumping stations. To this must be added river transport and fisheries, as well as river water quality in terms of human needs (Ahmad 1989). There is also the question of valuation: changing currency values against international standards, and inflation. Finally, there remains the ever present problem of data availability and reliability, including recognition of possible data manipulation, both at the village and the regional/national levels, with the intent on attracting disaster relief. In the present context it is necessary only to provide an outline of the current situation; the discussion will focus on Bangladesh again, rather than Gangetic India.

Between 1872 and the 1991 census the population of Bangladesh (or the area occupied by the present independent polity of Bangladesh) increased by a factor of about 5.5. From 1872 to 1951 this increase was gradual, doubling over the 80-year period. After 1951 there has been a rapid acceleration and the total population tripled in the 40-year period, 1951–91.

The pattern of population increase has been predominantly one of increasing density on the flood plains, the most fertile areas of the country. With this increase there has been a corresponding reduction in the size of land holdings and, therefore, a significant increase in the number of people vulnerable to excess flooding. It is the beels and swamps, that previously served as natural storage for flood waters, that have witnessed a process of occupation of 'new land' *per se*. This extension of agriculture also implies that there is much less flexibility for coping with flood hazards. As holdings become smaller the traditional strategies for coping with high-magnitude flooding have been significantly constrained. Hofer and Messerli (2002) conclude from this reasoning that, even if the flooding patterns are shown to have remained identical to those of a century ago, with no increase, the need for increased attention to the vulnerable people remains acute. The number of people and the amount of property at risk has definitely increased and will continue to do so as long as the population continues to multiply.

Rogers *et al.* (1989) cautioned against over-reaction by the development and aid agencies. They concluded that there was no evidence for a trend toward more extensive physical flooding downstream from the Himalaya. In this general sense there is broad agreement with Alford (1992b) and Hofer and Messerli (1998, 2002). Rogers *et al.* (1989) go on to make a number of insightful recommendations, perhaps the most promising (and unconventional) being exploration of the possibilities for exploitation of renewable underground storage. This reflects the increasing attention being paid to the magnitude of water storage in the extensive sub-surface aquifers and the rapid spread of tube wells. At this point their discussion goes beyond the scope of the current chapter.

Flood perceptions of the local people

Hofer and Messerli (2002) make a significant contribution to an understanding of how the local people in Bangladesh perceive their situation. The following findings were derived from analysis of a large number of questionnaires and village/household interviews. The local people who were interviewed did not consider monsoon flooding as a particularly serious hazard compared with other problems that they have to face. The demands from governmental authorities and donors alike for highly expensive and hard engineering solutions to the perceived hazards of flooding have been tempered over the last several years. However, the need for open exchange of hydrological and climatic data remains urgent.

Hofer and Messerli (2002, Chapter 7) reported on an extensive socio-economic study that was an integral part of their overall investigation of flooding in Bangladesh. Fieldwork was carried out in three test areas of rural Bangladesh: Bhuanpur in Tangail District, Serajgonj in Serajgonji District, and Nagarbari in Pabna District. The area covered by fieldwork in each district corresponds to a half circle 15–20 km in diameter with the diameter aligned along the river course. The three large sections were chosen to ensure that areas far from the main rivers were included. Villages located behind flood protection embankments were investigated, as well as those between the embankments and the river, and islands within the river (*chars*). A total of 148 interviews were undertaken, one for each village.

It became apparent that lateral river erosion was rated a far more serious hazard than flooding. To support this perception, there are many instances of rapid and dramatic lateral erosion by the main rivers. On several historic occasions major rivers, such as the Brahmaputra, the Ganges, the Tista, and the Sun Kosi, have completely abandoned their channels and taken a new course over long sections. An impressive example of the instability of an alluvial river course is provided by Winkley *et al.* (1994). They report that between 1973 and 1988 the Arial Khan River developed three meander cut-offs along a 35-km stretch shortening the river course by 18 km. Two new meanders were also formed. When such processes occur, or when a river cuts away several hundred metres of land on one bank, a large number of people lose all of their possessions; the most disastrous loss, apart from their actual lives, is their land. As landless poor, their long-term outlook is extremely bleak. In the case of actual inundation by monsoon flood waters, after a period of what may be extreme discomfort, the flood waters withdraw and the land remains, often replenished by a new layer of nutrient-rich sediments.

The questionnaires also revealed that a below normal monsoon flood is regarded as much more serious than high-magnitude flooding as it leads to dry season water shortage and drought. In recent years the winter season crops have become more important than those of the traditional monsoon season, thus further exacerbating this hazard. Finally, cyclones and tidal surges, especially for those living in the more southerly locations of Bangladesh, are much more devastating than high-magnitude monsoon flooding.

Hofer and Messerli (2002) proceed to examine the attitudes toward flooding exhibited by engineers, politicians, and journalists. While this goes far beyond the purposes of the current discussion, suffice it to say that perceptions of flooding by these groups of stakeholders rarely parallel those of the villagers at risk. Nevertheless, the varying perceptions of this group of stakeholders have been changing somewhat in recent years, although there is a tendency for them to revert to the earlier standard demands for costly infrastructural solutions during and immediately following high-magnitude flooding, such as occurred in 1998. The reader is referred to Hofer and Messerli (2002, Chapter 7) for the details of this fascinating discussion. To emphasize one aspect of the contrasts in perception, however, the two extremes of news media reporting that refer to the same 1998 flooding are quoted from that discussion:

> FLOODS IN BANGLADESH – Catastrophe in Asia's house of the poor: hundreds of deaths, millions without shelter.
>
> (an amalgam of Western Press reports)

> Have no fear: The children are enjoying diving in the River Jumuna.
>
> (Bangladesh *Daily Star*)

Conclusion

The new detailed information and research analysis accumulated by the Bernese–Bangladeshi research group should effectively terminate the dispute over whether or not land-use change in the Himalaya is the cause of increased magnitude and frequency of flooding in Bangladesh and Gangetic India (Hofer 1993, 1998; Hofer and Messerli 2002). In this respect it reinforces the earlier more tentative, or more narrowly focused discussions by Ives and Messerli (1989), Bruijnzeel and Bremmer (1989), Rogers *et al.* (1989), Alford (1992b) and others. Similarly, in a recent large compendium on inland flood hazards (Wohl 2000) there is repeated emphasis on highland–lowland hydrological relationships, although little precise analysis. Nevertheless, there is a specific statement to the effect that 'most of the severe floods in Bangladesh (1974, 1987, 1988) appear to be influenced by heavy rains and runoff from the Brahmaputra and Meghna basins, not the Ganges' (Hirschboeck *et al.* 2000: 50). These authors explain that rainfall patterns in the upper Brahmaputra and Ganges watersheds appear to have less influence on Bangladesh flooding than the relationship between local torrential precipitation and the 'climatologic and hydrologic factors that characterize the lower sections of the river systems'.

First and foremost, the perception that late-twentieth-century flooding in Gangetic India and Bangladesh has increased in magnitude and frequency is found to be insupportable. Annual variations in the physical extent of flooding certainly occur. The eastern sections of the Ganges–Brahmaputra–Meghna basin appear to have experienced a slight increase in flooding during the course of the century. This coincides with an increase in monsoon season precipitation over the same period although neither 'trend', if that term is justified, is statistically significant. Nevertheless, there is no comparable trend in the western section of the basin. Furthermore, flooding in Bangladesh usually coincides with below normal flooding in Uttar Pradesh and Bihar, upstream on the Ganges flood plain.

More tenuous, but nevertheless reasonably well supported by the available data, is the argument that flooding in the foothills of the Nepal Himalaya, or in the Garhwal and Kumaun Himalaya, does not propagate downstream along the Ganges. Even within the Nepal Himalaya, Alford (1992b) has demonstrated that there is no substantiated coincidence of high-magnitude flow on the mountain rivers.

There is a positive correlation between the regional patterns of monsoon precipitation receipts and the regional patterns of flooding. Thus it is possible to conclude that Hamilton's intuitive, and at first glance simplistic, explanation that 'it floods in Bangladesh when it rains in Bangladesh' is undoubtedly correct in a general sense. Nevertheless, rainfall incidence on the lower Ganges flood plain and on that of the Brahmaputra in Assam and its transfer across the international frontiers into Bangladesh must be added as a qualification. Thus, given the enormous quantities of monsoon precipitation receipts across the entire area of Bangladesh, Assam, Bihar, and West Bengal, and especially on the Meghalaya Hills (Cherrapunji), it is difficult to understand how land-cover changes in the Himalaya could have been cited to explain flooding patterns far downstream. The 1974 quotation credited to Eckholm is intriguing in this context – 'But I read several newspapers every day and have followed the accounts of many devastating floods over the last few years, and I have discovered that the news accounts *never mention deforestation as a cause of the flooding*' (Müller-Hohenstein 1974: 131 – my emphasis). It would appear that the discourse has come full circle.

This examination of the causes of flooding in Bangladesh, and by inference, on the Ganges flood plain, is only one, if crucial, step in the dispute. For the most part, Himalayan deforestation itself has been in progress for over two centuries. In many regions it peaked in the early twentieth century. In the last 20–30 years, in many areas of the Middle Mountains, and especially in Nepal, forest cover has actually expanded. The impressive agricultural terraces, both bari and khet, but especially the khet, that have replaced much of the original mountain forest and were originally portrayed as the *cause* of increased landsliding, soil erosion, and accelerating monsoon runoff, have been shown to be effective aids to soil conservation and water management (see previous chapters).

There is no doubt that the actual losses in lives and property on the Ganges and Brahmaputra flood plains have mounted significantly over the last hundred years or so. The increase, however, is due to the rapid population expansion and extensive development of infrastructure that has augmented the level of vulnerability. Even in this instance, and regardless of the pictures of disaster and human tragedy that the news media appear compelled to emphasize, the perceptions of risk held by the politicians and journalists are often out-of-phase with those of the rural people. Their way of life has evolved hand-in-hand with the annual flooding cycle. Nevertheless, there has been a serious increase in degree of hazard as their numbers have multiplied. This calls for more assistance regardless of the debate over the causes of flooding. In this context, Brichieri-Colombi and Bradnock (2003) have put forward a valuable proposal for bringing into use water from the Brahmaputra that currently is 'wasted' as it flows unused into the Bay of Bengal. However, they emphasize that without international cooperation no effective scheme for regional basin management will evolve. At a very different level, Paul (2003) discusses the increasing pressures for flood relief to assist the poor in the context of experience derived from the responses by government and NGOs to the 1998 floods.

Ultimately, it is necessary to return to regional physical geography. At the mega-scale, it is maintained that little has changed in the Ganges–Brahmaputra–Meghna hydrological system over many thousands of years. This remarkable highland–lowland interactive system from Himalayan crest to depths several kilometres below sea level in the Bay of Bengal is responsible for creation of the landscape of Bangladesh and the lower Ganges and Brahmaputra flood plains. The livelihoods of several hundred million people, mainly subsistence farmers, have been adapted to the vicissitudes of the system. While there are many justifications for assistance, over-dramatization and misidentification of the causes of vulnerability serve negative rather than positive ends. As firm and simple as the various steps in this discussion may now appear to be, misreporting, exaggerated reporting in the popular media, incautious scholarship, and harsh government policies inimical to the well-being of minority mountain peoples remain. Why is this so? This will be deliberated in subsequent chapters.

6 Mountain hazards

The figure of a woman, representing the Motherland, has lowered her head, dropped her hands, and is frozen in eternal grief. Behind her marble frame, in the depths of the Yarhich River valley, lie the drab hills of Tajikistan's largest grave.

(The Tragedy of Khait, Alexander Yablokov 2001)

Introduction

Very steep slopes, frequency of violent weather, and ongoing tectonic activity, these characteristics may lead to the perception that mountains are dangerous places in which to live and travel. Earthquakes, landslides, rockfalls, ice- and snow-avalanches, and floods, these too are part of the mountain image.

In the Himalayan region three specific facets of mountain hazards need to be identified. The first is that one form of catastrophic process often triggers another. Thus, landslides, rockfalls, and avalanches are frequently initiated by earthquakes, and these secondary factors may be responsible for much more damage than the initiating agent. Similarly, landslides may be triggered by heavy rainstorms and may themselves dam entire valley floors, thereby creating ephemeral lakes that eventually burst through or over the retaining dam to become devastating floods. Glaciers may advance from tributary valleys to block trunk valleys and create unstable lakes that also may drain catastrophically causing downstream flooding. Fortunately, vulcanism, another conspicuous mountain hazard, does not have a significant presence in the Himalayan region.

The second facet to be considered is the location of people and their habitations in relation to unstable slopes and other dangerous sites. Vulnerability is often coincident with poverty; it is the poor who frequently live in badly constructed homes located in less secure sites; and among mountain populations there is a high proportion of poverty.

The third facet is closely related to the second. Following a mountain disaster, poor weather may combine with difficulty of access so that rescue and relief responses may be greatly delayed or obstructed. Distance from major population centres, the primary locations for search and rescue and supply of relief materials, or poor communications that are a common feature of mountains, or both, inhibit timely response.

A further consideration is that human interventions on the mountain environment frequently exacerbate pre-existing slope instabilities thereby accentuating vulnerability to mountain hazards. For instance, poorly designed and badly constructed and aligned roads began to permeate the Indian Central Himalaya in the years following the border war of 1962 with China and this has resulted in a significant increase in losses from landslides and debris flows (Haigh 1984; Valdiya 1985) (see Chapter 4).

More recent large-scale engineering developments, such as hydro-electric installations and large dams, especially those located in vulnerable areas, have greatly increased the potential for serious losses. One of the long-lasting controversies in this context, for example, has been India's construction of some of the world's highest multi-purpose dams in areas of suspected high seismic risk; many of these have been located along the contact zone between the Himalaya and the foothills and flood plain. Furthermore, collapse of an artificial dam due to an earthquake would cause the release of a large volume of water on unsuspecting and unprepared downstream populations and infrastructure.

These and other hazards are the topic of this chapter. Some of the actual hazards, such as avalanches and glacial lake outburst floods (*jökulhlaup*), are unique to mountains and their adjacent piedmonts. Others, such as landslides, earthquakes, and floods, are not, although they occur with greater frequency in mountain regions than elsewhere.

The processes introduced so far are generally described as *natural hazards*. It is worth emphasizing that such processes become risks, or threats, to humans and their property only when there is potential for direct contact. Thousands of ice- and snow-avalanches and rockfalls occur annually in the Himalaya and adjacent mountain systems unseen by humans, or seen from afar as dramatic mountain phenomena. It is these events to which the term *natural hazard* is most appropriate. The term *mountain hazard* (Dow *et al.* 1981; Ives and Messerli 1981) was introduced during the earliest stages of the UNU mountain programme because it was recognized that many of the catastrophic processes referred to above may be augmented by human interventions on the landscape. Thus, the original name of UNU research in the Himalaya was the *Mountain Hazards Mapping Project*, and the term is retained here.

The intent of this chapter is not to provide an exhaustive discussion of the mechanics and distribution in time and space of these complex geophysical processes, nor of their comprehensive impacts on human activities in the Himalayan region. For this, the reader may wish to survey sections of the very large literature available. A much more limited objective of using specific examples from the region as case studies is adopted although, where relevant, reference will be made to some of the more technical literature. For an overview of the inter-relations between rapid slope processes and associated phenomena and human affairs the reader is referred to the outstanding work of Hewitt (1997) and Valdiya (1998).

Visitors to mountains, as well as the permanent inhabitants, become subjected to many of these mountain hazards. To this must be added accident and illness on steep slopes and at high altitude and the associated problems of rescue and evacuation. Although this topic is of considerable importance it will be considered only in relation to tourism and its impacts (Chapter 7).

There are other threats to mountain inhabitants and their environment that exceed in destructive force the processes referred to above as *mountain hazards*. These begin with ethnic and gender discrimination and accelerate all the way to guerrilla activity, insurgency, active warfare, and genocide. They will be introduced in Chapter 8.

Earthquake hazards

Thousands of earthquakes of varying intensities and with equally varying impacts on human life and property have occurred throughout the Himalayan region. The intense seismic activity results from the Indian tectonic plate thrusting beneath the Tibetan Plateau at a rate of about 45 mm/year and rotating slowly in an anti-clockwise manner (Sella *et al.* 2002, quoted by Bilham 2003). Bilham explains that, because of complexities in the structural units along the western, northern, and eastern boundaries of the Indian plate, such velocity is not directly observable across any single fault system. He stipulates that there is a potential slip at about 1.8 metres/century available to drive large thrust earthquakes beneath the Himalaya. Under such conditions earthquakes associated with a slip of six metres, for instance, cannot occur before a lapse of at least 300 years between major seismic events. Survey measurements across the subcontinent indicate that, as with many continental plates, very little contraction occurs (less than 5 mm/yr) and that the origin of earthquakes there is likely attributable to flexure of the Indian plate as it is bent down beneath the Himalaya. Based on these observations, Bilham (2003) provides a compilation and discussion of the historical record of earthquake occurrence for the Himalaya and the subcontinent.

There is extensive documentation of major, and many more minor, earthquakes, but the record is incomplete even though events were recorded in the Mahabharata, for instance, as long ago as 1500 BC. Bilham concludes that the record of major events may be complete only for the last 200 years, but it is certain that many large events occurred prior to 1800 that have not been recorded. Furthermore, a general absence of surface ruptures implies that many historic earthquakes occurred on faults that, so far, have not been mapped. A corollary to this is that many hundreds of subsurface faults are potentially awaiting reactivation for which there is no current geological evidence. In the Himalaya, the largest fault is invisible and lies 10–15 km beneath the surface. It is responsible for the slippage of the Indian subcontinent during Great Earthquakes (M> 8.0[1]) and is the active fault closest to many cities and engineering schemes constructed throughout the Himalaya (Figure 6.1).

These observations are of critical importance for estimation of relative earthquake hazard in the context of future security of large-scale infrastructure, especially hydro-electric dams, in the Himalaya. Bilham nevertheless supports the hypothesis of 'seismic gaps' that leads to the notion that a major earthquake (M=8.0, or greater) may be imminent along several large sections of the Himalayan front. The incomplete history of the region's earthquakes makes it difficult to estimate the timing of seismic activity. Nevertheless, the amount of slip in future earthquakes, should they occur, can be calculated with reasonable certainty. In some places, such as western Nepal and Kumaun, as much as nine metres of slip may be stored in elastic energy in the Himalaya, sufficient to drive a Great Earthquake.

The Tehri Dam

The controversy surrounding the planning and construction of the Tehri Dam is introduced in some detail because it epitomizes the overall issue of large-scale construction in seismically active mountain regions. Planned as one of the world's largest and highest rock-filled structures, it has been a centre of conflict in India for several decades. The controversy is multi-faceted. The project involves negative consequences, such as the drowning of thousands of hectares of fertile farmland and many villages, and the forced evacuation of about 100,000 people, as well as the positive aspect of provision of an enormous addition to northern India's electricity capacity. Thus there are two distinct components to the controversy. The one to be discussed here is the earthquake hazard in relation to the design of the dam and the disagreements between government and non-government experts on the adequacy of the design for withstanding a large earthquake. By implication, this issue is relevant to several other equally massive and contentious structures. The human and socio-economic aspects will be discussed in Chapter 8.

The Indian authorities that have pushed for construction of the Tehri Dam have been accused of secrecy and misrepresentation since the Uttar Pradesh Planning Commission gave

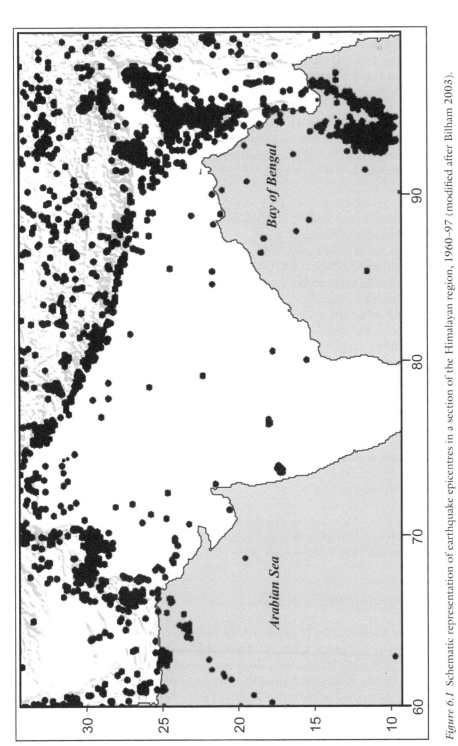

Figure 6.1 Schematic representation of earthquake epicentres in a section of the Himalayan region, 1960–97 (modified after Bilham 2003).

approval for construction to begin in 1972. Because of concern over this lack of transparency, given the potentially very high magnitude of the hazard, a major workshop was organized 15–16 January 1993 in New Delhi. It was co-sponsored by the Wadia Institute for Himalayan Geology, Dehradun, the Friedrich Naumann Foundation, Germany, and Development Alternatives, New Delhi. Professor Vinod K. Gaur (1993) edited the book arising from the papers presented at the workshop. In the preface, N.D. Jayal, Indian National Trust for Art and Cultural Heritage, draws the following conclusion:

> the response from some of the leading seismologists and earthquake engineers of the world was gratifying, and their unanimous conclusions . . . based on the learned scientific papers presented by them at the workshop, will . . . lead to a better understanding and respect amongst the public and policy makers for scientific truths.
>
> (Jayal 1993: vi–vii)

He goes on to refer to the 'scores of high dams being recklessly planned in the geologically and ecologically fragile Himalaya'.

Considerable detail is presented here on the seismic and technical engineering aspects of the Tehri Dam based on the book (Gaur 1993). This is because it introduces remarkable insights into both the earthquake hazard itself and the highly sensitive political and social issues that surround it. The Tehri Dam controversy is the *cause célèbre* of the problems of large-scale infrastructural development within and closely adjacent to the Himalaya. It is a region that holds some of the world's greatest potential for harnessing water power that is urgently needed to ensure progressive development, especially of the densely populated plains to the south of the mountains.

In the introduction to the book, Professor Gaur refers to the 2,000 km long arcuate shape of the Himalaya. He states that:

> The four great Himalayan earthquakes of this century, each of a magnitude 8.5 or greater and accompanied by about 200–300 km long rupture of the detachment plane, leave one in no doubt that the intervening segments constitute seismic gaps which are most likely now close to rupture, each capable of producing equally great earthquakes within the next 100 years or so.
>
> (Vinod Gaur and I now believe that the magnitudes are 1897:M=8.1, 1905: 7.8, 1934: 8.1, and 1950: 8.5, smaller than stipulated hitherto. The 1905 rupture was 150–190 km long and the 1897 event was not in the Himalaya but beneath the Shillong Plateau.)
>
> (Roger Bilham pers. comm. 24 April 2003)

In a recent evaluation of Himalayan seismic hazard Bilham (2003) concludes:

> Despite the diverse quality of the data in the past two centuries, we can be sure that we are not missing any great event since 1800. This permits us to estimate the minimum slip potential that has accumulated along the Himalaya since the last great earthquake [Figure 6.1]. We divide the central Himalaya into ten regions with lengths roughly corresponding to those of the great Himalayan ruptures (~220 km). With a convergence rate of 20 mm/yr along the arc, six of these regions currently have a slip potential of at least four metres – equivalent to the slip inferred for the 1934 earthquake. This implies that each of these regions now stores the strain necessary for such an earthquake.

Moreover, the historic record indicates no great earthquake throughout most of the Himalaya since 1700, suggesting that the slip potential may exceed six metres in some places. A case can be made that no great earthquake has occurred since 1505 in Kumaun and western Nepal, giving a slip potential of 9.5 m or M=8.2.

The site of the Tehri Dam falls within one such seismic gap and is less than 15 kilometres from the primary fault line. The dam is 261 metres high and 575 metres across, situated in the foothills of the Garhwal Himalaya on the Bhagirathi River close to the town of Dehradun.

Despite the inadequate seismic record of the Himalayan region, there was little disagreement among the participants of the 1993 conference that areas along the Main Boundary Thrust on either side of the dam site have experienced 'Great Earthquakes', magnitude of M=8.4–8.7, in the last 100 years. Furthermore, it was agreed that the sections of the Himalayan front within the seismic gaps would also experience a Great Earthquake at any time within the next 100 years. Bilham, however, now believes that these estimates are too high. His most recent studies suggest that the 1885 'Kashmir' and the 1905 'Kangra' quakes were only M=7.0 and 8.0 respectively (pers. comm. 24 April 2003). Regardless of these more recent estimates, it is believed that those sections of the Himalayan front within the seismic gaps could experience a Great Earthquake within the next 100 years. Thus, the question remains: is the Tehri Dam designed to withstand such a cataclysm? Valdiya, in the introductory chapter to Gaur's book (1993b: 1–34), provides an extensive overview of Himalayan tectonics and recent crustal movements. The later chapters take up the technical questions in detail.

The contributors to Gaur's book emphasized the lack of good quality earthquake data, and absence of acceptable experimental determinations to characterize ground behaviour during an earthquake. There was discussion about peak ground acceleration and the relatively low figure for this vital component of engineering risk assessment adopted for the actual Tehri Dam design. Bolt (1993: xvii) clarified that the concept of Effective Peak Ground Acceleration that he had earlier proposed had been misused by the Tehri Dam earthquake engineers to arbitrarily reduce the value of estimated Peak Ground Acceleration by 50 per cent. This is an example whereby the planning engineers had lowered dam specifications significantly so that a lower figure for cost of construction was derived. Thus the risk of failure was increased.

Another example of disagreement over dam design criteria is introduced to emphasize some of the background to the challenge presented to the authorities concerning this massive undertaking. Iyengar (1993: 138–44) pointed out that the design and safety evaluations had been based upon the incorrect assumption that, during an earthquake, the structure in question would behave as a two-dimensional unit. He argued that it would respond rather as a three-dimensional unit, given its location in a narrow, steep-sided valley. This led to the conclusion that the fundamental seismic frequency of the dam would be 1.8 times higher than the two-dimensional value of 0.51 Hz assumed by the designers, with consequent accentuated seismic response.

A third design issue, introduced here, relates to the dynamic response analysis conducted by the Soviet consultants who had been part of the Soviet Union's aid to India. In any large structure, computer simulated responses for displacements begin to increase significantly at about 8 seconds after the initial shock (Finn 1993: 131) and reach a maximum at 12 seconds. However, the computer simulations in this case had been terminated at 12 seconds and so there was an assumption that the level of shaking would decrease thereafter, thus creating an illusion of safety. Finn argued that during a Great Earthquake, of say M=8.7 magnitude involving a rupture length of 200–300 km, the maximum intensity of shaking attained at 12 seconds would continue to stress the structure for an additional 20–30 seconds. This implied that a realistic response spectrum was required that would be compatible with sustained ground shaking for at least 30–40 seconds.

Although Finn (1993) concurred that the construction design for the Tehri Dam did contain various defensive measures, there was a unanimous conclusion that significant uncertainty remained. This led to the conviction that, from a scientific, non-political, point of view, all was not well. It implied that work on the high dam should be suspended until more thorough assessments could be made, despite the large sums of money already spent.

There remained an equally important consideration that, while not part of the scientific and engineering assessment, it most certainly lay behind it. The plans would necessitate the eventual displacement of more than 100,000 people, and the Government of India's previous record involving the forced movement of people has been heavily criticized. The earlier design estimates for cost–benefit, including siltation rates and projected longevity of the reservoir, were of dubious value. Several examples of longevity estimates for a series of dam projects have been shown after completion to have been optimistic and irresponsible (Ives and Messerli 1989: 217–19). High dams appear often to be foisted onto an unsuspecting public with promises of greater benefits than ever materialize. In the Himalaya, it is now well substantiated that engineers have almost invariably grossly underestimated rates of siltation by ignoring the bedload component. Furthermore, there has been a lack of competent measurements of sediment load in suspension or solution. Certainly, no reliable data were available prior to finalization of the Tehri Dam specifications. This section will be concluded with a quotation from one of the workshop participants:

> There is always a trade-off between life safety and cost and between life safety and economic benefit to the population. Decisions on what are reasonable degrees of conservatism seem best treated if the situation is analysed in the broadest probability context, with all available data scrutinized and all the various hypotheses debated.
>
> (Bruce A. Bolt 1993: 91)

While public debate based on all available data is not practical where millions of people, very many poor and illiterate, are involved, the Tehri Dam project has become a crisis of confidence. Furthermore, large dams invariably have been constructed for the benefit of vested interests in the lowlands while the mountain inhabitants, whose good environmental management is essential for the protection of the public investment, rarely receive any of the benefits. There is also the need to consider that the weight of the dam itself, and especially of the huge amount of water accumulated upstream in the resulting reservoir, can itself induce earth movements.

Other Himalayan region high dams and seismicity

Despite the lack of public and expert confidence in the Indian government's efforts to complete the Tehri project, Soviet engineering design has succeeded in construction of high dams even more massive than the Tehri Dam, also in mountain areas subject to high risk seismic activity. Figure 6.2 and Figure 6.3 are photographs of the Nurek hydro-electric power station and dam, Tajikistan, taken in 1987. The dam is more than 300 metres high and is reputed to have withstood a 7.9 Richter scale earthquake. A sister dam, the Rogoun Dam, further up the Vakhsh River, was well into the construction phase in 1990 and was originally

Figure 6.2 The Nurek Power Station, Tajikistan, on the Vakhsh River. The dam, located immediately off the right margin of the photograph, was built in part by volunteer Communist Youth labour and, at 300 metres, was reputed the highest rock- and earth-filled dam in the world. It is reported to have withstood an earthquake of Richter-scale 7.9 intensity. Photograph, July 1987.

Figure 6.3 The Nurek Dam, an earth- and rock-filled structure. Close to the Afghanistan frontier, for several decades it has been heavily guarded against the possibility of sabotage, July 1987.

planned to be about the same height. Construction was suspended in 1991, not because of seismic activity but by civil war followed by a long and continuing period of guerrilla activity.

The Pamir have been depicted as the most seismically active mountain region in the world. Due to inaccessibility, both geographical and political, no major infrastructure has been undertaken so far, except for the Nurek and Rogoun dams in the western foothills. However, there have been large earthquakes that have left their mark on the landscape and on the nervous systems of the local inhabitants and government authorities. The most significant occurred on 10 July 1949 and resulted in the loss of more than 28,000 lives. It was an earthquake-triggered rockfall and landslide that obliterated the regional centre of Khait and many villages. Another earthquake-triggered landslide occurred during the winter of 1911 damming the River Murghab and causing the formation of Lake Sarez that, by 1999 had a volume of about half that of Lake Geneva. These are only two of many incidents, but they will be discussed in detail in the following section on landslides.

Glacial lake outburst floods – *Jökulhlaup*

Jökulhlaup is an Icelandic term literally meaning 'glacier leap'. It has entered much of the glaciological literature because the phenomenon of catastrophic release of large volumes of water from beneath glaciers was recognized and studied more than three centuries ago in Iceland (Thorarinsson 1960). In fact, the term is much older than that and is part of the Icelandic vernacular that has been adopted into the technical literature. The term most commonly used in the Himalaya is *glacier lake outburst flood*, or simply *GLOF*, deriving from some of the first descriptions of such events in Nepal by Galay (pers. comm. 7 July 1986). Groups of mountain people have clearly long been aware of such phenomena as they have their own terminology (for instance, the Sherpas of the Khumbu use the word *tshoscrup*).

Early knowledge of such catastrophic drainage from lakes ponded behind glaciers derives from exploratory travels in the western Himalaya and Karakorum (e.g. Mason 1935) and the major instances have been catalogued by Hewitt (1964, 1982). Most of these events occur when glaciers in tributary valleys advance to block the trunk valley causing the main river to pond against the glacier dam. The size and volume of the accumulating lake depends upon the height and strength of the ice dam. Eventually the growing volume of water either lifts the ice dam hydrostatically, over-tops it, or drains by enlarging subglacial tunnels and crevasses. In almost all cases the ensuing release is very rapid. The strength of the ice dam, and hence the volume of ponded water needed to break the dam, determine the size of the resulting flood that could be of considerable magnitude and the devastating effects felt far downstream. A large jökulhlaup is one of the most catastrophic events in nature and, in extreme cases, may cause extensive damage extending over several hundred kilometres downstream of the initial outburst. More modest floods may result from the drainage of smaller lakes that accumulate along the margins of valley glaciers. They may drain regularly on an annual basis, or even several times each year. Regardless of the magnitude of the event, however, the route taken by the flood/debris flow is extremely hazardous. This is the classic jökulhlaup – the catastrophic drainage of a glacier-dammed lake.

Although this form of jökulhlaup is still quite common, its incidence has been much reduced because of the progressive climate warming that has occurred worldwide over the last century, and especially since about 1970–80. Glaciers, in general, have thinned and retreated in recent decades. Nevertheless, this widely accepted generalization is controversial to a degree. Alford (pers. comm. 24 April 2003) reports significant glacier advances in the Karakorum of northern Pakistan in recent years. Above the Karakorum Highway, for instance, the Gulmit and Passu glaciers appear to have active ice at their termini. In the upper Shimshal Valley, the Khurdopin Glacier advanced across the valley to form an ice-dammed lake shortly before 2000. The Yukshin Gardin Glacier, immediately to the east, also advanced across the valley to become contiguous with the Khurdopin Glacier. Further down-valley, several other glaciers were also advancing, and in the Ghizer, east of Gilgit, the Karambar Glacier, that had been the cause of a recent jökulhlaup, was again advancing. Alford visited Shimshal with Galeeb Kuchra in October-November 2000. He presented an internal report to Focus Humanitarian Assistance (Washington DC) on the Shimshal jökulhlaup event that occurred on 10 June 2000 and made recommendations for possible future hazard assessment and mitigation. These glacier advances in the Karakorum are probably the result of glacier surges and are not necessarily related to recent short-term worldwide climate warming but rather to a much longer-term glacier dynamic.

The process of glacier recession in other parts of the Himalaya has witnessed the formation of innumerable small lakes caused by the infilling of the depression between the outer moraines of many glaciers and the retreating ice front. The more recent prominent end and lateral moraines of Himalayan glaciers were formed during the so-called Little Ice Age (Grove 1988), or Neoglacial Period (1500–1850). There have been numerous studies of glacier mass balance and climate change throughout the world and the reader is directed to a very useful general reference for more detailed information that lies beyond the scope of this chapter (Benn and Evans 1998). Here the focus will be on the relatively recent development of moraine-dammed and supra-glacial lakes in the Himalaya that has been recognized as a threat to humans and their infrastructure only in the last 20–30 years.

The altitude of most of the end and lateral moraine systems of the Little Ice Age is close to, or slightly above, the lower regional limit of permafrost (perennially frozen ground). This relationship was identified by Müller (1959) as early as 1956 during the successful Swiss mountaineering expedition to Everest and Lhotse. It is an important observation for the current discussion because many of the moraines that serve to hold up the evolving glacial lakes are (or were) frozen and/or contained ice cores. This made them highly impervious barriers to the release of ponded melt water. Furthermore, the pattern of Himalayan glacier melting has led to the formation of lakes, not only between the receding ice fronts and the end moraines, but also by amalgamation of many small melt ponds on the glacier surfaces. Most lakes are probably a combination of the two modes. The current climate warming trend is also causing the lower altitudinal limit of permafrost to rise so that the ice cores of many of the end moraines have begun to melt, thus reducing their ability to retain the glacial lakes that are forming behind them.

There are earlier accounts of glacial lake discharges in the Nepal and Bhutan Himalaya (Gansser 1966, 1970; Buchroithner *et al.* 1982; Fushimi *et al.* 1985). Cenderelli (2000: 73–103) reports 11 floods from the collapse of moraine dams in Nepal and Tibet since 1935. Nevertheless, it was the catastrophic drainage of the moraine-dammed lake, Dig Tsho, in the Khumbu on 4 August 1985 that eventually set in motion governmental reaction. This was because the flood waters and masses of sediment that they incorporated destroyed a nearly completed hydro-electric power plant some 12 km below Dig Tsho (the Thame mini-hydro-electricity plant, a project developed with Austrian government assistance). In addition, three or four Sherpa lives were lost, and 30 houses, many hectares of scarce agricultural land, and 14 bridges were destroyed. Nevertheless, the Thame plant was rebuilt close to the same site with Austrian funding and engineering assistance and production began in 1994.

The Dig Tsho glacial lake disaster, Khumbu Himal, 1985

The disaster occurred during the final stages of the UNU hazards mapping study in the Khumbu, and Zimmermann and Vuichard returned to the Khumbu in September 1985 to survey the effects of the flood immediately following receipt of the news from Kathmandu. The following summer a detailed report was prepared on the event, sponsored by ICIMOD (Ives 1986; Vuichard and Zimmermann 1986, 1987). Victor Galay, a hydrological engineering consultant with the Nepal government, provided valuable assistance.

As part of the evaluation of the Dig Tsho disaster a strong recommendation was made for systematic mapping of glaciological hazards throughout the Himalayan region (Ives 1986). Intermediate and more detailed studies were proposed at selected locations, such as other parts of the Khumbu, the Arun River watershed to the east, and the Rolwaling valley to the west, where examination of remotely sensed imagery pointed to the existence of potentially dangerous glacier lakes.

Imja Glacier, Khumbu Himal – potential disaster

The development of 'Imja Lake' is used here to illustrate the rapidity of formation of glacial lakes in the Nepal Himalaya. Quite extensive detail is available because of the serendipitous acquisition of vital photographs, thus making 'Imja Lake' the first of the Nepalese Himalayan lakes to receive detailed attention. The lake's very rapid formation (approximately 1972–1984) indicates how quickly potentially hazardous situations can arise.

The baseline for the study was derived from the work of Fritz Müller who spent eight months in the Khumbu in 1956. He made extensive use of Schneider's photo-theodolite as well as conventional cameras, taking many hundreds of photographs. After his untimely death in 1979 the UNU project obtained access to much of his photographic record, although his black and white negatives had been lost. Following the Dig Tsho jökulhlaup of August 1985 a review of Müller's photographs uncovered several of the Imja Glacier. These show the glacier abutting the Lhotse Glacier with its lowermost section covered by surface moraine. A few small melt ponds are apparent amongst the morainic debris, but certainly no lake. Barry Bishop (pers. comm. 12 October 1987) had photographed Imja Glacier from the summit of Ama Dablam in 1963 and confirmed that no lake had existed on that occasion. Additional photographs taken in 1971 by Fushimi (Fushimi *et al.* 1985) show approximately the same situation. Japanese glaciological expeditions onward from 1971 produced a full record of the progressive amalgamation of a series of small melt ponds amongst the surface debris to eventual creation of 'Imja Lake'. Hammond (1988) and Watanabe *et al.* (1994) give detailed accounts of this development. By 1984, air photographs obtained by Bradford Washburn (pers. comm. 14 January 1986) depict a considerable lake some 0.54 km^2. Kawaguchi (pers. comm. 7 September 1992) provided the remarkable oblique air photograph looking up the length of the glacier on 4 November 1991 (Figure 6.4). Subsequent fieldwork under the leadership of Watanabe includes a detailed assessment of lake evolution and its associated potential hazards up to 1994.

In 1992 the approximate dimensions of the lake were: area, 0.69 km^2; maximum length, 1,450 m; maximum depth, 99 m. The lake was then expanding eastward by about 5 m/yr, and very slowly westward. Bottom melting between 1991 and 1997 was about 2 m/yr (Mool *et al.* 2001a: 144).

The potential hazards associated with the lake's rapid formation (Hammond 1988; Watanabe *et al.* 1994) have since proved to have been somewhat over-stated. Given the lake's

Figure 6.4 In 1956 Fritz Müller photographed the Imja Glacier the lower part of which was completely moraine covered, except for a few small ponds. This photograph taken by K. Kawaguchi on 4 November 1991 shows the rapidly expanding supra-glacial 'Imja Lake', at this stage about 1.1 km long and 90 m deep. The end and lateral moraines are ice-cored. The lake continues to expand eastward into the glacier front and to increase in depth. Whether or not it will produce a *jökulhlaup* cannot be predicted. It is located in the Khumbu Himal immediately south of Island Peak and Nuptse.

growth rate and rapid increase in depth, however, it was not unrealistic to alert the Nepalese government authorities and public to a possibly dangerous situation, particularly since several tens of kilometres of the main trekking route to the Mount Everest base camp, Sherpa summer settlements, and parts of permanent settlements in the lower Imja Khola Valley would be at risk if an outburst flood occurred.

Subsequent and more detailed investigation (Watanabe *et al.* 1995) has revealed that the frontal morainic zone of the Imja Glacier is underlain by extensive dead ice and a small lake overflow channel has formed. As the lake expands eastward, largely because the lake waters are melting into the glacier front, the lake level is falling. Thus, while the lateral moraines are extremely unstable, with debris falling into the lake continuously, future attention must be focused on the lake level and the rate of melting of the frontal dead ice which alone separates lake waters from the end moraines. Melt-water from the Imja Glacier and from the winter snow accumulation will continue to supply water to the lake, ensuring that an indeterminate degree of hazard will remain.

Rolwaling Valley and glacial lake Tsho Rolpa

While the research on the Imja Glacier was under way, another moraine-dammed lake, Tsho Rolpa in the Rolwaling Valley to the east of the Khumbu, was recognized as a potentially serious threat to settlements in the valley below.

Fears of the danger posed by Tsho Rolpa increased in July 1991 when the moraine dam supporting nearby Tsho Chubung collapsed. The glacial lake outburst flood that followed damaged six houses in the Sherpa village of Beding and drew attention to the much more formidable Tsho Rolpa, the level of which was seen to be close to spilling over the end moraine crest. The lake itself had formed progressively over the period 1956–1991 and was more than 1.3 km long by 1991. The mapping of its development shows a similar pattern to that of 'Imja Lake' (Mool *et al.* 2001a: 149). Several agencies and individuals responded to the Sherpas' request for assistance: the Water and Energy Commission (WECS), HMG Nepal; Michael Damon, IIAS & ES, Enchede, Netherlands; ICIMOD; and various engineering consultants, students, and NGOs. The lake was carefully surveyed, the stability of the retaining end and lateral moraines reviewed, and a series of attempts were made to lower the surface level of the lake. Grabs and Hanisch (1993) formulated a technique for lowering the level of glacier lakes artificially by the application of syphons. Wavin Overseas, Netherlands, the leading manufacturer of plastic pipe in Europe, designed a siphon system for Tsho Rolpa and installed three siphons that functioned successfully for 14 months in 1995–1996 and intermittently until 1999 in this cold, high altitude and severe environment at close to 5,000 metres above sea level.

In 1996 reports from the ongoing Tsho Rolpa lake survey indicated the possibility of an imminent jökulhlaup that was estimated to reach and impact the Khimti hydro-electric power station located 75 km downstream. This prompted the government to commission the Bhutwal Power Company to model the anticipated jökulhlaup (Mool *et al.* 2000a; Rana *et al.* 2000). A further survey of the Tsho Rolpa end moraine in May 1997 reinforced the impression that a disaster was imminent, which led to the evacuation of most of the Rolwaling inhabitants. Police and military camps were set up as part of an attempt to provide an early warning of danger. No flood occurred and the local people returned to their homes. It became apparent, subsequent to the alarm, that official reaction to the potential threat had been excessive and the evacuation had been unnecessary (Dipak Gyawali pers. comm. 21 November 2003; see Chapter 10).

Additional siphons were installed on Tsho Rolpa, although the rate of lowering of the lake level barely exceeded the inflow of melt-water. Subsequent measures, also financed by the government of The Netherlands, have involved cutting a notch through the end moraine protected by a coffer dam. Release of lake water via the notch was scheduled for the spring of 2000. It had been estimated that a period of 12 days overflow at about 0.4 million cubic metres per day would be sufficient to lower the lake by three metres. In fact, 5.3 million cubic metres of water were released over a period of 14 days. A new maximum lake depth was established at 140 metres by echo-sounding following the lake lowering in July 2002 and there were plans for construction of a mini-hydro-electricity plant during 2003.[2] The next step was to be the cutting of a much deeper channel parallel to the test notch, designed to achieve a further lowering of the lake surface by 17 metres. The total 20-metre lowering was considered to be a permanent solution (Mool *et al.* 2001a: 185). In effect, this was completed by December 2002 with funding of nearly US$3 million from The Netherlands and a contribution of more than US$100,000 by HMG Nepal (many of these details are by pers. comm. Pradeep Mool, 14 April 2003).

Arun III Hydro-electric Project

As these threats from glacier lake outbursts received attention, the World Bank decided to review the potential glacial hazards facing its plans to promote construction of a major hydro-electricity project on the middle section of the Arun River in Nepal to the east of the Khumbu. The Arun, which has over 90 per cent of its 27,000 km² watershed in Tibet (China) has been appraised for its potential to produce a very large hydro-electricity output. Initial planning was for the construction of a 'cascade' of power stations along the river. Arun III, one of the components of the multi-stage project, was the choice for the first phase of construction. Since the run-of-the river potential electricity capacity for Arun III alone, estimated at 402 MW, exceeds by many times the amount Nepal as a whole would ever likely use in the immediate future, the prime purpose was for sale to India to acquire much needed foreign earnings for Nepal.

From the beginning, the Arun 'cascade' project has been heavily criticized for its assumed massive ecological impact – for instance, a high quality road is needed over a distance of 118 km to the site of Arun III. Furthermore, as the sale of electricity to India, as the sole feasible purchaser, was the main justification for the undertaking, strong objections were raised both from within Nepal and internationally. Nevertheless, the World Bank in the early 1990s was determined to proceed, with strong support from the Nepalese government.

The growing awareness of the potential danger of glacier lake outburst floods that had been prompted by the Khumbu mountain hazards studies and Galay's consultancy with the Nepalese Water and Energy Commission (WECS) led to a significant reconnaissance of glaciers and glacier lakes in eastern Nepal and neighbouring Tibet. A joint Sino-Nepalese study of the upper Arun watershed in 1991 identified over 100 glacier lakes, most of which were in Chinese territory. Also in 1991, WECS, with technical support from the Japan International Cooperation Agency (JICA), conducted field studies of several glacier lakes within Nepal. Second to Tsho Rolpa, the 'Lower Barun Glacier Lake', 55 km upstream from the Arun III construction site, was identified as the most serious potential hazard. The lake volume was estimated to be about 28,000 m³ and it was calculated that any major breach of its moraine dam would produce a flood wave that would reach the Arun III site within about an hour.

This rapid accumulation of information about the potential hazard induced the World Bank to organize a meeting of experts in Paris (27–28 April 1995) to review the situation. The government of Nepal was represented as well as the engineering firms involved in the design of Arun III and institutions that had provided technical advice concerning glacier lake problems. Observers from the anticipated major donor governments, Germany and Japan, were also present. To avoid any confusion during the ensuing two-day hearings, the discussion was rigorously restricted to examination of the potential threats to the proposed structures from mountain hazards and especially the drainage of glacier lakes. Consideration of ecological impacts and social and economic aspects was excluded.

The participants had been provided with an October 1994 advisory report presented to the Ministry of Water Resources, HMG Nepal, by Wolfgang Grabs, Teamleader, Snow and Glacier Hydrology Project. The report indicated the uncertain situation surrounding glacial lake hazards and Arun III at that time. For instance, it referred to the 'unsatisfactory state of institutional cooperation' and noted that the 'Snow and Glacier Hydrology Project cannot take the responsibility for any consequences that arise from a Glacial Lake Outburst Flood'. The report also expressed concern that no co-ordinated efforts were under way to ensure prevention or mitigation of glacier flood hazards, nor to prepare any flood hazard maps of most endangered areas in the valleys below known glacier lakes. Nevertheless, one of the results of the April 1995 meeting was to substantially correct this situation of institutional neglect.

The unanimous recommendation from the April meeting, however, indicated that there was no reason to delay the start of construction on the Arun III project in the narrow context of the potential for glacial lake outburst. It was agreed that the accompanying proposal for a glacier hazards survey, with plans for the possible lowering of the level of the 'Lower Barun Glacier Lake', could be accommodated within the lead time that would be provided by completion of the access road. It was recommended, therefore, that construction proceed. However, within two months of this recommendation, the German and Japanese governments withdrew their support because of ecological and socio-economic considerations and the World Bank concluded that the project had serious economic deficiencies. Thus Arun III was shelved indefinitely. Nevertheless, planning activities have since been reactivated and Asian Development Bank funding was approved in June 2002. Gyawali (2001: 66–87 and 148–52) provides an assessment of the politics behind the Arun III Project and this is discussed more fully in Chapters 9 and 10.

Other potentially dangerous glacial lakes

Acknowledgement of the potential danger of up-stream glacier lakes to the proposed Arun III Project was followed by concerted efforts to determine the nature of jökulhlaup hazards across a large area of the Himalaya. Yamada (1998), with support from JICA, published a survey of glacier lakes and their outburst flood potential for eastern Nepal. It included some spectacular oblique air photographs of many of the lakes under consideration. Watanabe *et al.* (1995) investigated glacier lake flood events in the Bhutan Himalaya. And a major survey and glacier and lake inventory of the entire Nepal and Bhutan Himalaya was completed by Mool *et al.* (2001a, 2001b). They documented 3,252 glaciers with a total area of 5,324 km^2 in Nepal and 677 glaciers with a total area of 1,317 km^2 in Bhutan. A surprising total of 2,323 glacier lakes were listed for Nepal, of which 20 were considered to be potentially dangerous. Similarly, 2,674 glacier lakes were recorded in Bhutan. Although many of them are very small, 24 were considered potential threats. These two extensive inventories (Mool *et al.* 2001a, 2001b), with numerous photographs, technical drawings, and innumerable factual details, have been published by ICIMOD with financial and technical assistance from the United Nations Environment Programme (UNEP). Unfortunately, availability of the inventories has led to an over-dramatized assessment, predicting millions of deaths during the twenty-first century as a result of glacier lake outburst floods (Pearce 2002). This will be discussed again in greater detail in Chapter 10.

The phenomenon of glacier lake outburst floods in the Himalaya and surrounding mountain ranges has finally attracted governmental and media attention. While the major classic jökulhlaup has long been recognized in the Karakorum and adjacent areas (Mason 1935; Hewitt 1964, 1982) the extremely rapid development of glacial moraine-dammed and supraglacial lakes over the last few decades has become an event of considerable importance and is being slowly, but progressively, taken into account throughout the region. Undoubtedly, small mountain villages and individuals have suffered from this form of mountain hazard over the centuries, although the traditional siting of villages has usually taken such hazards into account whenever possible. What has put the process into its present-day perspective is the continued expansion of modern infrastructure into a mountain system that only a few decades ago was isolated from world events.

Landslides and rockfalls

The terms *landslide* and *rockfall* are used in a variety of ways and each has several synonyms. *Landslide* is applied as a general term throughout the Himalaya and is often used to include debris flows (Caine and Mool 1982), mudflows, and other forms of movement of mass on slopes under the influence of gravity. *Rock avalanche* is often used as a synonym for *rockfall*. Rockfalls and landslides often grade into each other; furthermore, jökulhlaup and other suddenly released bodies of water have the ability to pick up debris so that the incorporated material may far exceed 50 per cent of the total volume.

Any discussion on landslides and rockfalls must take into account the enormous range in the magnitude of individual occurrences – from the collapse of entire mountain slopes that may entail more than 10 km³ of material to small soil slips of a few cubic metres on the front edge of agricultural terraces. Several triggering mechanisms that prompt the downslope release of rockfalls and landslides have been identified. They range from earthquakes, to torrential rain storms, to periodic warm weather melting of interstitial ice on unstable rock walls, to the eventual automatic release of masses of rock on high angle slopes that have been inherently unstable since slope over-steepening by glacial erosion during the last Ice Age. Several triggering mechanisms may combine and human intervention will frequently destabilize a slope, or accentuate instability that is present naturally. For example, rapid and poorly planned road construction has been held responsible for thousands of landslides (Haigh 1984; Valdiya 1985, 1987).

The world's largest recorded rockfall occurred in the Langtang Himal, Nepal, probably more than 25,000 years ago (Heuberger *et al.* 1984). The volume may have exceeded 10 km³. The resultant landforms were so vast that visiting scientists walked over them mapping glacial moraines on their surface without recognizing the main underlying feature. Heat generated by friction along the gliding plane was sufficient to fuse rock material to form *frictionite*. Heuberger *et al.* (1984) estimated that between 60 and 75 per cent of the originally deposited material was removed by glaciers during the last major glacial period. The assumed minimum age of more than 25,000 years, at first glance, may suggest that such events are scarcely relevant in terms of today's human occupation and development of the Himalayan region. None the less, gigantic events of this order of magnitude, while frequently very hard to detect and almost impossible to predict, are characteristic of these mountain areas of maximum relief.

Cenderelli (2000: 73–103) cites Mason who in 1929 reported the collapse of a 300-metre high earthquake-triggered landslide dam in the upper Indus Gorge. This produced a massive flood wave that extended for more than 100 km down the Indus, destroyed more than 100 villages, and killed several thousand people.

Hewitt (1988) reported three catastrophic rockfalls that deposited about 20 × 10⁶ cubic metres of debris on the Bualter Glacier in the Karakorum, northern Pakistan, between 29 and 31 July 1986. The first and largest fell almost 1,500 m vertically and travelled 4.8 km horizontally, covering more than 4.0 km² of the glacier surface. Five months later, the glacier suddenly increased its rate of flow from less than 1 m/day to more than 7 m/day and it advanced 2 km. Maximum velocity of the rockfall upon impact with the glacier surface was calculated at more than 440 km/hr. As in this case, many such catastrophic events cause no direct damage or loss of life, but their magnitude and unpredictability represents a serious hazard.

Such high magnitude, long recurrence interval events not only transform the mountain landscape but they may significantly change the downstream sediment load and fluvial characteristics of the rivers for long periods. In cases where a main valley is effectively blocked by landslide deposits, a lake will eventually form upstream of the dam and itself become a major hazard. One of the most spectacular combinations of mountain hazards is the Usoi landslide dam and the formation of Lake Sarez in the Pamir of Tajikistan. During the winter of 1911 an earthquake triggered a massive rockfall that completely blocked the Bartang Valley and impounded the Murghab River, tributary to the Pianj–Amu Darya system which ultimately flows into the Aral Sea. The total volume of the deposits is estimated at 2 km^3 with a maximum height above the original valley floor of 550–700 m. By 1999, Lake Sarez, which had formed behind the dam, was more than 60 km long with a maximum depth in excess of 500 m. The dam is the highest, natural or artificial, in the world. Its name, Usoi Dam, was taken from the village that was totally destroyed during the event. The lake level continues to rise at about 20 cm/year and about 50 m of freeboard remains between the lake surface and the lowest point on the crest of dam. Its volume is roughly half that of Lake Geneva. Thus in 1997 several questions arose: since the lake at 3,200 m lies in the heart of the Pamir, a highly seismic zone, what is the likelihood of another rockfall hitting the lake surface and sending a wave to over-top the dam? Could the dam itself collapse under the accumulating pressure of the ponded water? Alternately, as drainage has penetrated the lower section of the dam to re-supply the Murghab River, what is the chance that it would continue to enlarge its channel and cause the dam to collapse? Computer simulation by staff of the United States Army Corps of Engineers developed a worst case scenario that envisaged a catastrophic flood/high speed debris flow extending 2,000 km to the Aral Sea and placing at risk five million people and much property and infrastructure in four different countries – Tajikistan, Afghanistan, Uzbekistan, and Turkmenistan (see also Chapter 10). In June 1999, under the leadership of the UN International Decade for Natural Disaster Relief (IDNDR), a multi-disciplinery team in co-operation with the Government of Tajikistan visited the region of Lake Sarez to assess the hazard. Although the assessment effectively dismissed the worst case scenario, nevertheless, the possibility of an earthquake-induced landslide or rockfall hitting the lake surface remains (Alford and Schuster 2000). In such an event, a wave could over-top the dam crest and, while not likely to cause it to break, would still threaten to annihilate 32 Pamiri villages below it in the Bartang Gorge that extends 120 km to its confluence with the Pianj River on the border with Afghanistan. Electronic monitoring of the lake level and the surrounding mountain slopes and installation of an early warning system is planned and competitive bidding for the project is under way.

Several comparable events have occurred in the Pamir in the undetermined distant past. These can be detected from the abandoned shorelines of former lakes and the remnants of landslide dams that have collapsed. There are also historic reports of immense rockfall/landslide catastrophes that did not form large lakes. Perhaps the most dramatic is an event that occurred on 10 July 1949 in northeastern Tajikistan (Yablokov 2001). Early July 1949 brought unusually heavy rainfall throughout the northern Pamir. Then, between 7 and 9 July, a series of earthquakes caused numerous small rockfalls and landslides. On 10 July a giant earthquake occurred (magnitude greater than 8.2 Richter); it is estimated that 250 million cubic metres fell off the top of Borgulchak Mountain and, within minutes, travelled

12 km to obliterate the district centre of Khait and the many villages in its path. An estimated 28,000 lives were lost (Figure 6.5 and Figure 6.6). Remarkably, the Soviet Union kept the disaster secret and it was not until more than 50 years later that the details of the tragedy were fully reported (Yablokov 2001). This event was a giant rockfall that extended as a land-slide, or probably a high-speed debris flow, that was triggered by heavy rains and a massive earthquake.

Fort and Peulvast (1995) report many major mass movements throughout the Himalayan region ranging from the Pamir, Kunlun, and Ladakh to the Annapurna–Pokhara area of central Nepal. The Annapurna–Pokhara event is especially interesting as the present-day townsite of Pokhara and Lake Phewa and much of the surrounding area were totally transformed. Fort and Freytet (1982) and Fort (1987, 2000) estimate that this event occurred about 500 years ago and interpret it as 'the indirect consequence of an earthquake that destabilized the steep High Himalayan Front of the Annapurna Range (>7,500 m)'. The resulting ice avalanche/rockfall/debris flow from the west face of Annapurna IV (7,524 m) deposited an estimated 4 km^3 of material in the intramontane basin of Pokhara at 900 m, burying the original landscape beneath an average mantle 100 metres thick (Fort pers. comm. 3 October 2003).

Li (1994) has compiled a catalogue of major landslide events in China dating back nearly 4,000 years, the world's longest record. While many hundred of these occurrences were located well beyond the broader Himalayan region, many occurred in Sichuan and Yunnan provinces. Most appear to have been triggered by earthquakes, or by torrential rainstorms, or by a combination of the two. Of particular note is the 1786 Kangding–Luding earthquake (Richter ~ 7.5) which caused a massive landslide that dammed the Dadu River. After ten days the accumulating lake waters over-topped the dam resulting in a gigantic flood that extended more than 1,400 km downstream. The earthquake accounted for 400–500 lives, but the flood swallowed up more than 100,000 people. Li (1994) makes a conservative estimate that between 1954 and 1990 at least 140–50 people in China were killed each year by landslides. Property damage has been correspondingly high.

To this point only the high magnitude events have been discussed. At the other extreme of the impact spectrum, the soil slips on the outer edges of agricultural terraces, although numerous, do not constitute an immediate danger and so are not discussed further. Inter-mediate between the two extremes, however, are the frequently discussed landslides (debris flows). While they are landslide/mass movement events, the usual triggering mechanism is either a torrential rainstorm, or several days of sustained heavy rain, or the two combined, and so they are included in the next section on torrential rains (as reported in Chapter 4, earthquakes, especially when they occur during periods of heavy rain, will produce numerous landslides simultaneously).

Torrential rainstorms/snowstorms

Periodic torrential rainstorms are characteristic of monsoon climates; they are also a common feature of mountains (Barry 1992). The Himalayan region presents many examples. Two of their peculiarities are the unpredictable nature and the very local occurrence; areas a few kilometres apart may be subjected to entirely different precipitation patterns, ranging from 'normal' to 'extreme' and 'catastrophic'. They also occasionally occur 'out-of-season'; for instance, after the assumed end of a monsoon season or prior to its onset. Moreover, very heavy downpours occasionally hit semi-arid areas such as the Hunza Valley in northern Pakistan.

Figure 6.5 The tragedy of Khait, Tajikistan Pamir. On 10 July 1949 a combination of heavy rains and an 8.2 Richter-scale earthquake caused half of the upper part of Mount Borgulchak to collapse and fall into the valley below. This view shows the upper section of the collapse as it appeared in October 1991.

The enormous local relief of the Himalayan region leads to great differences in the amounts of precipitation received by windward and rain-shadow slopes. As most weather stations are located on the valley floors, recorded receipts are assumed to be unrepresentative and the higher slopes and ridge crests are believed to experience much heavier amounts. Furthermore, precipitation at the higher elevations may be in the form of snow. In turn, this will have a significant effect on the geomorphic impacts of any rainstorm because runoff will be delayed. These highly varied characteristics are illustrated in the series of examples of catastrophic downpours that are described below.

Rainfall amounts in strictly local areas may exceed 600–800 mm, or even 1,000 mm, over a two-to-three day period. Such huge amounts and high intensities produce large numbers of landslides (debris flows and mudflows). When such events occur after periods of 'normal' rainfall, especially late in the monsoon season when soils and weathered slope materials are already saturated, the effect can be devastating, with heavy loss of life and property. This process will be first illustrated by describing the October 1968 Darjeeling rainstorm.

Between 1400 hrs on 2 October and mid-afternoon on 5 October 1968, 630 mm rain was recorded at the Darjeeling weather station. The station officer, based on his practical experience, indicated that much higher amounts, possibly twice the recorded amount, must have fallen on the surrounding and more exposed ridge crests. The storm had affected a

Figure 6.6 The giant rockfall from the summit of Mount Borgulchak travelled as a high-speed debris flow/landslide to eliminate the district centre of Khait 12 km below, and many intervening small villages. Within minutes, 28,000 lives were lost. This view, looking down-valley across the former site of Khait, shows the hummocky landslide deposits on the lower valley floor, October 1991.

considerable area of the Darjeeling Lesser Himalaya and was the heaviest recorded since 1950 (Ives 1970; Starkel 1972a, 1972b; Froehlich *et al.* 1990). On the afternoon of 5 October, close to the end of the storm, it was reported that innumerable landslides released, seemingly almost simultaneously (Figure 6.7). The 50-km highway from Siliguri on the plains to Darjeeling had been cut in 92 places; many of the cuts were small, but several were massive (Figure 6.8); the famous narrow gauge railway was almost completely destroyed; much track was eliminated and rolling stock was knocked down the steep slopes (Figure 6.9). During the International Geographical Conference symposium held in Darjeeling the following month an attempt to count the number of landslides failed because of sheer numbers, so a best estimate of 'at least 20,000' was made. The major rivers of the area had been in spate, the main road bridge across the Rangeet River, tributary to the Tista, had been destroyed. Where the Tista and other smaller rivers emerge from the mountains onto the flood plain large aprons of coarse alluvium had been deposited and several villages wiped out. Froehlich *et al.* (1990) determined from the local weather records that such torrential downpours occur between two and four times a century and act as the dominant geomorphic process for slope development in the region.

Starkel and Basu (2000) have compiled the results of over a quarter century of landslide research in the Darjeeling area since the 1968 catastrophic rainstorm. This confirms the broad

Figure 6.7 The Darjeeling rainstorm of October 1968 was a 50 to 100-year event and produced thousands of landslides. Many were small, yet entire mountain sides were mutilated and only foot travel was possible for many weeks. Within a few years, however, most of the landslide scars had been obscured by natural revegetation, November 1968.

aspects of the previous discussion. However, some significant additions emerge and, in particular, Starkel and Basu place much more importance on the effects of deforestation than is proposed in Chapter 3. They provide a useful overview of the Darjeeling situation as follows:

- Extensive forest clearance began in the mid-nineteenth century but was much reduced during the mid-twentieth century.
- Catastrophic rainstorms, which they define as downpours of 1,000–2,000 mm in 2–4 days, occurred in 1899, 1950, and 1968.
- Mean annual precipitation varies from 2,000–2,500 mm on the plains to the south, to 4,000–6,000 mm on the mountain front, declining to about 2,000 mm north of Darjeeling town.
- On the more exposed ridges annual precipitation may exceed 8,000 mm.
- The rate of soil erosion on deforested areas in 1968 exceeded that within the forests by one or two orders of magnitude; revegetation of landslide scars occurred within 5–15 years.
- Land use/land cover is classified as: forest, 38 per cent; agriculture, 37 per cent; and tea gardens, 22.5 per cent.

Figure 6.8 Darjeeling Highway, West Bengal, India. In October 1968 the Darjeeling Himalaya experienced a rainstorm of over 1,000 mm in a three-day period. Thousands of landslides were released and the 50-kilometre highway between Siliguri on the Terai and Darjeeling was cut in 52 places. Here one of the major landslide breaks is undergoing massive and expensive repair, November 1968.

They conclude that there is a tendency toward recent acceleration in deforestation, in large part due to the rapid population growth. They propose that reforestation, control of water discharge courses, elimination of excessive mining of low-grade coal on steep hillslopes, and reduced construction of multi-story buildings on unstable slopes are essential if future severe environmental and economic consequences are to be avoided (Starkel and Basu 2000: 155).

In central Nepal between 19 and 21 July 1993 a rainstorm of magnitude 540 mm/ 24 hr with maximum hourly intensity as high as 70 mm/hr hit the Kulekhani watershed (Dhital *et al.* 1993). The storm produced a landslide intensity of 47 per km^2 and siltation of the reservoir reduced its design life from 60–70 years to 10 years (see Chapter 4). A comparable, highly local rainstorm in the same general area in 1981 (Carson 1985) caused dozens of landslides on a single mountain slope.

Hewitt (1993) has provided a detailed account of a much more widespread rainstorm that affected a large area of the western Himalaya and the Karakorum. Gilgit, situated in the semi-arid upper Indus Gorge, received 53 mm of precipitation on 9/10 September 1992; Chilas, a little further downstream, received 94 mm, Abbotabad received 390 mm, Murree 326 mm, Srinagar 104 mm, and Jammu 146 mm. The last four are located on the southern flank of the western Himalaya and, while the totals were much higher than those of Gilgit and Chilas, the latter receipts were much more unusual in proportion to mean annual totals. For this to occur in September was especially unusual and the geomorphic impacts in the

Figure 6.9 The famous Darjeeling narrow gauge railway closely follows the highway for much of its length. The October 1968 rainstorm caused almost total destruction with much of the line washed away and rolling stock precipitated hundreds of metres down the mountain side, November 1968.

upper Indus and headstream valleys, including the Braldu, were devastating. Referring to the Braldu road, Hewitt (1993: 373) remarked '[s]ince the road is a geomorphological nightmare, only the completeness of its destruction, from end to end in a few hours, is remarkable'. Most of the erosion was concentrated below 3,000 m, the reverse of what is usually observed. Snow came down to 4,100 m.

Hewitt relates that Finsterwalder (1960) assumed that transhimalayan incursions of the summer monsoon were rare (one in 50 years). More recently, as Hewitt (1993) indicates, this type of event has affected the valley floors below 3,500 m four times in the period 1984–92. The 1992 torrential rainstorm emphasizes once again the considerable increase in economic and material losses in mountain regions as road systems and associated infrastructure continue to expand.

Several other hazardous phenomena that occur in most mountain regions, including the Himalaya, have not been discussed. Floods often accompany landslides and debris flows, as indicated above, and it is frequently difficult to differentiate the effects of mass movements, floods heavy with sediment, and water damage. Major flooding in the context of Bangladesh, however, has been discussed separately in Chapter 5 because it is one of the critical aspects of the eight-point scenario of the Theory of Himalayan Environmental Degradation and so will not be considered further here. Snow and ice avalanches are very frequent at the higher elevations and have been widely publicized by the news media primarily in relation to mountaineering accidents. This aspect of mountain hazards is considered sufficiently specialized that it will not be introduced here. Nevertheless, the example of untimely heavy snowstorms will be introduced because, with the rapidly accelerating development of trekking tourism throughout much of the region, this phenomenon has become a significant hazard. A case history of a particularly tragic event that engulfed a large section of the Nepalese Himalaya will be used as an example.

Heavy rain, with snow at the higher elevations, began on 9 November 1995 and extended across a large section of the Himalaya from east of Kangchenjunga westward to beyond Mustang. Snowfall totals in excess of three metres on the major trekking areas were reported together with torrential rainfall at lower levels (Frankel and Roberts 1995). As November is renowned for its mild conditions and especially for its normal lack of precipitation, and is therefore a prime trekking and mountaineering season, it is not surprising that hundreds of trekkers and mountaineers and equal numbers of porters and guides were affected. A minimum of 61 deaths, including 22 foreign trekkers, was recorded, the majority in Sagarmatha National Park.

Many lives were lost to debris flows/landslides released by the torrential rains, although snow avalanches accounted for the largest number of deaths. The storm persisted for 36 hours leaving hundreds stranded, unable to make progress on foot through the heavy accumulations. Yaks as well as humans suffocated. By 12 November the government had organized a massive helicopter search and evacuation mission. During the subsequent four days 238 trekkers and 279 Nepalese were evacuated. Alton Byers (pers. comm. 6 October 2003) was undertaking repeat photography south of Mount Everest when he realized that a serious storm was approaching; he managed to walk to safety through the rapidly accumulating snow. In the *Newsweek* account of the snow storm Byers was quoted as saying 'I predict this will be seen as one of the most remarkable events in the history of mountain rescue' (Frankel and Roberts 1995: 17).

The human losses on such occasions are magnified by the surprise nature of the event. Many of the trekkers who were trapped for several days by the deep snow had only lightweight clothing and equipment. Porters and guides accompanying tourists in those periods that are regarded as fair weather months would not be equipped to survive such inclement conditions for long. Although the storm must be classed as a very rare event, the loss of lives was unprecedented.

Conclusion

Following the Rio de Janeiro Earth Summit of 1992, and especially after the United Nations declaration of 2002 as the International Year of Mountains, there has been a rapid growth in literature promoting or debating 'sustainable mountain development'. This expression has different interpretations, but one aspect of 'development' is the expansion of infrastructure, especially roads and hydro-electrical facilities. The economy and social circumstances of much of the Himalaya region are being rapidly transformed by such developments. The approach to mountain geography is being revised as the relatively inaccessible mountain regions are being integrated on a global scale by progressively increasing access. Nevertheless, this chapter, in describing mountain hazards, indicates that accessibility is relative. The Karakorum Highway, for instance, owes its existence to military/political considerations (Kreutzmann 1991, 1993). Trade, tourism, and other non-military reasons alone would not have succeeded in connecting Islamabad and Xinjiang by a major highway nor have provided the resources to keep it open.

The magnitudes of the large and rare events, such as the Langtang rockfall or the three rockfalls of 1986 that Hewitt (1988) reported hitting the Bualter Glacier in the Karakorum, are so great that they are hard to visualize. The recurrence intervals of such phenomena cannot be calculated. They are rare in both time and space, yet they do happen, as in 1949 when 28,000 people and an extensive area of land were engulfed in the vicinity of Khait, Tajikistan. Many lives and much property are lost annually to occurrences of intermediate and lower magnitude mass movements. Moreover, the construction of large-scale infrastructure, such as high dams, in locations perceived to be hazardous, cause serious social and political problems. However, the long recurrence intervals of the high magnitude events and the minute area affected by them in proportion to the total area of the Himalayan region should serve as a caution against the tendency to over-dramatize their overall effect. Nevertheless, mountains do present a series of hazards that are obstacles to development, and their effects increase in magnitude as the processes of modern development penetrate further into mountain regions. Yet, as serious as this overall assemblage of mountain hazards may be, it pales into relative insignificance when the hazard of human conflict is introduced. This will be the function of Chapter 8.

7   Tourism and its impacts

Something lost behind the ranges . . .
                    (Rudyard Kipling)

Introduction

Tourism has been described as the world's largest and fastest growing industry. It is estimated by the World Tourism Organization that the industry will employ 250 million persons and promote over 10 per cent of total capital investment by 2010. Mountain tourism in its various forms will account for about 20 per cent of the total (WTTC 1999). Religious pilgrimage, drawing travellers to the sources of sacred rivers and the abodes of gods, has a long history in the Himalaya, and in some areas continues to have important economic and ecological impacts. Over the last half century, mountaineering, trekking, and mass tourism have developed dramatically in the Himalayan region. In some respects these newer forms are homologous with traditional pilgrimage; certainly the secular inspiration derived from recreation in high places is comparable to the spiritual inspiration associated with pilgrimage. There is an important difference, however: modern tourism in the Himalaya, unlike traditional pilgrimage, necessarily entails 'development', fundamental change in livelihoods and social structures of the impacted communities.

In the 1960s the initial rapid expansion of international tourism after the end of the Second World War was thought to be a possible solution for some of the world's more intractable problems. It was hoped that tourism would reduce inter-cultural misunderstanding, or at least facilitate the transfer of wealth from affluent visitors to poor inhabitants in many destinations (IUOTO 1963; Hinch and Butler 1996; Price *et al.* 1997). In those early halcyon days, tourism was welcomed as a benign means of creating economic opportunities in backward destinations at comparably slight cost, as their very attributes – traditional culture and pristine nature – needed only to be recast as amenity resources. It was anticipated that the demands of the tourist would create local jobs and a market for handicrafts, leading to increased economic independence, rise of local living standards, enhanced cultural pride, and self-determination. The more optimistic proponents even anticipated that it would promote increased attention to nature preservation.

The discussion so far encompasses all forms of tourism, not just that related to mountain destinations. It also emphasizes international travel, and the statistics often quoted by the World Tourism Organization lump together as 'tourist industry' everything connected with travel. Nevertheless, it serves to indicate that tourism, however defined, has acquired a critical mass, sufficient to effect change on a large scale.

Certainly many benefits have accrued, although it has been increasingly argued over the last two or three decades that more harm than good is being done and doubts are replacing the early optimism. Expectations have not been realized. Typical downsides of tourism development have included environmental damage, disruption of mountain cultures, economic 'leakage', and a widening gap between the local poor and the not-so-poor. Local property values are inflated so that indigenous people are frequently forced out of their homes. In many instances, tourism has aggravated gender asymmetries, as women tend to be restricted to the poorest paid of the tourism-related jobs and often may not even maintain control over the slender wage earnings that their labour produces (Sicroff and Ives 2001). Additionally, women often have to carry a heavier workload as local farm labour is frequently relinquished by their menfolk. Nevertheless, there have also been some impressive benefits associated with this growth.

This introductory general discussion refers primarily to what became known (derogatorily) as mass tourism. In response to increasingly voluble criticism, tour operators and scholars alike have promoted alternate varieties such as *adventure tourism, nature tourism, ecotourism, trekking tourism, low-impact tourism*, and so on. The sales promotion to attract tourist clients today often stresses a concerted approach to minimize environmental impacts; the cliché 'sustainable ecotourism' has become standard terminology.

Hard on the heels of the green vanguard, promoters of unreformed mass tourism have co-opted the rallying cries of the politically correct. This 'green-washing' has led to much hand-wringing on the part of those alternate operators who hoped to gain a marketing advantage that would amortize the added costs of eco-sensitivity. Recently, a variety of labelling and certification schemes have arisen. However, these schemes are impractical, since the cost of inspection and the introduction of uniform standards mean that family-owned enterprises in remote and impoverished destinations could not aspire to certification (Sicroff and Alos Alabajos 2001).

The development of tourism has been constrained by three factors: political openness, physical accessibility, and financial investment. The first two factors have tended to operate synchronously while financial investment is progressively attracted as initial access becomes more secure. Mountaineering ventures, however, are an exception. Attempts to reach the highest summits can be regarded as both the forerunner of Himalayan tourism and as a special case. Official permission for access to the high peaks in areas still strictly closed to foreigners was obtainable on an expedition-by-expedition basis long before trekkers were able to approach the base camps. When the Tajik Pamir were controlled by the Soviet Union, it was only a rare foreign mountaineering expedition that was permitted to attempt the high summits and it was even rarer for a scientist to journey with Soviet scholars into the high mountain valleys. Nevertheless, the best known political/mountaineering events are those associated with Mount Everest (Chinese – Chomolungma; Nepalese – Sagarmatha).

Prior to the Second World War, before Himalayan tourism was considered practical, the so-called British route from Darjeeling to Everest's northern approach through Tibet was periodically negotiated with the Dalai Lama. After Communist Chinese control was extended across Tibet this route was denied. Coincidentally, Nepal opened its frontiers, facilitating the rush of Swiss and British teams to Everest and Lhotse via the Khumbu Glacier and South Col. In fact, practically all the major summits of the Himalaya and Karakorum were reached within the 1950s, from K-2 and Nanga Parbat in the west (Pakistan) to Cho Oyu (Nepal) and Kangchenjunga (Nepal/India) in the east. Trekking tourism did not reach these areas until several years later. Lhasa and the northern approach to Everest remained closed until 1980, which date also saw permissions for attempts on other Chinese summits, including

Namche Barwa, the Chinese Pamir, and an increasing number of summits in the Hengduan Mountains, such as Gongga Shan (western Sichuan). Warfare in Afghanistan and civil war in Tajikistan placed temporary restraints on parts of the Hindu Kush and Pamir. Although access to Afghanistan, Kashmir, and Bhutan is still limited, the Himalayan peaks attract hundreds of expeditions each year.

Physical accessibility has been progressively enhanced as highways and roads, together with jeep trails, have penetrated most of the region's mountain ranges and almost all areas have been declared 'open' to foreigners, most recently Mustang and Dolpo in north-central Nepal. The lack of roads had not hindered the growth in mountaineering activity since the long 'walk-in' to base camps had been regarded as a necessary part of acclimatization for the so-called 'death zone' at extreme altitude. Nevertheless, the growth of Himalayan tourism of all genres has been greatly accelerated by the extension of road and air access. The availability of helicopter charter services has become increasingly important.

The third control, financial investment, has come predominantly from sources in the industrialized countries outside the mountain regions. Even with the recent increases in investment from within India, China, and Pakistan, the new stakeholders are primarily low-landers. This aspect of tourism development has deprived indigenous people of control over development and minimized their share of the economic benefits.

The influence of mountaineering on the development of mountain tourism has been primarily indirect. The flood of books recounting heroic ascents incited an unprecedented level of fascination with the region. Herzog's *Annapurna* (1952) is reportedly the most popular mountaineering book of all time, having sold 15 million copies in 50 languages (*Economist* 24–30 May 2003). Such books also publicize the numerous religious and architectural attractions and, above all, have elevated the Sherpas of Khumbu into a world class icon. Trekking tourism followed on the heels of John Hunt, Edmund Hillary, and Tensing Norgay. Yet it must be emphasized that, by the beginning of the new millennium, the number of tourists who confine their visits to Kathmandu and some of the famous hill stations of the British Raj, such as Darjeeling, Simla, and Nainital, was greater than the combined influx into the Annapurna Conservation Area and Sagarmatha National Park. Similarly, the number of visitors to sacred sites overwhelms the combined total of the trekkers to all other Himalayan destinations: for example, Badrinath, in the Garhwal Himalaya, receives a half million pilgrims and secular tourists per year. It is also significant that the last two decades have witnessed a prodigious growth in domestic tourism, especially in India and China. While foreign exchange remains extremely important, the volume of domestic tourist traffic overshadows the number of foreign visitors by orders of magnitude so that, in terms of negative impacts on environment and local cultures, domestic tourism has probably become the more serious threat.

These and other aspects of Himalayan region tourism will be discussed in more detail below. Nevertheless, the topic is complex due to varying local conditions, national policies, and the timing of developments. A series of type examples will be presented and discussed. Several examples are taken from Nepal (pp. 146–55) because it is one of the most heavily impacted countries of the Himalayan region. Next, development of tourism in the Lijiang region of northwestern Yunnan, China, will be outlined (pp. 156–65). Westward from Nepal, tourism development in the Kulu Valley of Himachal Pradesh will be discussed (pp. 165–8). The evolution and set-backs to mountain tourism in Pakistan's Northern Areas and Northwest Frontier Province will be considered (pp. 168–70). Finally, the case of Tajikistan's Pamir will be examined as a region with considerable potential for development of tourism, the initiation of which was suspended due to civil war and remains obstructed by continuing political instability (pp. 170–1).

Mountain tourism in Nepal

Nepal can be considered the prime example of the development of mountain tourism in the greater Himalayan region. Over a horizontal distance of little more than 100 km the land rises from less than 150 m above sea level on the Terai to above 8,800 m, not only representing the highest point in the world, but also the greatest difference in height over a specific horizontal distance. The gazetting of national parks and other forms of protected areas in Nepal has a long history in comparison with other Himalayan tourist destinations. By the end of the millennium there were 17 national parks and protected areas in Nepal. Together they occupy 16 per cent of its total area. The *modus operandi* implemented in the establishment of these parks runs the gamut from the controversial expropriation of indigenous lands at Rara Lake to what has (rightly or wrongly) been perceived as a model for co-operative management in the Annapurna Conservation Area Project (ACAP).

Rara National Park is one of the smallest of the parks in Nepal's system, and it was also the first, in 1975, to be officially established. Located above Jumla in western Nepal, Rara National Park was created at a time when the contemporaneous United States/New Zealand model provided the standard for defining a national park. This model assumed that national parks should comprise scenic locations with no permanent human habitation. To this end the Nepalese military was dispatched to forcibly eject two small ethnic minority groups, Thakuri and Kamis, who had used the area around Lake Rara for generations. The two groups occupied separate villages, each numbering 150–200 persons. Some provision had been made for their relocation although opinions about their subsequent well-being are mixed. Warren Smith (pers. comm. 21 October 2003), who was responsible for construction of the new park warden's quarters in 1975, feels that the episode should not be regarded as an unmitigated tragedy given the extremely marginal condition of the indigenous people at the lake. However, this assessment is made without reliable information about the subsequent disposition of the evacuees.

Sagarmatha National Park was gazetted in 1976 by the Nepalese government, supposedly to protect the environment of the Mount Everest/Khumbu Himal region. The park was subsequently declared a World Heritage site. The lead-up to this auspicious event included serious official consideration for the forced evacuation of an appreciably larger ethnic group, the Sherpas, than the unfortunate residents of the Lake Rara area. Fortunately, the mountaineering Sherpas, as 'tigers of the snows', had already acquired worldwide admiration and the entire community had attracted some powerful supporters. Eventually the park was established and the Sherpa settlements accommodated by a system of exclosures. Nevertheless, the sense of intrusion onto traditional Sherpa lands by a sometimes insensitive central authority has remained, even if in reduced intensity, into the twenty-first century.

Any examination of the development and impacts of mountain tourism in Nepal must reflect the several different modes of operation that have evolved and that function simultaneously. The selection for discussion, therefore, includes Sagarmatha National Park, the Annapurna Conservation Area, the Arun/Barun National Park and Conservation Area, the Kangchenjunga park and conservation area, and the most recent developments in Mustang.

Sagarmatha National Park/Khumbu Himal

Sagarmatha National Park is located in the northern section of Solu Khumbu District in northeastern Nepal. Its northern limit, the apex of which is Mount Everest, runs along the Himalayan crest and forms the international border with Tibet, China, for a distance of about

40 km. This cluster of the world's highest mountains is deeply dissected by three glacially carved valleys whose rivers converge to form the main stream of the Dudh Kosi that flows almost due south across the park's southern boundary, and for the further 10 km to the mountain airstrip of Lukla, before entering a narrow gorge section on its way to the Terai lowlands and the Indian border. The entrance to the park lies at about 2,800 m. Khumjung is the traditional Sherpa 'capital', but the Sagarmatha National Park (SNP) headquarters and most of the commercial activity is concentrated at Namche Bazar (3,440 m), a burgeoning town now spilling out of its amphitheatre-like bowl high above the confluence of the Bhote Kosi and the Dudh Kosi. This provides it with a spectacular mountain view (Figure 7.1). Namche is at the juncture of all the main trails that, with the advent of trekking tourism, have become the primary trekking routes, thus ensuring the expansion of its traditional pre-eminence in long-distance trade and its emergence as the region's leading commercial centre.

Figure 7.1 The Saturday bazaar at Namche, Khumbu Himal, with mountain backdrop. Porters and tradesmen walk up from the low hills and the Terai to this commercial and festive event. Photograph by Bruno Messerli, April 1979, showing the traditional market. More recently many imported goods, including electronics and mountaineering equipment are offered for sale to tourist visitors as well as to local people.

Following the heroic mountaineering accomplishments of the 1950s that included the Tilman, Shipton, Houston reconnaissance expeditions and the British and Swiss successes on Everest and Lhotse, tourism hardly stirred for the next dozen years, with only 20 reported visitors to the Khumbu in 1964 (Naylor 1970). The numbers exceeded 5,000 by 1980, 17,000 by 1996/97 (Nepal 2000), and 27,000 by 2001 (Byers 2002). By the 1990s the number of trekkers per year exceeded by three or four times the number of permanent Sherpa inhabitants. To the total number of visitors must be added some 80,000 non-Sherpa support personnel from outside the Khumbu – porters, guides, cooks, and traders at the weekly bazaar. An additional uncounted group includes hotel and restaurant workers and farm labourers recruited from south of the park and as far afield as the Terai and distant hill regions of Nepal. The farm labourers have become indispensable because many of the Sherpa villagers committed themselves to the tourist trade and had insufficient time to tend their own farms.

The great majority of the trekkers have the Mount Everest base camp or nearby Kala Pattar promontory as their prime destination, although from the 1980s onward an increasing number have begun walking into the tributary valleys of the Dudh Kosi, especially the Gokyo valley leading to Cho Oyu. Bjønness (1980), Fisher (1990), Brower (1990, 1991, 1996), Stevens (1993, 2003), and Brower and Dennis (1998) have documented the significant changes that this rapid growth in the traffic has wrought on the culture and way of life of the Sherpas. Byers (1986, 1996, 2002) studied the impacts of these changes on soil erosion and vegetation cover. Fisher (1990), who worked with Sir Edmund Hillary from the beginning of the school and hospital building programme in 1964 and co-ordinated the establishment of Lukla airstrip that facilitated Khumbu's tourist boom, has provided an anthropologist's reflections on the changes that have occurred. Nepal *et al.* (2002) have published an extensive assessment of the evolution and dynamics of tourism in the Khumbu and comparisons with similar developments in the Annapurna Conservation Area (much of this and the following section have been taken from Nepal *et al.* 2002).

It is hardly surprising that the Sherpa way of life has undergone fundamental change given the remarkable growth in the numbers of mountaineer and tourist visitors. Construction of hotels, restaurants, shops, and trail-side tea houses has seemed to expand exponentially. Most of the tourism infrastructure, as well as a large proportion of the Kathmandu trekking companies and even airlines, is owned and operated by Sherpa families, so that there has been little of the economic leakage so typical of development in remote destinations.

Although affluence has increased, there has been some criticism of the widening economic gap between the rich and poor. This is, in part, a reflection of original disparities between rich traders and poor subsistence farmers who had little land and few domestic animals. It also derives from the position of the villages in relation to the main trekking routes, with the valley of the upper Bhote Kosi being the most remote and, therefore, deprived of the economic benefits that have influenced the growth of the more accessible areas. Furthermore, as the Bhote Kosi leads to the Nangpa-la[1] and into Tibet, the valley remained officially closed to visitors until the early 1990s, which further marginalized the area in comparison with that traversed by the main route to Everest. Nevertheless, the living standards of the poor are substantially better than 40 years ago.

The number of scholarly publications on the Khumbu and Sagarmatha National Park is immense; only a few especially relevant ones can be mentioned here. Much of the early discussion on the environmental impacts of tourism has been melodramatic; it includes the highly controversial claims of forest and general environmental degradation of the 1970s and early 1980s in contrast to more objective accounts of the relatively intact and stable landscape below the upper treeline and the previously overlooked potentially serious damage to the shrub/heath alpine belt above. The many papers by Byers (also discussed in Chapter 3) provide the most authentic account, strongly supported by an extensive array of replicate photography (see Figures 3.1–3.4) from the mid-1950s to the present (see also Stevens 2003). Similarly, initial over-reaction concerning the prospects for disastrous geomorphic events likely to result, in part, from excessive visitor impacts, is most authoritatively discussed and counteracted by Byers (1986, 1987c), Zimmermann *et al.* (1986), and Vuichard and Zimmermann (1986) (see also Chapters 3 and 4). The special case of actual catastrophic processes – the sudden outburst of glacier lakes (see Chapter 6) – is examined by Ives (1986), Vuichard and Zimmermann (1986, 1987), and Watanabe *et al.* (1994, 1995).

Tourism began to have significant impacts on the life and culture of the Khumbu as recently as the late 1970s, beginning along the main route from Lukla, via Namche (Figure 7.2), to the Everest base camp and progressively extending throughout the Khumbu. Within a quarter century the standard of living of the Sherpas has improved immeasurably: education; health; diet; housing; and general material comfort so that, as an ethnic entity, they have become one of the most affluent groups in the entire Himalayan region.

There have also been negative impacts. One of the most frequently reported is the vast amount of refuse – that left behind by visitors and that generated by the changing life-style of the Sherpas as they become more 'Western' and adopt a 'throw-away' mode of living. The waste disposal problem has led to the designation of the Khumbu as the 'garbage trail to the Mount Everest base camp'. The condition of the camp itself has been heavily criticized

Figure 7.2 Above Namche some of the world's most famous mountain scenery comes into view. The relatively isolated village of Phortse (3,840 m) nestles within stone field walls on the terrace while Ama Dablam (6,856 m) forms the right-hand skyline. The summit pyramid of Mount Everest appears on the upper left. Photograph by Alton Byers, May 1984.

although it appears that the most serious situation relates to the upper slopes of the mountain and the South Col which has been described as the 'world's highest garbage dump'. Over the last 15 years or so there have been strenuous efforts to effect a clean-up. Yet, according to Junko Tabei (2001, 2002), the first woman to reach the summit of Everest and director of Himalayan Adventure Trust of Japan (HAT-J), an organization dedicated to preserving mountain environments, these efforts are hardly keeping pace with the inflow of new waste. The melt-out of human waste on the thinning and retreating Khumbu Glacier poses an accelerating water pollution hazard downstream. Nevertheless, various reports from Namche in 2003 describe a remarkable degree of cleanliness and tidiness and indicate that, with the exception of specific sites, such as the Everest base camp, the garbage issue has been greatly exaggerated.

The current status of mountaineering, the process that set these changes in motion, is far removed from that of the 'golden age' of the 1950s. During the 29 May 2003 Golden Anniversary celebrations of the Hillary/Tensing first ascent, some 130 mountaineering teams were competing to reach the summit. A helicopter crashed near the base camp the day before the anniversary, killing three; the wreckage was subsequently removed by helicopter. It has been deplored that today almost any ordinarily fit person with US$60,000 to spend can be pushed-pulled to the summit. Furthermore, the death and injury of young Sherpa males in mountaineering accidents is a substantial loss to Sherpa society. Sir Edmund Hillary made a personal plea to HMG Nepal to close the mountain for some years, or at least severely restrict the number of expeditions allowed at any one time. It is to be doubted that even Hillary will have much influence when each expedition permit puts at least US$70,000[2] into the coffers of the Nepalese government.

Far more influential in slowing the caravan to the summit has been the Maoist Insurgency that closed the Khumbu to visitors during 2002. Despite the Maoists' insistence that tourists are not to be targeted, many have been roughed up and subjected to extortion for protection money, either by Maoists or by opportunist bandits masquerading as Maoists. The shattering impact of political and military turmoil is a reminder of the vulnerability of all remote destinations to disturbances that interrupt the flow of foreign visitors. This was emphasized by Stevens (1993) who pointed out that by 1990 the economy of the Khumbu depended to the extent of 90 per cent on smoothly operating tourism. Although there were no reported instances of violence in the Khumbu during 2002, the disruption was serious. Under such circumstances, of course, it is not the Sherpas who have suffered most; the heaviest blow has fallen on the porters and seasonal hotel labour who have no financial reserves and no safety net of any kind. However, after several months of cease-fire tourists were returning in the spring of 2003 and, according to Brot Coburn (pers. comm. 19 October 2003), all flights into Lukla were fully loaded with tourists in September/October. Nevertheless, the resurgence of Maoist activity following the 2003 Dasain religious festival in October throws the situation into renewed uncertainty across the whole of Nepal (see Chapter 8).

Michael Thompson, in a published interview with Kanak Mani Dixit, the editor of *Himal*, (*Nepali Times* 28 March–3 April 2003) deplored the present state of mountaineering and argued that 'if you do more with less, the benefits are long-term'; that cell phones should be deposited at the park entrance; and that if helicopters, and especially TV crews, are present, it is not mountaineering (Thompson climbed on the southwest face of Everest with Chris Bonnington in 1972). This, of course, is a purist point of view; cell phones, for instance, will eventually replace standard telephone communications, the Sherpas need to be able to communicate as much as anyone else, and the cell phone has been a major factor in many recent rescues from remote locations worldwide.

While the word 'Sherpa' has come to be synonymous with 'porter', the truth is that portering was a significant part of the local economy only in the early stages of tourism development. The majority of the early mountaineering porters were from poor families; portering was beneath the social aspirations of relatively affluent Sherpa families. Some Sherpas still take work as elite high-altitude porters and guides, but most have found other ways to profit from Khumbu tourism. In fact, the most remarkable characteristic of tourism development in Sagarmatha National Park is that it has been founded on local initiative. The traditional Sherpa entrepreneurial experience as long-distance international traders between Tibet and India, together with the degree of affluence that this provided, enabled the initial construction of lodges (small hotels) and tea houses. The early Sherpa involvement in mountaineering expeditions also added to their managerial expertise. When the opportunity arose, they were ready to invest in and operate trekking companies, competing successfully with international tour operators in the globalized travel market.

The advantages of ethnic homogeneity, sense of independence (going to Kathmandu in the 1950s was referred to as 'going to Nepal' (Fürer-Haimendorf 1975)), and confidence in their culture, so important in the Sherpa transition to tourism over the last 40 years, were tempered by the central government's control of park management and by the presence of a military base at Namche.[3] Insensitive efforts to control Sherpa traditional access to their natural resources, and especially the attempts to eliminate domestic goats from the park (Stevens 1993, 1997), have ensured continuing friction (problems of park management are discussed more fully in Chapters 9 and 10). Death and injury of many young Sherpa mountaineers and prolonged absences on mountaineering expeditions, together with the withdrawal of many males from farming activities into tourism, have increased the burden carried by the women who are responsible for both their own and their partner's work. It has become necessary to relieve the burden on women by hiring farm and hotel labour from beyond the boundaries of Solu Khumbu. However, it can be argued that the recruitment of labour from far afield is a positive development in that it provides wage labour and spreads the benefits of tourism development in the Khumbu (Seth Sicroff pers. comm. 28 November 2003).

A further environmental consideration is that, despite the more recent refutation of claims that Khumbu forests were fast disappearing (Byers 1996, 2002), the great increase in demand for construction timber and wood fuel has certainly affected forests somewhere – especially south of the park in the Pharak area (Stevens 2003; Figure 10.1). This is additional to the degradation of the shrub juniper communities in the alpine belt above treeline (Byers 2002). The single account of deliberate forest destruction in the park is the elimination of trees around the park headquarters by the military as security against Maoist assault (Deegan 2003) (see Chapter 10).

Conflict over management policies persists. The most effective garbage clean-up operation, for instance, after the problem was highlighted by outside 'clean-up' expeditions (McConnell 1991; Tabei 2001), has been the establishment of a successful commercial enterprise by a Sherpa NGO that charges fees for garbage collection and disposal. This underscores the advantages of local initiative and management, a perception that led to two of the resolutions approved at a recent mountain conference held at Namche (May 2003: Bridges-PRTD): a proposal to establish a hotel association in Solu Khumbu in an effort to mitigate cut-throat competition; and a proposal to establish a programme that would facilitate the sharing of local expertise in tourism development with other host communities in remote mountainous destinations.

One especially interesting project aimed at reducing wood fuel consumption has been the construction of the Thame hydro-electric facility under an Austrian Aid programme.

The first attempt ended in disaster when the 95 per cent completed plant was destroyed on 4 August 1985 (see Chapter 6) by a glacial lake outburst flood (GLOF). At that time criticism was levelled at the choice of site, an unstable debris cone with the conspicuous glacial lake hazard directly upstream (Ives 1986). Yet the hydro plant was rebuilt close to the original site and came into operation in 1995. The International NGO, Eco-Himal, instructed local residents in maintenance and management, and the privately incorporated Khumbu Bijuli Company now supplies electricity to eight villages. However, there are negative aspects even to this commendable effort. Because of the high cost of electricity, the plant is operating well below capacity. The presence of electric power lines traversing the landscape of a World Heritage Site would appear an anomaly, although this is not a problem that is unique to the Khumbu. Finally, how safe in the long term is the potentially vulnerable site? Nevertheless, as the hydro-electricity plant was built to reduce pressure on local forests, the power lines could be considered a reasonable trade-off; at least, within Namche itself the power lines are underground.

In summary, extensive changes to the Sherpa community have accompanied the development of mountain tourism and mountaineering. The original prognostications of environmental disaster have been demonstrated to be gross exaggerations. Yet environmental deterioration is occurring and requires a more proactive response. An auspicious recent initiative is a project, supported by The Mountain Institute, ICIMOD, and the relevant regional governments, to encourage integrated planning for development and nature conservation of the region on all sides of Mount Everest (Sherpa *et al.* 2003). In terms of cultural repercussions, the picture is clearer. Societies evolve or die. The overall well-being of the Sherpa community has greatly improved and numerous jobs are being provided for seasonal workers from far across Nepal. Some of the increase in wealth has been used to rebuild monasteries and cultural treasures. Sherpa traditions have survived. Most important, the change is driven by the Sherpas themselves. They created a Sherpa myth that, like a charm, has drawn tourists and supporters. With extraordinary sagacity, they set priorities that both appeal to donors and also safeguard their future: education and (more recently) cultural and natural conservation. They held onto the economic reins and created a money-machine that is the cornerstone of Nepal's economy. Despite the loss of control entailed in park status, the Sherpas are masters of their destiny. The Khumbu probably represents the best case scenario of tourism development in the Himalayan region.

Annapurna Conservation Area Project

The Annapurna Conservation Area Project (ACAP) was established in 1986 with a pilot project operating in the village of Ghandruk. That year it is estimated that over 36,000 visitors entered the area. In 1989 entry fees were imposed and nearly 37,000 visitors were registered.

The conservation area is more than twice the size (5,300 km²) of Khumbu and the Pharak region to the south (2,300 km²) and supports a much larger population (118,000 in 1999). Furthermore, the ACAP population is much more diverse than that of the Khumbu. It consists of Brahmins, Chhetris, Gurungs, Magars, Thakalis, and Tibetans, together with a number of occupational castes, such as Damai (tailors) and Kami (blacksmiths), while significant Newar minorities are to be found in the bazaar towns near the major rivers. In terms of range of elevations and vegetation types also, the Annapurna region is much more diverse than the Khumbu. It includes subtropical lowlands in the extreme south, and extends northward through a range of forest belts that give out onto alpine meadows, and finally to the glaciers and snow peaks of the Annapurna range. North of the main range the landscape, lying in the rain shadow, is arid; long, steep, bare slopes, cut by deep river gorges, are common.

The much larger area and more numerous resident population of the ACAP, together with its greater range of topography, attract many more visitors than the Khumbu. It is also more easily accessible from Kathmandu. While the conservation area was established ten years after the Sagarmatha National Park, the route known as the Annapurna Circuit, a complete circumambulation of the Annapurna range, had been an international attraction since the early 1970s. By 1980 the number of tourists exceeded 14,000, more than twice the number that visited the Khumbu in the same year. This increased to over 35,000 in 1990 and to over 54,000 in 1997/98 (Nepal *et al.* 2002: 39). Nevertheless, the types of impacts are similar, as are the benefits and costs. The main difference is that the Annapurna Conservation Area Project is largely independent of direct government authority. This is in part due to local resistance to formal park designation.

A turning point in Nepalese tourism development and conservation policy appears to have been the 1986 inauguration of the ACAP and its management by the King Mahendra Trust for Nature Conservation (KMT-NC), an NGO. The management structure represented a significant departure from systems in place in other Nepalese protected areas (such as Sagarmatha National Park) which are managed directly by the relevant government department. The KMT-NC, with a mandate to encourage local conservation activities, established the Annapurna Conservation Area Project. Support and co-operation from local communities within the conservation area is actively sought and efforts are made to accommodate their livelihood needs. One vital advantage obtained by the ACAP, in contrast to the other protected areas in Nepal, is that entry fees paid by visitors remain under the control of the conservation area management and are not absorbed by the national government. Funding is therefore available to finance activities in the area and covers about half the total budget of the ACAP. The other half is provided by the international donor community.

ACAP maintains a central office in Ghandruk and seven regional offices. In 1998 the staff numbered more than 200, a third of whom were female. This has facilitated direct village involvement in conservation projects, promotion of tourism at the village level, improvements in housing, drinking water access, and waste disposal, as well as training and education.

The success of ACAP has been remarkable, although some negative aspects have emerged. For instance, in many villages, the local organization committee has been taken over politically by the more affluent families who have pursued their own interests at the expense of their less wealthy neighbours (see Chapters 9 and 10). However, ACAP has assisted in the establishment of 159 village-level conservation and development committees, 288 mothers' groups, and 19 lodge management committees. Efforts to reduce the use of wood fuel and increase income from sale of handicrafts have followed. Village attractions, such as costumed dance performances and tourist overnight stays in the houses of local residents (as distinct from lodges after the pattern of the Khumbu and elsewhere) have broadened the economic base. Furthermore, the sharing of know-how gained from these successful experiments has encouraged similar activities in other parts of Nepal. Nevertheless, Nepal *et al.* (2002) report that 'it is sometimes difficult for the average visitor to understand where all the money goes – money which has been generated over the years from an increasing number of tourists'. The 1998 entry fee was NR2,000 (about USD 140).

Upper Mustang

In contrast to the successes of ACAP, one of the most recent developments, in Upper Mustang, has the appearance of a disaster. Upper Mustang is the area north of Kagbeni and its northern limit is the frontier with Tibet. It has lagged behind most of the rest of Nepal in terms of improvements in overall well-being. It has lost population because of out-migration due to deepening poverty that has also resulted in the slow decay of its renowned Buddhist monasteries. Initially a wealthy region, Upper Mustang suffered badly with the 1959 closure of the Tibetan frontier and the consequent loss of its salt trade. This was exacerbated by the central government's unwillingness to open Upper Mustang to foreign visitors. The regulations restricting foreigners were finally lifted in 1992 and the area became a northern extension of the Annapurna Conservation Area. However, the form of tourism development permitted by the government was severely limited. No lodge or tea house could be constructed by local people; tourist numbers were restricted to 1,000 per year; entry permits were very expensive on Nepalese standards (US$70/person/day with a minimum stay of ten days). Only groups accompanied by government liaison officers and trekking guides were permitted and all supplies had to be brought in from outside.

As compensation to the local residents, the government initially agreed to reinvest 60 per cent of the revenues generated by tourism in local improvements. On this understanding the KMT-NC established the Upper Mustang Conservation and Development Area. Thus, investments in community development, improvement in infrastructure, restoration of monasteries, and environmental conservation, had been anticipated by the local people. The underlying government policy appears to have been the phased introduction of tourism as the necessary infrastructure and support services were developed. Local entrepreneurship, with the construction of lodges and tea houses, so common in other tourist destinations, would then follow.

The outcome is now perceived as a breach of trust (Nepal *et al.* 2002). The initial 1992 reinvestment amounted to 42 per cent of the revenues, instead of the promised 60 per cent; this fell to less than 5 per cent by 1997, despite sustained rapid growth in tourist numbers. Revenues exceeded US$500,000 by 1996. The official upper limit of 1,000 tourists per year was disregarded and it is assumed that the government used the opportunity for its own financial advantage. Similarly, the tour operators based in Kathmandu and Pokhara made significant profits as they held a virtual monopoly over the people of Upper Mustang who were not permitted to establish competing operations of their own. Local residents had to contend with a 300 per cent inflation of commodity prices. Many families depend on what they can earn from tourists by posing for photographs and what their children could obtain by begging.

Other Nepalese ecotourism destinations

Two additional Nepalese protected areas illustrate different management regimes: the Makalu-Barun National Park and Conservation Area, and the Kangchenjunga Conservation Area. Both are recently established protected areas; Makalu-Barun was opened in 1991 and Kangchenjunga in 1998. Their management strategies, like most others before them, have been heavily influenced by international agencies and reflect the perceived success of the Annapurna Conservation Area and the continuing evolution of international conservation thought.

The Makalu-Barun Park and Conservation Area

The Makalu-Barun area has a twofold administrative system, a national park and a conservation area. Both are controlled by the Department of National Parks and Wildlife Conservation; the development involves close association with The Mountain Institute, West

Virginia, USA. The conservation area serves as a buffer zone for the park. It is 700 km² in area and extends approximately west-to-northeast in a narrow strip about 12 km wide along the southern boundary of the park. It supports about 40,000 subsistence farmers and its original forest cover has been heavily modified by intensive farming. Only a small percentage of its total area remains under subtropical forest although, as Zomer *et al.* (2001) have demonstrated, the reduction in forest cover between 1972 and 1992 was slight, yet there have been extensive shifts in land cover between different categories. The park itself is about 1,500 km² in extent. Much of its northern border is formed by the crestline of the High Himal extending southeastward from Everest and includes the 8,463-metre summit of Makalu. The northern section of its eastern border is contiguous with Sagarmatha National Park. Because this park and conservation area have been established so recently, little information on tourist impact is available (see Chapter 3 pp. 52–3).

The Kangchenjunga Conservation Area

The Kangchenjunga region was designated one of the World Wildlife Fund's (WWF) 'Global 2000' eco-regions and was declared a 'Gift to the Earth' on 29 April 1997. The following year a 1,650 km² section, subsequently expanded to 2,035 km², was officially declared as the Kangchenjunga Conservation Area (KCA), to be managed jointly by the Department of National Parks and Wildlife Conservation and the WWF. The primary objective is nature conservation; the area is extremely rich in endemic plants, including 24 of Nepal's 37 species of rhododendron and a significant number of endangered animals, such as the snow leopard, red panda, musk deer, and many species of birds. An ultimate aim is the establishment of a 'tri-national peace park' in collaboration with the two contiguous protected areas: Qomolungma Nature Reserve (Tibet, China) and Khangchendzonga Biosphere Reserve (Sikkim, India).

Trekkers penetrated what is now the KCA as early as 1988, although total numbers have remained modest. There were 87 in 1988, followed by a steady increase to 801 by 1999. The present plans call for an upper limit of 1,000 per year. The small numbers are explained by the area's remote setting in the extreme northeastern corner of Nepal, the limited availability of facilities, and the rather heavy annual precipitation (Gurung 1996).

The indigenous population numbers less than 5,000 and consists of Sherpa and Limbu, the two largest groups, and Rai and Gurung. They are basically subsistence farmers practising several forms of transhumance and also trading with Tibet. There are four main villages which are the focus of planned eco-tourism developments. Proceeds from tourism so far have made only a slight contribution to the economic base as most of the trekkers arrive in self-sufficient groups organized by Kathmandu-based trekking agencies.

Although the KCA is in the very earliest stages of its development, a survey of local residents by Müller-Böker and Kollmair (2000) indicated that 'nearly all of the interviewees had expectations that went far beyond the intended and economically feasible potential of the project'. They quote the highly critical conclusions of Pimbert and Pretty (1997) who claim that this type of conservation area management exploits local participation in order to achieve conservation goals set by outsiders (Müller-Böker and Kollmair 2000: 330). This kind of system has no mechanism to ensure the effective transfer of control to the local people.

Northwestern Yunnan, China: benign development or a pact with the devil?

The development of tourism at the extreme eastern extent of what is here defined as the Himalayan region is discussed because it provides an example where a remarkable potential for a truly appropriate tourism is being eclipsed by a juggernaut of mass tourism promoted by the government. However, the size of the region is so large that there may still be room – physical, economic, and political – for 'appropriate tourism' provided the political will and management expertise can be marshalled (Justin Zackey pers. comm. 5 October 2003). The following account is based on a series of personal visits to the Lijiang region between 1982 and 1995 (Messerli and Ives 1984; Ives 1985; Ives and He 1996; Sicroff and Ives 2001). The intent is to demonstrate the situation that prevailed in the late-1980s, to indicate the opportunities that existed at that time for appropriate tourism as a force for improving the well-being of the poor mountain minority people, and to describe how these opportunities are in danger of being overwhelmed by massive development.

During the first field studies in Lijiang County in 1982 and 1985 the entire area of northern Yunnan was closed to foreign visitors; domestic travel was also extremely limited (Ives 1985). Despite decades of heavy-handed and inconsistent policies emanating from the Communist Party, the life-styles of the seven distinct minority peoples had not departed significantly from their traditions. In 1982 a high proportion of the population was living in extreme poverty, below even the contemporary Chinese definition of the 'poverty line'. Lijiang Town, the ancient capital of the Naxi people, the dominant local minority, had a population of about 60,000 in the early 1980s. The 'Old Town' (Dayan) was well preserved. The main town, that had grown up around the ancient core, boasted two small hotels, both without running water; bathing was possible only in the town's public baths. A large section of the Jade Dragon Snow Mountains, which overlook the town and are sacred to the Naxi religion, had been designated a 'forest preserve', although commercial logging (run by the government) and illegal freelance logging were rampant throughout the county, including the preserve (see Chapter 8).

In June 1985 the State Council and Military Commission of Yunnan Province declared Lijiang Naxi Autonomous County open to foreign tourists. In December 1986 the Old Town (Dayan) was designated as a province-level historical and cultural attraction. During the following decade international tourism rapidly expanded from a trickle to more than 100,000 visitors in one year. Concurrently, domestic tourism has expanded, outpacing international tourism by an order of magnitude. World Heritage Site status came in 1997. According to figures released in 2001 by the Lijiang Tourism Bureau, more than 90 per cent of tourists were domestic.

Situated across latitude 27° North and with elevations ranging from 1,500 to 5,600 m, the landscape of Lijiang County is extremely varied. It includes rice paddies and lush meadows at the lower elevations, forested middle slopes steepening to the upper timberline at about 4,200 m, alpine meadows above, leading up to precipitous ice-fretted ridges and peaks that support Eurasia's southernmost glaciers. This majestic panorama is traversed by the spectacular course of the upper Yangtze (Jinsha Jiang) which climaxes in the 'Tiger Leaping Gorge'.[4] Here, for a distance of 25 km, the Jinsha Jiang has carved one of the world's deepest gorges along a major tectonic dislocation between the 5,500-plus metre summits of the Haba Xue Shan (Haba Snow Mountains) and the Yulong Xue Shan (Jade Dragon Snow Mountains) (Figure 7.3). Tiger Leaping Gorge, named for an instance in Naxi legend, has established a biological sequence from subtropical monsoon rainforest (including the upper

Figure 7.3 The high summits of the Yulong Xue Shan (Jade Dragon Snow Mountains, 5,596 m) rise above the northern end of the Lijiang Plain, northwestern Yunnan, China. Yu-hu, the village used by Joseph Rock as his headquarters for part of the period 1924–1950 lies on the lower left. The mountain massif rising above the farmland is now part of the Three Parallel Rivers World Heritage site, May 1985.

limits of banana plants), to alpine tundra and permanent snow and ice in a horizontal distance of barely five kilometres. The terrain and its extremely diverse flora once supported a comparably varied fauna, including deer, red panda, leopard, bear, tiger, wolf, fox, and many birds, among which are several species of colourful and rare pheasants endemic to the Jade Dragon Snow Mountains, as well as a variety of raptors.

The monsoonal climate is ameliorated by the altitude, so that summers are cool and pleasant, relative to much of China and Southeast Asia. The extensive topographical and biological diversity has provided habitats for many distinct ethnic minorities, including Naxi, Yi, Tibetan, Bai, Pumi, Mousu and Han. Thus the cultural diversity matches the biological diversity.

The Naxi insist that the pictograph writing system of their ancient Dongba culture predates the development of Han script (Figure 7.4). Regardless of the accuracy of this claim, the Naxi culture certainly has produced an assemblage of architectural gems, many of which survived the Cultural Revolution or have been repaired and rebuilt since 1980. These are concentrated in and around the Old Town of Lijiang. There are also numerous outlying temples and monasteries, well preserved traditional villages, manicured agricultural landscapes, and an array of minority cultures which, together with the mountain setting,

Figure 7.4 Section of a Naxi (Dongba) pictograph telling the legendary story of the 'Love-Suicide Meadow'. This written 'language' is claimed by the Naxi people to predate Chinese script. Today Naxi newspapers use the Roman alphabet, September 1994.

collectively provide the incentive for both mass tourism and appropriate tourism. Given that mass tourism has received high-level governmental priority since 1990, the primary question is whether any niches remain for appropriate tourism.

With the opening of the Lijiang region to foreign visitors in 1985, the tourism scene evolved from an extremely primitive destination at the end of a tortuous two-day bus or jeep drive from Kunming, to a boom-town with rapidly extending paved roads, four-star hotels, attractive restaurants, linked to the provincial capital of Kunming by an air-conditioned overnight bus service and a modern airport (Figure 7.5). When the Old Town was granted World Heritage status, there was a reduction in the rate of uncontrolled construction, although Lijiang is still surrounded by tasteless modern architecture that is spreading out across the Lijiang plain as so much urban blight. A catastrophic earthquake in February 1996 destroyed a third of the buildings in the Old Town and damaged most of the others, but following the destruction the county and provincial governments eliminated many of the non-traditional structures (damaged and undamaged) and took steps to re-establish much of the traditional beauty. This was a very positive and courageous response to a serious natural disaster.

Concurrently with the development of Lijiang City and the Old Town, much of the Jade Dragon Snow Mountains is supposedly protected as a nature preserve. Nevertheless, the small

Figure 7.5 In startling contrast to Lijiang's traditional architecture this ultra-modern airport structure was opened in October 1995. It provides a 30-minute link with Kunming, Yunnan's provincial capital, November 1995.

tectonic basins that run along the eastern foot of the mountains have become sites for elitist hotels, a pleasure park, and a golf course, the construction of which resulted in the forced evacuation of a Tibetan subsistence farming village. The integrity of the mountain nature preserve itself has been compromised by the construction of a chair lift and a large gondola system. The former opened in 1994 and provided tourist access to one of the more attractive subalpine meadows, steeped in Naxi tradition; the latter cuts a swath through the mountain forest belts to 5,000 m, well above timberline, providing instant access to snow fields, glaciers, and delicate alpine meadows (see Figure 2.7); a third lift has been constructed recently. The apparent motive here is to enable some of the more affluent millions of the city people of Southeast Asia and of China to experience an alpine environment. Finally, the integrity of the Tiger Leaping Gorge, a potential World Heritage site in its own right, has been challenged by construction along its entire length of a road-bed adequate for large tour buses. The route runs along the left bank, inside neighbouring Zhondiang County. This prompted Lijiang County to blast a competing route on the right bank as far as the main rapids. Scores of rickshaws now shuttle hundreds of camera-laden tourists each day from the bus parking lot along the flagged route to the look-out and back.

While domestic tourists comprised by far the greatest number of visitors, in the early 1990s it was the foreign tourists who comprised the vast majority that participated in *appropriate* tourism. Most of those undertaking the trek through Tiger Leaping Gorge were of 'Western' origin. They were the cohort who walked or bicycled to the more distant and off-road villages. Their presence in the old town of Lijiang provided a sense of *déjà vu* – a feeling of nostalgia for the Kathmandu of 30 years ago.

While increasing numbers of young Chinese (Han) were entering the mountains as trekkers, the original target market for the development scheme outlined below was the foreign sector. This was because of the importance of attracting foreign currency and because the kind of low impact backpacker tourism was an established style of travel among Western and 'Westernized' tourists and not undertaken by domestic travellers of that time. However, the very recent and rapid growth of interest in trekking by young Chinese identifies a new emerging group that should be encouraged.

A proposed trek around the Jade Dragon Mountains

The Jade Dragon Snow Mountains loom over the town of Lijiang (Figure 7.6). They are the sacred peaks of the old Naxi Dongba tradition, a recurring motif in Naxi lore. The combination of colourful folklore, beautiful and proximate mountain scenery, extensive remnants of ethnic minority cultures, temples and monasteries, and the Tiger Leaping Gorge on the far side of the range, provided a rich resource base for tourism development. These attractions are enhanced by the aura of the remarkable figure of Dr Joseph Rock, who lived and worked at the foot of the Yulong Xue Shan from 1923 until 1949 and has emerged as a kind of folk hero.

Figure 7.6 The Yulong Xue Shan towers over the old town of Lijiang, capital of the Naxi Autonomous County. As a sacred Naxi mountain it has had a major impact on the lives of the people. This impact continues today, but in its new form as a tourist attraction, October 1995.

The sacred mountain of the Yulong has a commanding place in Naxi folklore, comparable to that of the Tibetan Shambala (fictionalized as Shangri-la in James Hilton's *Lost Horizon*). It impinged more drastically on Naxi society after 1723 when the Qing Emperor Yongzheng tightened Han control over his minority subjects through the military governor in Lijiang. The political and social institutions of a previously open and matrilinial society were reshaped to conform with the Confucian values of the central government. Hitherto, teenage relations had been essentially unrestricted and 'love marriages' were more common than parentally arranged contracts. After 1723, among many other changes, pre-marital chastity and Han-style prearranged marriages were required; out-of-wedlock childbirth was severely deprecated. This led directly to suicide pacts among love-stricken Naxi young people, often in groups of up to 12 couples. According to Naxi beliefs, those who killed themselves following prescribed rituals in designated auspicious locales beneath the high peaks would attain blissful everlasting youth in a mystic valley beyond the mountains. For more than 200 years, and as recently as the 1950s, this tragic custom took a significant toll on the Naxi population. The most famous of the 'jumping-off' points is the subalpine meadow called *Yunshanping* (Spruce Meadow), now known to the burgeoning tourist trade as 'Love-Suicide Meadow' (Swope *et al.* 1997). It was largely to exploit this somewhat macabre resource that a chair lift was opened in 1994.

Tiger Leaping Gorge, itself, was opened officially in 1992. The traverse through the gorge involved a trek of about 30 km between Qiaotou, a small rough-hewn settlement in the south, and Daju, the northern trailhead. More than twice the depth of the Grand Canyon, but much narrower and with snow peaks on either side, Tiger Leaping Gorge trek could be completed in as little as two days, but there were many possibilities for side excursions, and the family-owned guest houses at Walnut Grove became well-known attractions.

The number of visitors to the gorge increased from about 8,000 in 1992 to 25,000 or more in 1995. The drop to about 13,500 in 1996 was probably due to the after-effects of the February 1996 earthquake, the epicentre of which was directly beneath the gorge and resulted in increased slope instability (Sicroff 1998). Efforts to blast the motor road through the entire route from Qiaotou to Daju resulted in a temporary closure after 1996. Since completion the number of visitors per year has sky-rocketed; the vast majority are domestic day-trippers and bus passengers from Lijiang.

Despite these early developments, the Jade Dragon Snow Mountains and the Tiger Leaping Gorge offered excellent opportunities for trekking tourism. There was an upper trail accessible in the early 1990s that today leads trekkers above the devastated lower route. In the past ten years, lodges and tea houses have sprung up along this route. Elsewhere the picturesque but impoverished villages needed only a modest amount of assistance to draw economic advantage from their cultural and topographic legacy. In circumambulating the mountain core in 1993, it was found that there were only a few sections of the route that needed to be improved for trekkers, but that the greatest need was for locally owned and managed lodges located at convenient intervals.

It was proposed that the trek could begin and end with the two villages, Yuhu and Wen Hai, at the southern tip of the Jade Dragon range. The western branch of the walk, from one to three weeks, would pass through the gorge, now diverting to the upper trail, while the eastern branch winds through low hills south of Daju and beneath the sheer eastern face of the range. Yuhu, a Naxi village within an hour's jeep drive of Lijiang Town, is the site of Joseph Rock's expeditionary headquarters of some 70 years ago (Rock 1924, 1947).

From Yuhu, a four-hour walk leads up to a wooded ridge and then descends gently to the attractive lakeside village of Wen Hai (3,100 m). The lake is actually seasonal and disappears down a limestone sinkhole during the winter dry season. In summer it reflects the

Figure 7.7 The village of Wen Hai lies at 3,100 m above the limit of rice cultivation. It is located on the southwest side of the Yulong Xue Shan and is the ideal starting point for mountaineering ascents of the highest summit. The farm in the foreground was acquired for the village co-operative in 1995 with the intent of setting up the first of a series of lodges and a trekking route to circle the mountain range, November 1995.

Jade Dragon. Wen Hai itself is a valuable tourist attraction and is the logical base for the easiest route to the highest summits in the range (Figure 7.7).

The trekking route would proceed northwestward from Wen Hai through a series of small and isolated Yi villages (Figure 7.8). This direction eventually leads downslope into the mouth of the Tiger Leaping Gorge. Once the gorge has been traversed and Daju attained, the eastern limb southward offers several variations via Yi and Tibetan villages. Side excursions include a long but technically unproblematic ascent of the northern summits to about 5,400 metres and an attractive day trip into a side canyon replete with waterfall and hanging glacier, that was virtually unvisited (as recently as 1995 at least) by anyone other than local shepherds (Figure 7.9).

The establishment of village co-operatives in Yuhu and Wen Hai was initiated in 1995 with assistance from the United Nations University/Ford Foundation research team. A farmhouse in each village was acquired for development as tourist lodges. The Governor and Party Secretary of Lijiang offered support and links were established between the village co-operatives and the Lijiang Co-operative Research and Training Centre. The latter was being assisted by a research group from Simon Fraser University, Canada, funded by the Canadian International Development Research Centre (IDRC). The objective was to

Figure 7.8 The proposed trekking route extended northwest from Wen Hai through a series of small Yi villages before winding down into the Tiger Leaping Gorge. The Yi women and children of the second village proudly display their colourful costumes which belie the extreme poverty of their situation. Appropriate trekking tourism would greatly improve the lives of these people, November 1995.

facilitate the development of co-operative management skills among Yuhu and Wen Hai personnel. But already in 1995 the growing impingement of mass tourism was threatening to overshadow these prospects. Would there be time to help the mountain villagers help themselves?

By 2002 the Nature Conservancy, which had begun co-operative work with the local and central government agencies, was attracted to the possibilities so apparent at Wen Hai. Under their guidance the original lodge proposed as a base for trekking was resurrected, local people were trained in management and spoken English, and a modest tourist destination was established. So far this represents a small fraction of the potential.

The question posed in the title of this section (benign development or a pact with the devil?) will not yield an absolute answer. Certainly 'benign development' is too optimistic. There is no doubt that on a local scale the disparity between wealth and poverty is being widened; this appears to occur wherever tourism expands rapidly in regions lacking sufficiently strong cultural and institutional structures. It is equally clear that the overall standard of living in northwestern Yunnan has improved dramatically since 1979; an accurate cost–benefit ratio will not be immediately forthcoming. Nevertheless, one does not need to probe very deeply to observe certain disconcerting developments.

One example is the 'Love-Suicide Meadow'. Here it was the extremely poor Yi villages nearby who seized the initial benefits from the tourist potential (see Chapter 8). As their tourist entrepreneurship collapsed through the development of mass tourism, two troupes

Figure 7.9 A trek around the Jade Dragon Mountains would offer many side trips of varying challenge, including an easy, if laborious, ascent of the northern summits to above 5,300 metres, and exploration of beautiful canyons such as this, complete with views of an avalanching glacier and the highest peak, October 1995.

of young Yi village dancers were 'imported' from further afield. The girls were paid a pittance to satisfy the curious tourists.

At the other extreme is the success of the revitalization of Lijiang's Old Town after the 1996 earthquake and its designation as a World Heritage site in 1997. The town is now a comfortable and culturally interesting destination. Nevertheless, the rise in property values has encouraged many of the original owners to sell their properties. They have been purchased by incoming entrepreneurs, mainly from Fujian, Sichuan and Guangdong provinces who, in turn, have established a host of boutiques, hotels, and restaurants that are perverting the original functions of the old buildings. The result is loss of authenticity and dimination of the exotic. Yet many of those who capitalized on their traditional dwellings have been able to buy apartments in the new town and now enjoy the benefits of running water and electricity.

The situation in Tiger Leaping Gorge is more serious. Such needless damage to a world-class attraction is indeed a 'pact with the devil'. The persistent economic distress of many poor villages, such as Yuhu and Wen Hai, suggests that there remains a possibility for benign tourism development in a mode that will lift the living standards of the local minority peoples while promoting cultural diversity and protection of the environment. There is still an opportunity for a village-managed trekking route around the Jade Dragon. Yet the onslaught of mass tourism continues. It will not stop in Lijiang but will move into the rest of northwestern Yunnan. The prospects for appropriate rural development appear to be fading. In fact, within the last six years neighbouring Zhongdian and Deqin counties have attempted to out-compete Lijiang County. Roads are being improved, hotels constructed, and Zhongdian already has an airport; ominously, the county has been officially renamed 'Shangri-la County'.

Meanwhile, the Chinese authorities have proposed to UNESCO that a vast area of north-western Yunnan, the 'Three Parallel Rivers National Park', be designated as a World Heritage Site. This was approved by the World Conservation Union (IUCN) for UNESCO in July 2003 following agreement by the Chinese authorities to extend the eastern boundary to include the Tiger Leaping Gorge and the Jade Dragon Mountains. How the gorge can be included in a World Heritage designation, considering the damage by road construction referred to above, will remain a problem. A far more serious challenge to future management of the newly designated Three Parallel Rivers World Heritage site is that it is home to 315,000 permanent residents, 36,500 of whom live in the various 'core zones'. In addition, there were over 200,000 tourist visitors in 2001, 90 per cent of whom were domestic.

In summary, three groups have been affected by these extensive and rapidly unfolding developments: the 'investors' and the government who take the lion's share of the profits; Lijiang town urban dwellers who benefit appreciably; and the vast majority, the rural poor, who see minimal gains. Even this group can be divided into a minority on the fringe of tourism, such as those trafficking at the 'Love-Suicide Meadow' and similar locations and those who managed to build small lodges along the upper Tiger Leaping Gorge trail, and the majority who have benefited very little.

The Kulu Valley, Himachal Pradesh, India

The Kulu Valley is the core of the Upper Beas watershed, a major headstream of the Indus originating in the Pir Panjal and the Greater Himalaya which separate the valley from Spiti and Lahaul. The main stream rises near the Rohtang Pass (4, 304 m); its tributaries, such as the Parvati and Sainj, occupy long valleys and enter the Beas nearly at right angles, creating a complex landscape that Coward (2001) refers to as 'the Kulu valleys'. The main river runs approximately 80 km from north to south along the geological strike between the Greater Himalaya and the Dhauladhar, or Lesser Himalaya, to the village of Larji where it makes a sharp right turn to cut through the Dhauladhar and debouch onto the Plains of Mirthal.

From the point of view of human habitation, Kulu is a narrow valley, its floor less than two kilometres wide, set deep in the mountains. Before 1950 the valley supported a popu-lation of about 50,000 that included three small villages, Manali, Kulu, and Bhuntar, and numerous hamlets and scattered farms. The economy was overwhelmingly dependent on subsistence agriculture. Mixed farming dominated, with a heavy dependency on animal grazing at all elevations up to the alpine meadows in summer, and access to forest resources, such as fodder, medicinal herbs, hunting. Farming was supplemented by trade wherever possible. According to Singh (1989), existence was marginal yet set in an environment of bucolic beauty and serenity. The unusually supportive British colonial administration, based on the so-called 'Anderson settlement' (Anderson 1894), had provided the inhabitants with documented rights of access to the natural resources of the valley. Transmission of these rights of usufruct over the generations has contributed significantly to the good management and preservation of the forest cover in comparison with many other regions in the Himalaya (see

Chapter 3). The Anderson settlement also has become a vital factor in the present-day disputes with the government of Himachal Pradesh (see below).

Expansion of the road network in the 1960s and 1970s and government subsidies for development of horticulture and construction of hotels and guest houses have together contributed to the remarkable expansion of tourism. Tej Vir Singh's (1989) book, *The Kulu Valley: Impact of Tourism Development in the Mountain Areas*, provides an exhaustive treatment of the growth and expansion of tourism, its causes and consequences, up to the late-1980s. Singh documents the growth, initially slow and then accelerating, in the number of visitors, principally from domestic urban sources such as Delhi and Calcutta. For Manali (1,830 m) alone the figures are: 10,000 in 1965, 40,000 by 1975, and in excess of 140,000 by 1985.

The attractions of the Kulu Valley, as perceived in the 1960s, were access to spectacular mountains, interesting architecture, and well preserved forests and wildlife, including Himalayan black bear, musk and barking deer, leopards, bharal, tahr, and ibex. The area was also rich in birdlife, with several species of pheasant and one of the world's only two nesting sites of the Western Tragopan. Early in the tourism development phase, as described by Singh, proposals for creation of a national park were introduced; the Greater Himalayan National Park has since been established in 1999 (see below).

The development of the road network to access the Rohtang Pass and to penetrate far up the main tributary valleys toward the Greater Himalaya has favoured development of a combination of sight-seeing tourism and active trekking. The great commercial success of the apple orchards, together with tourism and secondary occupations such as construction, resulted in total realignment of the valley's economy. Singh (1989) provided a detailed assessment of Manali's growth and prospects and lauded its far-sighted urban planning. He maintained that the impacts of tourism were more positive than negative, despite considerable leakage of profits. Nevertheless, he concluded that by the late-1980s the Kulu Valley had reached a critical point, and that the projected continued growth in tourism would cause increasingly severe problems, such as environmental degradation, urban overcrowding, and accentuated disparity between rich and poor.

Singh's (1989) questionnaires returned by the local people revealed a number of relevant factors. Tourism seemed to have encouraged conservation of the physical environment by introducing ecological awareness to the permanent residents and by promoting cleanliness. However, the abnormal rise in land prices prompted purchase by outsiders and inflation was causing increasing hardship for the poorer residents. The typical 'all-inclusive' tours organized from as far afield as Simla, Delhi, and Calcutta, seriously inhibited local enterprise. Of the large number of employment opportunities created, it was primarily the low-paid jobs that went to the local people. At time of writing (Singh 1989) 1,440 persons were directly employed in the tourist sector: 676 in accommodation; 265 in the trekking business; and 160 in transport.

The permanent valley population has obviously increased in phase with the developments described: 38,000 in 1881; 46,000 in 1911; more than 300,000 in 1991 (Coward 2001). Likewise, according to Coward, Manali had 363 hotels and guest houses by 1998, and Kulu 94. It is apparent that Singh's (1989) threshold has been far surpassed.

The contemporaneous growth of tourism throughout Himachal Pradesh was significantly assisted by the government's continued support. However, the collapse of tourism in Kashmir during the early 1990s, due to the serious political and military confrontation between Pakistan and India, produced a major acceleration in adjacent Himachal Pradesh. Coward (2001: 8) indicates that by 1995 there were more than 15 million domestic visits per year to the state, with Simla, the capital, and the Kulu Valley being the two most popular destinations. Kumer (1995) estimated that annual visits to the Kulu Valley had exceeded 1.5 million.

By this point, the situation facing the people had taken on several disturbing characteristics. The over-crowding, water pollution, inadequate waste disposal, deteriorating architectural standards, and general environmental damage, which Singh had predicted in the late-1980s, had arrived in force (Gardner 2003). Moreover, two additional developments appear to have combined to increase tensions and disparities between the rich and the poor, and between permanent Kulu residents and the government: the decision to enforce national park regulations that preclude all human access to the park's natural resources, and the government programme to promote construction of multiple run-of-the-river hydro-electric power stations (see Chapter 8).

The Greater Himalayan National Park had been designated in 1985 and formally incorporated in 1999. It occupies 754 km² and encompasses all or parts of 13 different forest divisions. Five of these are reserved, or strictly protected, forests and eight are Class II protected forests in which various traditional villager uses are allowed. For many years government bureaucrats, especially those of the Forest Department, had expressed strong convictions that grazing within the forests and on the alpine meadows above, together with collection of medicinal plants and fodder, was causing serious environmental damage. No official action, however, had been taken against the offending villagers and no scientific evidence had been mustered to support these claims. Local opposition responded that the traditional practices had been a major factor in ensuring that the environment was in such good condition. Certainly, the presence of villagers in the forests limited, if it did not entirely prevent, illegal logging. It appears to have been politically expedient prior to the 1999 establishment of the park for the state government to have ignored community rights issues stemming from the Anderson settlement of 1894 as well as the *de facto* rights that had evolved informally from long-term use subsequent to the settlement.

In a precipitous about-face, the government of Himachal Pradesh responded to a May 1999 decision of the Indian Supreme Court by declaring that all claims to traditional access rights in the area of the park had been settled (Coward 2001). The Court's decision was related to its judgement that throughout India the settlement of claims to traditional access rights to natural resources in what had become protected areas was unsatisfactory. Coward (2001) inferred that the Himachal Pradesh government response to formally incorporate the park at this time was a reflection of its determination to promote its own agenda. Moreover, the government's claim that settlements with the local people had been completed was strongly denied. The settlements supposedly would provide a combination of alternate grazing access and financial compensation. Not only was the process arbitrary, but much of the proffered alternate grazing land was impractical because it was too distant from the recipient's farms. Financial compensation was considered grossly inadequate, and no allowance was made for those whose claims depended on usufruct rights established subsequent to the 1894 settlement.

The next setback was that the government deregulated a large section of the park to permit construction of the Parvati hydro-electricity plant (see Chapter 8). Housing for 6,000 employees of the power plant would be constructed in the village of Sainj, which has a current population of about 2,000. Nearly all the new workers would be brought in from outside the valley, many from out-of-state. These would be added to the already considerable force of transient workers. The planned electric power would be for sale outside Kulu, much of it outside Himachal Pradesh. Thus local residents would receive little, if any, benefits from the developments and would have to contend with the many negative effects.

Saberwal (1999), Saberwal and Chhatre (2003), and Mehra and Mathur (2001) describe the Greater Himalayan National Park controversy in detail. The most flagrant development, according to them, is that in the first instance the government had used the threat of human pressure on the breeding grounds of the Western Tragopan pheasant as part of the justification for exclusion of traditional access to the park. This was then followed by declassification of part of the same area to facilitate construction of a hydro-electricity plant. The details of the political conflagration are beyond the scope of the present discussion and are available, albeit in a rather one-sided exposition, in the papers cited. The intention here is to illustrate yet another instance of neglect of indigenous rights, specifically of poor mountain people with little political power. In the case of Kulu, however, it is not simple neglect but also a combination of corruption and hypocrisy. It is unlikely that such pernicious developments can produce long-term benefits, either for the economy of Himachal Pradesh, or for the environmental integrity of the Greater Himalayan National Park.

Northern Pakistan

Following the Second World War, mountaineering access to the Karakorum and Western Himalaya proceeded simultaneously with that in other parts of the Himalaya. The only serious setback occurred as the conflict between India and Pakistan engulfed the Siachen Glacier area (see Chapter 8).

Although the number of mountaineering expeditions to northern Pakistan continued to increase throughout the 1960s and 1970s, it required the opening of the Karakorum Highway in 1978 to expose the area to a significant and annually accelerating increase in tourism (Aghar Haroon pers. comm.10 June 2003). As a specific example, Kreutzmann (2000) points out that the number of foreign visitors to Karimabad, historic capital of the Mirs of Hunza, increased from 302 in 1979 to 5,361 in 1985. This rate of increase characterized all of the Northern Areas of Pakistan and persisted until the end of the twentieth century. The influx of tourists to Pakistan as a whole, excluding nationals, also showed a spectacular increase, although the data are not disaggregated so that the share of the Northern Areas is not known. This influx of visitors increased from 320,000 in 1990 to 1,175,000 in 2000. The very slight drop to 1,023,000 in 2001 is explained by the fact that the majority of visitors usually arrive before 11 September (Shariff 2001/2002).

Certainly, after 11 September 2001 there was a general collapse in the tourist industry. The arrival of news media and military and related personnel had no beneficial effect on the Northern Areas, where tourism has remained at a virtual standstill ever since (Watanabe pers. comm. 14 September 2003).

Kreutzmann (2003) provides considerable detail on the early stages of tourism growth after 1978 and discusses the implications of the 11 September events and the subsequent war in Afghanistan. Focusing on the Hunza Valley, Kreutzmann (2003) explains that by 1995 over 90 per cent of all Hunzacutt settlements were linked to the Karakorum Highway at least by jeep road and that Gilgit had become a rapidly growing regional centre. The opening of the Khunjerab Pass provided further impetus to tourism by making it possible for foreign visitors to travel from Islamabad to Kashgar and further into China. The fact that the rapidly developing tourism services were not well integrated vertically encouraged a proliferation of small hotels, control of which, along with a range of associated jobs (porter, cook, guide, jeep driver, handicraft artisan, musician, and dancer), remained largely in the hands of local entrepreneurs (Figure 7.10). This opportunity had a double-edged effect: first, it ensured rapid acquisition of capital by these local entrepreneurs, which was available for

Figure 7.10 The Hunza Valley, northern Pakistan. Simple hotels built by local people provide small numbers of visitors easy access to traditional scenes such as this, but the onset of political instability and the threat of violence have brought about a collapse of tourism, September 1995.

investment in further development; and second, it drew significant numbers of males from their traditional farming, greatly increasing the burden of farm work on the women and old people. In both respects, there was an accentuation of a pattern that had developed over the previous 20 years, whereby young men who left the Northern Areas to find employment in Karachi or abroad returned with savings which they invested in non-farm businesses, including the construction of small hotels. Felmy (1996) and Azhar-Hewitt (1998) explore this process and conclude that even in the case of benign efforts to encourage local, village-level developments, there are unintended negative impacts on the position of women.

In the final years of the twentieth century, however, large enterprises such as the Serena Hotel chain, based in Islamabad and other 'down-country' urban centres, were beginning to construct four-star hotels. Combined with the increasing local competition throughout the Northern Areas, this development resulted in a decline in the percentage of occupancy of hotel rooms. Despite the beginnings of over-saturation, construction continued, much of it on borrowed money. Within days of 11 September 2001, tourism collapsed in Pakistan as a whole, and especially in the Northern Areas, producing an economic and social disaster. It makes little difference that travel to northern Pakistan is completely safe (Watanabe pers. comm. 14 September 2003; Kreutzmann pers. comm. 24 September 2003). Nor has the government's attempt to encourage mountaineering expeditions in 2002 and 2003 by reducing the mountain peak climbing fees by 50 per cent produced a measurable upturn (Aghar Haroon pers. comm. 15 April 2003).

Recent visitors have reported an additional problem regarding management of the national parks (Watanabe pers. comm. 14 September 2003). Local mountain residents complain that they are totally ignored by the authorities. The National Park authority and the regional IUCN and WWF centres appear to be pre-occupied with Marco Polo sheep and snow leopards. Not only is the livelihood of the local residents ignored, but there appears to be little appreciation of 'entire ecosystem management'. In some instances the actual park boundaries are unknown. Khunjerab National Park is a partial exception, but even here locally organized groups, such as the Khunjerab Villages Organization and the Shimsal Nature Trust are opposing the imposition of the national park system. This may seem surprising, in view of the efforts of the international staff of the IUCN and WWF to ensure the participation of the local inhabitants. During the planning stages they had arranged a series of village-level workshops to facilitate dissemination of information on the potential advantages of park establishment and to win the support of the local people. However, international organizations have only an advisory role in their relationships with national governments (Lawrence Hamilton pers. comm. 23 October 2003).

Presumably, as regional tensions subside, tourism will recover. Nevertheless, Kreutzmann (2003) cautions that the post-2001 collapse has emphasized the importance of maintaining the viability of the subsistence farming system and noted that future development efforts must avoid heavy dependency on a single sector, especially tourism.

Pamir Mountains, Tajikistan

The Pamir Mountains have long carried the enticing name 'Roof of the World' and evoke images of the Silk Road and the Great Game. Recently they have become yet another remote mountain destination for trekkers and adventure tourists. The High Pamir offer mountaineering challenges comparable to any mountain region in the world (Figure 2.10 and Figure 2.11). The long gentle ridges of the Western Pamir, with their extensive mountain vistas, the Pamir Plateau to the east, and the long deep gorges that dissect the mountain core offer innumerable trekking routes. After decades of severely restricted access, the area began to open up when Communist Party General Secretary Gorbachev introduced his new policy of *perestroika* in 1987. A trickle of Westerners arrived, and authorities in Dushanbe, the capital, began to plan national parks and environmental conservation projects. But the large immediate financial gains came from big game hunters willing to bribe senior government authorities for support to shoot snow leopards and Marco Polo sheep.

An addition to the UNU project on mountain ecology and sustainable development included a partnership between Yuri Badenkov of the Soviet Academy of Science and the Tajik Academy of Science to extend the ongoing research in the Himalaya and Hengduan Mountains into the Pamir. A reconnaissance was undertaken in 1987 and, recognizing the importance of assessing the Pamir Mountains in terms of Tajik development, the collaborators made proposals for long-term research. With the support of the newly independent Government of Tajikistan, the plan was set in motion.

One of the more extensive field journeys was undertaken by Stephen Cunha. Hitch-hiking on Russian or Tajik helicopters, he surveyed the Pamir Plateau, Khorog, Lake Sarez and the Pianj River Gorge. His work buttressed the proposal to include a large section of the Pamir in the UNESCO Biosphere Reserve Programme, complete with buffer zones and appropriate development of trekking tourism (Cunha 1994, 1995; Badenkov 1997). This coincided with, and was influenced by, recommendations from a group of Tajik government officials and businessmen who were promoting the establishment of a Pamir national park.

The diverse topography and great range of elevations of the Pamir provide habitats for many unique plant and animal communities embracing 114 endemic vascular plant species and several large mammals, some of which are on the endangered list: snow leopard, Siberian ibex, Markor, Marco Polo sheep, and Asiatic bear. They are also home for small numbers of Mountain Tajik people that include a significant Ismaili Muslim population. During the Soviet era many of the Mountain Tajiks were forced to migrate by the Red Army to provide cheap labour for the expansion of the irrigated cotton plantations in southwest Tajikistan (see Chapter 8). With the introduction of *perestroika*, many of the Mountain Tajiks returned from their enforced deportation and set about rebuilding their ruined mountain villages.

In 1991 the Tajik government designated the upper Shirkent Valley as the first national historic park, although no financial support was made available. In 1992 a Pamir park was tentatively approved. However, civil war rapidly engulfed the new republic, bringing development planning to a halt. Massive transfers of armaments and militia from Afghanistan worsened the crisis, which has accounted for at least 40,000 fatalities and over half a million refugees, many of them fleeing across the southern border into Afghanistan. There emerged a remarkable, if temporary and fluctuating, series of alliances: a coalition of old guard Stalinist politicians, collective-farm chairmen, and factory managers, on the one hand, opposed by Islamic fundamentalists, liberal intellectuals, and rural district leaders, on the other. There was the threat that Gorno-Badhakshan would secede, and chaos and destruction reined until a UN-brokered peace was negotiated in 1997. Nevertheless, serious unrest continued, exacerbated by the fact that some of the warring militia had not taken part in the peace negotiations. In 1999 sections of the upper Vakhsh valley, including the incomplete Rogoun Dam, remained in the hands of defiant militia. As of October 2003, the dam construction remained at a standstill; a weak economy, lack of foreign investment, and continued political instability are cited as the main reasons (Yuri Badenkov pers. comm. 22 October 2003).

Thus, the new era of mountain tourism in the Pamir remains on hold. Gradually, improvements in the political climate seem to be occurring. With Swiss assistance, a highway has been engineered along the Pamir section of the Silk Road. The World Conservation Union (IUCN) is supporting establishment of a major peace park that would integrate the Pamir, China's Taxkorgan Nature Preserve, Pakistan's Khunjerab Park, and a section of the Wakhan Corridor of easternmost Afghanistan. The Aga Khan is intent on building a new university at Khorog, the capital of Gorno-Badhakshan, to specialize in mountain development issues for the entire Central Asian region. Efforts are being made to install a remote monitoring and advanced flood warning system below Lake Sarez (see Chapter 6). Undoubtedly, once the Central Asian region stabilizes, adventurous travellers will learn that a magnificent new destination awaits discovery.

Conclusion

From the diverse cases examined in this chapter, it can be seen that tourism (whether mass, appropriate, or something in-between) bears mixed blessings. Despite the popular consensus that ecotourism can be predominantly beneficial, negative impacts are felt even in the best of circumstances. Certainly, for many subsistence farmers in remote mountainous areas, the opportunity to earn a little extra money as guides and porters has been a huge benefit. In many instances, these earnings have enabled families to enlarge their dwellings and take in paying guests, the first step in the climb to relative affluence. Inevitably, the widening gap between the local 'rich' and the local poor leads to friction and discontent. Yet the worst impact seems to stem from the manner in which government authorities and outside investors disrupt local traditions and livelihoods while siphoning off most of the proceeds.

Several scholars have identified the drive on the part of international and central government interests to establish national parks, World Heritage sites, and other forms of protected areas, as part of a Western elitist agenda that has often conflicted with traditional rights of usufruct. And abuse of this agenda by corrupt local management is not uncommon. Two egregious cases are Upper Mustang and Kulu. Sceptical visitors often wonder what happens to all the money that is collected from summit fees and park entrance fees. Little goes to enhance park management, and even less into the hands of local people.

Regardless, as Justin Zackey (pers. comm. 8 October 2003) has commented after a long period of field research in northwestern Yunnan, the issue of tourism, whether 'appropriate' or 'mass' cannot be approached ideologically. Little can be done except to protest 'worst cases', such as the needless degradation of Tiger Leaping Gorge or the unconscionable treatment of the people of Upper Mustang. Yet a far more traumatic situation arises when a region that has attained considerable material benefits and has become dependent on tourism is pauperized by sudden acts of war, terrorism, or local insurgency. The lesson emphasized by Hermann Kreutzmann following his experience in northern Pakistan is that every effort should be made in the future to prevent the recurrence of such vulnerability. When an economic monoculture develops, primarily or completely dependent on tourism, the local people are hostage to conflicts and economic disturbances around the world.

8 Conflict, tension, and the oppression of mountain peoples

And I would have her taught Geography, said Mrs Malaprop, that she may learn something of the contagious countries.

(Sheridan)

Introduction

Twenty-three of the 27 extant wars in the world-at-large are taking place in mountain regions, claimed Jacques Diouf, Director-General of FAO, in his keynote address during the UN launch of the International Year of Mountains in New York on 11 December 2001. This dramatic statement marked the end of a long period during which mention of warfare by mountain scholars and development agency personnel was conspicuous by its absence. The cry to 'save the Himalaya' that was made under the auspices of GTZ and UNESCO at Munich in 1974 was basically an environmental plea. During the period from about 1970 until the close of the millennium, despite a few conspicuous exceptions (Allan 1987; Hewitt 1997; Libiszewski and Bächler 1997; Ives *et al.* 1997), academic and applied mountain research and development personnel barely acknowledged conflict and underlying tensions. It appeared politically indiscreet to bring the unpleasantness of the horrors of warfare and oppression of minority peoples into the mountain agenda. In contrast, a large number of NGOs, and especially Amnesty International, have called attention to disasters brought about by conflict. The same is true of the United Nations High Commission for Refugees (UNHCR) and UNICEF.

Armed conflict, as perceived as conventional war with combatant armies, is only one component. To this must be added defensive military stand-offs, guerrilla insurgencies, forced migrations and growing numbers of both international and internal refugees, expropriation of land for major construction or for establishment of national parks, and pervasive discrimination against the poor, the underprivileged, and the politically marginalized minorities. All these forms of oppression occur throughout the Himalayan region. Furthermore, conflict and tension are the greatest obstacles to achieving the major objective of the International Year of Mountains – sustainable mountain development.

In retrospect, there is a good case for arguing that the very Theory of Himalayan Environmental Degradation represented a form of oppression of the mountain minority poor. For the purpose of this argument, 'oppression' may be regarded as the widely embraced notion that it was the poor mountain farmers who were responsible for extensive clear-cutting of mountain forests and hence for inaugurating the assumed downstream despoliation. In many areas this has been used, and is still being used, to justify restrictive measures that have negatively affected their livelihood.

This chapter, therefore, will focus on the under-reported and frequently overlooked elements of conflict that are pervasive to the entire region. The widely publicized forms of violence, such as the recent overthrow of the Taliban government in Afghanistan by military might and highly sophisticated weaponry, will not be discussed in detail since these events are before us through the news media almost on a daily basis.

The underlying tension across the Himalayan region stems, in large part, from the prevalence of poverty that is brought into sharp focus as favoured local elites attain affluence. The poor mountain people become more aware of the 'outside' world with the progressive penetration of globalization and all facets of information technology. And despite some exceptions, there is an apparent inability to alleviate the condition of the poor. A further important consideration is that large-scale construction, usually with significant foreign aid, has frequently exploited mountain resources for the benefit of institutions and population centres beyond the mountains (Gyawali 2001, 2003). It also ensures enrichment of local elites, international consultants and contractors, and the providers of financial loans that are often equal to a large percentage of GNP; the local people rarely benefit directly and often undergo forced evacuation of their homes.

In Nepal, certainly, it can be argued that the Maoist Insurrection (1995–present) has drawn sustenance from the widespread poverty, coupled with rampant corruption and governmental ineptitude. The same may be said of the various armed insurrections and local unrest that has characterized most points eastward of Kathmandu. Moreover, religious and ethnic tensions, together with various attempts by central authorities to exert control, are underlying co-determinants. Starr (2001) is one of the few who have tried to assess the causes of the recent worldwide and seemingly simultaneous eruption of wars, protests, and destabilization in mountain regions. He points out that the popular explanations for violence in mountains – traditional warlike nature of mountain peoples, extreme ethnic and religious differences, imperially-imposed frontiers, centre versus local, and cultural issues – fail to take into consideration the previous long periods, often over many generations, when extremely disparate mountain groups lived together in harmony. He attributes the upsurge in violence of the last several decades to a combination of growing poverty, both relative and absolute, and the increasing awareness of the mountain people's position in relation to developments in the lowlands. As market forces penetrate the mountains, local cultures are undermined economically while the increasing effectiveness of modern communications arouses awareness of the apparent neglect and the expropriation of their resources. Starr (2001) writes, for instance, that '[m]ountain people's contact with commercial centers in the lowlands is great enough to bring HIV into their world, but not great enough to bring any treatment for it'.

Furthermore, the spread of malnutrition among formerly thriving mountain communities has reached such a level that in many areas they experience food deficits for up to six months of the year (Chapter 9), leaving the people with no choice but to migrate to the cities to seek low-paid labour or charity. This increasing contact with the cities also provides insights into the scale of corruption and wastage found there. Once poverty and malnutrition reach a critical stage, then mountain people are susceptible to the influence of radical political or religious groups who may sponsor violence. Each instance of regional mountain violence has its own distinctive trigger. Nevertheless, the situation has reached such crisis proportions that more sustained efforts are urgently needed to recognize what is happening and to attempt solutions. Starr believes that this requires working with the local people to improve their condition at the village level while at the same time the political tensions are addressed. The conventional approach has been to seek solutions to the political problems before tackling the poverty and this has met with very limited success. Starr cites the very successful strategy of the Aga Khan Foundation in the northern mountains of Pakistan that has focused on small-scale village-level improvements and the encouragement of self-help, an approach that is now being expanded into Tajikistan and Kyrgyzstan.

Nevertheless, some local struggles, such as the long-lasting insurrection in the Chittagong Hill Tracts of Bangladesh and the Nagaland aspirations for independence in northeast India, may soon find tentative resolution. Others, such as the repression inflicted by the Royal Government of Bhutan on the statistically significant Lhotsampa Hindu minority, have been largely 'concealed' until very recently, or at least have received little effective attention from agencies of the United Nations or bilateral aid organizations.

The intent of this chapter, therefore, is to bring to the fore the wide variety of different forms of conflict and oppression. However, it is not intended to attempt an exhaustive treatment. Examples have been selected to demonstrate the full range of conflict, oppression, and discrimination that underlies the social order of the region. At one extreme the lack of awareness of local minority aspirations leads to unthinking and inappropriate application of official policy that overrides local self-help efforts. This is illustrated by the callous, but probably unwitting disruption of the efforts of Yi people in Yunnan to profit from small-scale tourism centred on the 'Love-Suicide Meadow' (pp. 188–9). The plight of the hill people of northern Thailand is in large part attributable to the Thai constitution that deprives them of legal rights as human beings; they are also used as a convenient vehicle through which prejudicial treatment can operate (pp. 189–90). The Chakma and other minorities of the Chittagong Hill Tracts of Bangladesh were initially pushed aside as impediments to obtaining vast sums of foreign financial aid to develop and to distribute hydro-electricity nationally (pp. 175–7). The troubles of Nagaland appear to derive, in part, from British wartime and pre-Indian Independence promises of autonomy for the Naga people (pp. 177–8). Bhutan seems determined to reduce its ethnic and religious minority by a policy of severe repression while it poses internationally as a model of humane treatment of all people (pp. 178–83). The actual armed conflict along the Siachen Glacier can be described as a surrogate war between two large regional powers – India and Pakistan (pp. 185–7). The destructive civil war in Tajikistan was in part a legacy of the Soviet Union, complicated by its proximity to Afghanistan and the region-wide tensions between resurgent Muslim fundamentalists, liberal nationalists, and Stalinist hard-liners seeking to retain power (pp. 190–1). The large-scale international and intercontinental conflict in Afghanistan, and its ramifications along the Kashmir 'Line-of-Control', are acknowledged for their very serious destabilizing threat, but will not be discussed in any detail. Nevertheless, conflict and underlying political tensions based on ethnicity, religion, poverty, and, above all, the inequitable exploitation of mountain resources that is particularly disadvantageous to women, all of which are grounded in a complex history, appear to stand as the primary obstacles to sustainable mountain development.

Chittagong Hill Tracts

The Chittagong Hill Tracts comprise a large stretch of uplands in the southeastern extreme of Bangladesh, connected to the main lowland core by a narrow corridor a few kilometres in width between the Bay of Bengal and the southern border of the Indian state of Tripura. Ridge crests rise to between 450 and 900 m in the south and between 300 and 600 m in the north, separated by mostly narrow north–south valleys. The highest land is in the southeastern and northeastern frontier regions and exceeds 1,000 m. An estimated 80 per cent of the hill tracts is described as hilly or mountainous with steep slopes. The climate is similar to the rest of Bangladesh, except for a slight amelioration due to the increased elevation. Total area is about 13,000 km², or 10 per cent of the entire country. In 1991 it supported a population of about two million, more or less equally divided between indigenous peoples and Bengali settlers. The Chakma comprise by far the largest percentage of the indigenous peoples, although there are nine different minority groups.

From 1900 onward the Hill Tracts were administered by a special form of British Imperial management that was incorporated first into the jurisdiction of the former East Pakistan and subsequently into that of Bangladesh. In practice, it continued in effect formally until 1989 and, informally, until 1998 (Blaikie and Sadeque 2000).

The indigenous peoples have been principally shifting cultivators (*Jhummias*, or swiddeners), while the Bengali settlers produced irrigated rice and other crops. Over time, however, there has been a merging of actual farming types, and members of each group practise some of the traditional farming methods of the other. *Jhumming* was first drastically curtailed in 1880 when the British administration classified almost 25 per cent of the entire Hill Tracts as reserved forest. This set in motion increasing pressure on the land by shortening the period of forest fallow within the slash-and-burn cycle, although it did not reach critical levels until population increase accelerated after about 1950. A second major event was the completion of the Kaptai Dam in 1963. Funded by the World Bank and USAID, the Kaptai Dam submerged 650 km² of prime agricultural land and this eventually led to the militant reaction that consumed the region for more than 20 years.

The advantages of the Karnaphuli hydro-electric project, with the Kaptai Dam as its centre piece, were originally seen as providing a large amount of electricity, increasing fish production in the reservoir, improving water transportation, and bringing more land under irrigation. In practice, the production of 230 MW electricity, amounting to 12.5 per cent of Bangladesh's total power production, was totally outweighed by the ensuing insurgency and environmental damage. It has been estimated that 40 per cent of the most productive land of the Hill Tracts has been submerged. The increased pressures on the remaining land further reduced the length of the forest fallow cycle leading to declining soil fertility and reduction in crop yields. About one-sixth of the total indigenous population of about 100,000 had been promised financial compensation and new land. The attempts at resettlement were so inadequate, however, that many fled to India as refugees, while others formed the Shanti Bahini guerrilla organization and initiated armed insurrection.

Following severe loss of life and property, the long period of insurgency was brought to a close on 2 December 1997 (Shelley 2000: 107) with the signing of a peace treaty between the Bangladesh government and the Parbatya Chattagram Janasanghati Samity (PCJS), the political wing of the Shanti Bahini. Before this accord was achieved, 25 separate meetings between three successive Bangladesh governments and the PCJS had taken place. While the problem began before Bangladesh independence in 1971, all governments involved must bear responsibility for the conflict. Blaikie and Sadeque (2000: 35) point out that, while the construction of the dam 'could well be justified on the grounds of overwhelming national need', the total insensitivity to the needs and way of life of the local hill people led to severe socio-economic impacts and 'a policy disaster of the highest order'. The resentment re-invigorated the search for identity among the hill people under the then existing authority of East Pakistan. Displacement of a large number of ethnic minority people, accompanied by absolute neglect of their welfare by the various governments, led to the vigorous and prolonged armed insurrection.

In the early 1970s the long period of suffering and neglect of the Chittagong hill people emerged as a prolonged armed conflict against the central government. The government responses, driven by principles of state security, resulted in severe environmental impacts, frequent displacement of both hill people and Bengali settlers, and attempts at resettlement that were highly unfavourable to those involved. Progressively, the state appropriated resources vital to the livelihood of the local people, and those forced off their traditional lands by the hydro-electric project were classed as criminals for opposing the actions of the government. Many plains people were settled by the government on lands previously occupied by the hill people, further inflaming ethnic and religious conflicts. Moreover, the conflict has cost Bangladesh, one of the poorest countries in the world, considerable expenditure and loss of life.

The peace treaty of 1997 was facilitated by an eventual stalemate in the fighting and by a change in government in Dhaka. The Chittagong Hill Tracts were accorded quasi-separate status with a large element of self-rule and the hill people hope that they will be able to exert more control over environmental policy (Blaikie and Sadeque 2000). Uncertainties remain, however, and more time is required before any degree of success can be determined. The apparent impressive progress indicates that accord can be established with ethnic hill minorities if an appropriate approach can be found. Shelley's (2000) report on the Chittagong Hill Tracts represents an official government contribution to an international country review and, therefore, may be overly optimistic. At least it does not attempt to disguise the extent of the policy disaster.

Regardless, the 1997 Chittagong Hill Tracts Accord must be viewed as a promise for action rather than as a solution. As Roy (2002) has indicated, little actual progress has been made in terms of active resolution of grievances. The 1997 Accord established various legal and administrative instruments, including a new Ministry of Chittagong Hill Tracts Affairs with extensive minority representation. Major changes have been made to the structure of the land administration system. For instance, no grant, settlement, transfer, or acquisition of land in the Hill Tracts is to take place without the consent of the relevant Hill District Council. Additionally, the district councils are to share authority concurrently with village headmen and other revenue officials. Yet, as of June 2002, nothing had been accomplished in terms of the functioning of the new arrangements. Roy (2002: 36) insists that reforms are needed 'that can at least partly address the problems of corruption and circuitous official procedures that have bred corruption, and have been the bane of landless people. . . .' These are only a few of the difficulties facing the Chittagong Hill Tracts. Unless the various levels of government can tackle them efficiently and convincingly, the prospects for renewed political agitation, and worse, remain to threaten collapse of the Accord.

Nagaland and the northeast frontier of India

The Naga people number about three-and-a-half million and have traditionally occupied the extremely rugged and isolated land on India's northeast frontier, overlapping into northernmost Myanmar and southwest China. Historically they had a reputation for ferocity, even head-hunting, and have been cited along with the Pashtuns as one of the only two tribal groups that the British in India were unable to pacify. They are mountain people dependent upon a mix of swidden agriculture, settled rice and maize cultivation, supplemented by faunal and floral forest products. Despite their fierce independence they were apparently converted to Christianity during the last half of the nineteenth century. During the Second World War they provided critical support to the British struggle to contain the Japanese incursion into India. The famous British Field Marshal, Viscount Slim, is reported to have remarked, 'Many a British and Indian soldier owes his life to the Nagas, and no soldier of the Fourteenth Army who met them will ever think of them but with admiration and affection' (Slim, W.J., Viscount, *Defeat into Victory*, 1981, pp. 334–5).

Following the Second World War the British Government considered retention of Nagalim (the preferred name of the Naga people) as a Crown Colony, although the process was delayed because the Nagas insisted on complete independence. With Indian Independence in 1947, despite promises to the contrary by both Mahatma Gandhi and Jawaharlal Nehru, independence for Nagaland became a lost cause. Even as late as 1952, Prime Minister Nehru made the statement to the Lok Sabha, 'We want no people in the territory of India against their will.' Yet in 1954 the Indian Army began a campaign of subjugation of the Nagas that has lasted for almost 50 years. The Nagas claim that they have been subjected to ruthless military repression from 1954 onward with intermittent cease-fires and peace negotiations. A cease-fire was effected on 6 September 1964 and negotiations continued for two years before breaking down, with renewed fighting. The Nagas aver that the repression involved excessive use of force against civilians, the burning of villages and crops, extra-judicial executions, and the arbitrary arrest and disappearance of Naga leaders.

The territory of the Naga people extends into the Indian states of Assam, Arunchal Pradesh, Manipur and Tripura, and northernmost Myanmar, as well as the Indian state of Nagaland itself. This complicates aspirations for independence. Also the Nagas are intermixed with several other minority groups. Furthermore, they have been subjected to especially severe and widespread atrocities by the Myanmar military. And over the years, it has been estimated that the fighting with India has resulted in the loss of over 125,000 lives and countless injuries. Following a recent five-year cease-fire, peace talks began again on 3 January 2003. Luingam Luithui, who was exiled from India in 1995 for documenting human rights violations, and accepted by Canada as a political refugee, was granted a visa for travel to Delhi to take part in peace talks with other Naga leaders (*Ottawa Citizen* 31 January 2003).

As in the case of other similar struggles for independence, it is very difficult to obtain verifiable information. Furthermore, it must be emphasized that part of the efforts by the oppressed side to attract international attention may involve exaggeration. Similarly, the government accused of unjustifiable repression often tends to minimize its severity internationally and frequently accuses the insurgents of being terrorists. Nevertheless, the 50-year repression of the Nagas, as with all of the acts of subjugation and repression in the Himalayan region, has undoubtedly caused heavy loss of life on both sides and extensive destruction to infrastructure and the environment. The extreme isolation of the territory claimed by the Nagas, its dissection by international frontiers, and the almost totally restrictive access for independent foreign visitors have all served to place this mountain struggle in a vacuum that has attracted little Western attention.

Bhutan: a kingdom of peace – and strife

Bhutan has been described by many observers as a model for Himalayan cultural preservation and environmentally sound development. Essentially, this is a myth that has been sustained by active governmental propaganda, leading to naive perceptions in the West illustrated by the writings of Kean (1997) and Lumley (1997). A critical examination of the basis of the present conflict may help to unveil the misinformation and increase awareness, and indirectly to alleviate the danger of a wider conflagration that could be devastating for all participants.

The institution of monarchy in Bhutan is less than a century old. Aided by the British, Ugyen Wangchuck was installed as hereditary king in 1907. Three years later a treaty was signed to give Bhutan control over its internal affairs, leaving the British Raj responsibility for defence and foreign affairs. India assumed the British role in 1947. From its formative years until recently, the Kingdom of Bhutan was widely regarded in the West as a remote, largely closed, Himalayan paradise where peace, gentility, and happiness prevailed. This small country of less than 50,000 km² and a population of little more than one million posed as a fairly homogeneous polity of Mongoloid/Tibetan ethnicity and Buddhist religion. Its policy of seclusion has been a major factor in the creation of its image as an exotic mountain land of peace and enchantment.

Bhutan became a member of the United Nations in September 1971. It was assumed that the abrupt increase in the official population count from 600,000 to over one million was a device to qualify for UN membership, although it may have been a simple artifice of lack of adequate census-taking instruments. Membership in the international family of nations was conceived as a measure of maintaining its autonomy between its two powerful neighbours – India and China. In 1975, however, it was unable to send a delegation to the UNESCO MAB–6 regional meeting held in Kathmandu because India refused overland travel permits for the Bhutanese delegation and, at that time, no air link existed.

Upon gaining United Nations membership, UN agencies and several bilateral aid agencies established offices in Thimphu, the capital of Bhutan, although development aid had been initiated earlier by India and that source continued to predominate. Infrastructure, especially roads and hydro-electricity projects, received priority attention, although most of the power to be generated was destined for sale in nearby lowland India (Dhakal 1987). The national development policy was deliberately aimed at cautious and slow measures to ensure minimum impacts on the country's rich environment and culture. The present king, H.M. Jigme Singye Wangchuck, carefully protected and furbished the notion that his kingdom was a unique abode of kindliness where the expression 'gross national happiness (GNH) is more important than GNP' became an effective propaganda symbol. There were few reliable scholarly publications, the notable exceptions being Aris (1979), Rose (1977), Karan (1967, 1990), and White (1990). Official publications (Bhutan 1992) and tourist-oriented large format releases give the impression of a culturally homogeneous country. For instance, in referring to the country's demographic origins that include several distinct ethnic groups, the National Planning Commission stipulates, 'However, they have completely assimilated into the Buddhist derived culture of the northern and central parts of Bhutan. The settlers in south-western Bhutan are predominantly Hindus. The bulk of them have immigrated more recently' (Bhutan 1992: 15).

The government's claim of homogeneity has been widely supported by Western popular writing and by the news media until recently. This, however, is a misrepresentation of the highest order. While the National Planning Commission document on environment and development (Bhutan 1992) was under preparation for presentation to the UN Conference on Environment and Development (UNCED), these 'recently' immigrated Hindus (the Lhotsampas of Nepalese descent) were being brutalized by torture, rape, and imprisonment, and more than 80,000 of them were forced to flee the country (Dhakal and Strawn 1994). While several thousand of the refugees dispersed into nearby Indian territory, the vast majority, with Indian assistance, were moved to eastern Nepal where they have survived in refugee camps for the last 12 years (Dhakal 2000a, 2000b; Ives 2002).

With support from UNHCR, the Government of Nepal, and several humanitarian NGOs, the refugees have anxiously awaited determination of their fate while Bhutan and Nepal cavil over their status. India remains aloof while the donor countries to Bhutan continue their aid and seemingly do little to assist the refugees except to mildly suggest that financial support for His Majesty's government may be delayed. The following account summarizes the refugee situation over the last decade.

The Bhutan ruling elite surrounding the king, identified as Drukpa, began exerting pressure during the 1980s to achieve a form of homogeneity in the population as a whole. In 1985 a Citizenship Act was promulgated, containing inherent prejudices against citizens of Nepalese descent and predominantly Hindu religion. Most of these, the Lhotsampas, originated from Nepal and had migrated into southern Bhutan mainly at the beginning of the twentieth century, although some date their settlement several decades earlier. The Citizenship Act may have been a reaction to the realization that the already large Lhotsampa ethnic minority were reproducing more rapidly than the Drukpa (actual reliable numbers are not available, but there have been claims that the Lhotsampa already represented a majority, or almost a majority). Restrictions were imposed, including elimination of education in the Lhotsampas' own language (Nepali) and the enforced wearing of the Bhutanese national dress. This traditional costume derives from a high-altitude culture and is extremely inconvenient in the warm moist summer climate of the lower altitudes inhabited by the Lhotsampas.

Peaceful protests during September and October 1990 resulted in a wave of severe repression. The police and army were sent in and thousands of people were arrested, many were tortured, and frequently forced to sign so-called 'voluntary migration forms' (signature of such forms, often forced on illiterate people, legally deprives them of Bhutanese citizenship). Lhotsampa leaders were imprisoned, or else went underground fearing assassination, and eventually fled the country.

The flow of refugees began in 1991, starting with hundreds each day and peaking to thousands daily by mid-1992. By 2000 there were more than 97,000 Bhutanese refugees housed in seven refugee camps under the auspices of the United Nations High Commission on Refugees (UNHCR) in the Jhapa and Morang districts of eastern Nepal. It is estimated that another 10,000 live in Nepal outside the refugee camps and that 30,000 have settled in India. Almost one-fifth of Bhutan's population is in exile (there may be many more if reliable demographic data were available); on a per capita basis, Bhutan is the greatest source of refugees in the world.

When people become refugees they lose everything: that which was inherited and that which they had created by their own labour. Because the majority of the Bhutanese exiles are poor rural subsistence farmers it is extremely hard for them to line up for their daily rations; neither is it easy for them to fill long days with inactivity.

There are strong leaders in the refugee camps and an impressive organization evolved. In each camp classes are offered in useful skills, such as tailoring, typing, weaving, and shoe-making; there are language classes in both Dzongkha, the official language of Bhutan, and Nepali. Thirty per cent of the camp dwellers are children of school age and their education has been better than that of those who remained in Bhutan. According to Dhakal (2000a), in a review of camp activities: 'Today the majority of the refugee population is literate and is able to take advantage of the opportunities offered.' But what are the available opportunities?

Nepal was the reluctant recipient of the refugees and immediately sought accommodation with Bhutan in the hopes of effecting repatriation. A Joint Ministerial Committee was set up and has met intermittently since 1993. Upon Bhutan's insistence and Nepal's subsequently regretted agreement, it was decided to classify the refugees into four categories: (1) bona fide Bhutanese citizens who had been forcibly evicted; (2) Bhutanese citizens who had voluntarily emigrated; (3) Bhutanese with criminal records; and (4) non-Bhutanese people.

This agreement has become the means of interminable obfuscation by Bhutan. For instance, to prove actual citizenship, refugee families were required to produce legal documentation, yet the Bhutan authorities had attempted to destroy all personal documentation that they could find when the refugees were leaving Bhutan.

After eight rounds of talks at the Joint Ministerial level, agreement was reached to interview all refugees to effect the fourfold categorization. The process began in April 2001. Difficulties immediately arose. The procedure was incredibly slow and was made extremely inconvenient for refugee families. Any outside observer would question the integrity of the process. A strong indicator of this is that even the UNHCR personnel were not allowed to serve as third-party observers.

By December 2001, 12,500 refugees from the Khudunabari camp had been interviewed. As the process of verification extended into 2002, however, several critical, if not directly related events unfolded. First, most of the Nepalese Royal Family was massacred, including the king and queen, and a new king, H.M. Gyanendra Bir Bikram Shah Dev, came to the throne amidst widespread public unrest. Second, the Maoist Insurgency within Nepal continued to expand and resulted in further government instability. Then the 11 September 2001 attacks on New York and Washington unleashed the spectre of large-scale international terrorism, ushering in unprecedented military action in Afghanistan. This in turn led to the near destabilization of Pakistan and further tension and military brinkmanship between India and Pakistan and intermittent conflict along the Line-of-Control in Kashmir.

All of these developments have certainly combined to overshadow the now relatively minuscule problem of the Bhutanese refugees who continue to languish in eastern Nepal. However, the problem is by no means insignificant to the refugees themselves, nor to their host country, Nepal. Furthermore, the maintenance of the refugee camps has already cost the international community in excess of US$100 million.

The most recent information received from the refugee camps in Nepal indicates that the fourteenth Joint Ministerial Committee meeting in Thimphu was concluded in May 2003. The Committee had determined that only refugees in Category 1, 'bona fide citizens who were forcibly evicted from Bhutan', would be allowed to return home as fully designated Bhutanese citizens. However, this group included less than 3 per cent of the 12,096 refugees of the Khudunabari camp who had been processed in 2001 and 2002. Another 2 per cent were classed as criminals; 70 per cent were placed in Category 2, those who emigrated voluntarily, and they must wait for two years for their cases to be reviewed, and 24 per cent were declared 'non-Bhutanese'. There are still 600 refugees from the Khudunabari camp and the vast majority of the refugees in the six remaining camps to be processed. In practice, 95 per cent of the people have been rendered stateless.

> At the end of September, a frustrated Ruud Lubbers, the head of UNHCR, announced that since the agency cannot monitor the return of the refugees, it would not promote repatriation.
>
> (*The Economist*, 25–31 October 2003: 38–9)

This represents a serious criticism of the Government of Bhutan's persistent refusal to tolerate third party observation in any of the procedures relating to the refugees; third party observation is one of the important functions of the UN agency.

Objective documentary information is contained in the 1 April 2003 'Country Reports on Human Rights Practices 2002' released by the United States Bureau of Democracy, Human Rights and Labor, Department of State. This provides a grim record, both of the refugee situation and of the living conditions of the Lhotsampas who remained in Bhutan following the 1990 crack-down and who have been subjected to continuing discrimination, intimidation, and physical abuse. The report also brings to light the steadily increasing threat of destabilization stemming from the expansion of militant activities in northeastern India and the increasing use of Bhutanese territory by guerrilla groups for shelter from the Indian army.

When the extent of the mistreatment perpetrated by the Bhutanese government, both to the 100,000 refugees and the Lhotsampas who remain in Bhutan, is considered, it is surprising that a militant response has not erupted from within the Lhotsampas. Certainly within the camps, refugee leaders have persevered to induce their young people to adhere to a course of peaceful protest. Nevertheless, 12 years in the refugee camps has inevitably produced serious tension. Thus, on 9 September 2001 R.K. Budathoki, President of the Bhutan People's Party, was murdered by members of a student group calling itself Akhil Bhutan Krantikari Vidyarthi Sena (All Bhutan Revolutionary Student Force). The group is believed to have links with the Nepalese Maoist Insurgency. Given these developments and the apparent inability of India to contain the various militant groups on its own territory (National Democratic Front of Bodoland, the United Liberation Front for Assam, and the Kamtapuri Liberation Organization, fighting for a separate state within North Bengal) there is a very real potential for extension of the conflict.

The growing tension in Nepal and Bhutan is made the more sensitive by the Maoist Insurgency in the former country and by ongoing guerrilla activity across northeast India. As the confrontation between the Maoist insurgents and the Nepalese government continues (see pp. 183–5 below), reports as recent as May 2003 indicate that a new militant group, calling itself the Bhutan Communist Party (Marxist–Leninist/Maoist) has been formed with the objective of overthrowing the Bhutan monarchy.

It is tempting to compare the Bhutan situation of 2003 with that of the Chittagong Hill Tracts of ten or 20 years ago. However, the militant Hill Tracts groups were at least operating largely within a single country – Bangladesh. The repression in Bhutan can so easily erupt into militancy throughout Nepal, Bhutan, West Bengal, Assam, and Arunachal Pradesh. Moreover, this is a region on the contact edge between India and China – and the Indian demarcation of its northeastern border (McMahon Line) is not accepted by China.

The foregoing account is based in part on reports from Amnesty International and a book by Dhakal and Strawn (1994) that provides extensive details on the development of the conflict. In contrast, the Bhutan country report (Lhamu *et al.* 2000: 137–70) reads as a government exercise in propaganda. Despite its great detail, no mention is made of the different ethnic groups that make up the Kingdom of Bhutan. It reads as if Bhutan is a unique paradise of equality, serenity, and joy. The population is described as 'small, isolated, and homogeneous communities who share firm common beliefs and a common identity' (Lhamu *et al.* 2000: 148). The authors explain the Kingdom's underlying principle of Gross National Happiness (GNH) whereby the 'individual is placed at the centre of all development efforts, recognizing that people have material, spiritual, and emotional needs'. The final insupportable claim is a reference to 'sustainable development' which is seen as a means to ensure that the benefits are:

> . . . shared equitably between different income groups, genders, and regions, in ways that support social harmony, stability, and unity and contribute to the maintenance of a just and compassionate society.
>
> (Lhamu *et al.* 2000: 151)

The donors, by their silence, give the impression of support for policies involving brutal and inhuman treatment of a large proportion of a country's population. This cannot fail to increase the level of tension and frustration and encourage, rather than discourage, conflict throughout the Himalayan region. Above all, India must be held responsible to a considerable degree, both because it is the major source of assistance and because of its treaty influence over Bhutan's foreign policy.

Nepal and the Maoist Insurgency

Nepal, like many developing countries, has been slow to emerge from totalitarian control: initially by hereditary monarchy (1769–1854); next by hereditary family prime ministership (1854–1951); and then by an only slightly circumscribed hereditary monarchy that pandered to democracy by using the term 'panchayat democracy' (1951–1990). During the final period a quasi-representative parliament was in place, although actual power appeared as a shared commodity between the royal palace and a confusing array of international and bilateral aid agencies. This combination fostered the impression that development for the benefit of the poor majority could be imposed from above. Discontent among most sectors of society led to the overthrow of the panchayat democracy and its replacement in 1991 by a Westminster-style parliamentary system.

The net results of post-1950 'development' have been some impressive infrastructural projects, the emergence of a middle class, extensive freedom of the press, and an enormous expansion of tourism that has certainly brought positive gains to particular sections of society and the national economy. On any world conservation scale, Nepal must rank near the top, having declared almost a fifth of total territory as national park and nature reserve. Furthermore, recent adjustments in the forest policy have led to the establishment of more than 12,000 Forest User Groups (FUGS) (Piers Blaikie pers. comm. 23 June 2003) and a significant increase of village management of local forest land (but see Chapter 9 pp. 202–3). Regardless of many positive aspects, however, and according to the standards of international assessment, the majority of Nepal's people appear to have become relatively and absolutely poorer. This does not take into account the pattern of remittances to some villages, off-farm wage employment, both in-country and abroad, and the extensive informal economy (Gyawali 2001: 66–86). Gyawali (pers. comm. 23 November 2003) has also pointed out that many of the centres of Maoist Insurgency coincide with the places that have experienced major donor interventions.

The latest political adjustment of 1991, although with initial high expectations, has led to disenchantment, mainly on account of inept government resulting from fractionalization between and within political parties, and continued corruption.

In the first government under the new system, elected in 1991, the Nepali Congress Party held a large majority and the Unified Marxist/Leninist (UML) Party formed the main opposition with 69 seats in the 205-seat parliament. A radical left party, the United People's Front (UPF) gained nine seats. According to Chitra K. Tiwari (2002), in a published lecture presented to Williams College, Massachusetts, on 1 March 2002, the result has been 'political chaos'. Thapa (2003) reviews the origins and current status of the insurgency, and while his viewpoint remains critical of the government, it is much less so than that of Tiwari. Thapa believes that the increased literacy and external contacts of the mountain youth, coupled with limited opportunity for advancement, led to increasing demands and frustration. This was exacerbated by the apparent nonchalance of national-level politicians and the Kathmandu intelligentsia to the initial Maoist violence in the far west. The elite appeared to regard the violence as something that was best handled as a law-and-order affair to be put down by the police (Thapa 2003: 88–9).

Tiwari (2002) states that the popular expectations of the 1990 People's Movement had not been seriously addressed by the new democratic government:

> Corrupt individuals of the previous regime whom the people wanted to see in jail were rewarded with political appointments . . . Corruption, smuggling, embezzlement of public funds, bribery and kick-backs in the allocation of development projects gradually infested the system.
>
> (Tiwari 2002)

Public demonstrations, organized by the UPF, were severely repressed by armed police in 1992 and 16 unarmed people were killed. Following this initial period of unrest the UPF splintered as the mid-term elections of 1994 approached. One group participated in the elections and obtained a single seat. The larger group emerged as the Nepal Communist (Maoist) Party, boycotted the elections, and opted for a militant response.

The catalyst for what the Maoists refer to as the *Maoist's People's War* was a small rural protest in Rolpa District in western Nepal in 1995 that was brutally crushed by armed police. There were widespread reports of rape, torture, and indiscriminate killing of several hundred peasants, followed by the flight of several thousand to the jungles of the western Terai. From early 1996 the situation deteriorated, with repeated Maoist attacks on district police stations, and the conflict progressively widened to embrace almost the entire country.

The Maoist movement in Nepal has been described as the country's first indigenous insurrection and it spread across most of the country's 75 districts. The claim by the Maoists to have raised 15,000 hard-core fighters backed by over 40,000 militia, half of whom are women, is probably exaggerated. Nevertheless, it is reported that the insurgency and the government's response have resulted in more than 5,000 deaths, with over a thousand of these occurring in the final months of 2001 and early 2002. Failure of the government's attempts to quash the insurrection has been attributed in part to the initial sympathetic support given to the Maoists by many of the rural people, coupled with the government's repressive actions. There have been several attempts to obtain a peaceful settlement. The Maoists made three primary demands: abolition of the monarchy; election of a constitutional assembly; and formation of an interim government, that would include Maoists, with its main purpose being to supervise national elections.

On 26 November 2001 the government declared the Maoist Party a 'terrorist organization', presumably seeking international support in the immediate post-11 September period. A state of emergency was declared and the army ordered to crush the insurgents. Despite some early government successes, the insurgents remained highly effective in the more remote areas of the country, and the loss of life on both sides, and many innocent people in between, continued to rise, each side blaming the other for the reckless killings.

In addition to the direct losses due to the fighting and atrocities, Nepal has suffered severely in financial terms. The international and bilateral aid agencies have been reluctant to begin new projects. Tourism, Nepal's main source of foreign exchange, was reduced precipitously in 2001 and 2002, and the government was forced to drain 25 per cent from its development budget. Financial reserves also fell with the decline in business activities.

Tiwari (2002) contends that the government's decision to seek a military solution from the start was a serious mistake. He believes that economic measures aimed at improving the lot of the poor should have been put into place, but then partially contradicts himself by explaining that an essentially corrupt government would never be able to assuage the people. On the other hand, once the insurgents turned to indiscriminate killing of school teachers and other innocent people, much of their early sympathetic support was forfeit.

A cease-fire was negotiated on 29 January 2003 and the first round of talks took place in Kathmandu on 27 April. The Maoists presented their three primary demands, except that they agreed to have the proposed abolition of the monarchy settled later by popular nation-wide vote. Their contributions to the talks had been taken as indication of their seriousness in desiring a settlement. The government delayed for several months to respond and to propose a date for a second round of talks. In the meantime, the United States had put the Maoists on their 'watch list' of 38 secondary-type terrorist organizations. The United States ambassador to Nepal was reported by *The Himalayan* (6 May 2003) to have stipulated that they would be removed from this list when they were able to demonstrate their relinquishment of violence. This, apparently, has been Washington's response to the murder of two security guards (Nepali) at the United States embassy in Kathmandu and concern that the Maoist pre-29 January 2003 activities were targeting US interests. The talks were eventually resumed on 17 August 2003. However, they failed to produce a settlement and were broken off. By September 2003 violence had once more erupted and, after a brief lull to accommodate the major autumn religious holiday of Dasain, resumed in late October.

Tourists had begun to return to Nepal in the spring of 2003. The 29 May 2003 celebrations commemorating the 1953 first ascent of Everest had attracted large numbers, both in-country and from abroad. However, by October the situation had deteriorated again.

The past nine years of unrest and conflict have emphasized a major problem: development aid to Nepal has been far from successful despite nearly a half-century of efforts and expenditure of some hundreds of millions of dollars. There have been many successful aid projects, but these overwhelmingly have been the ones, usually small in scope, that have incorporated the input of the local people and their active participation. The larger enterprises, especially those involving infrastructure, have done little to help the poor people, who frequently have become losers rather than beneficiaries. Although attitudes among the donors have changed appreciably over the last two decades, the fact that a major insurgency, dependent at least initially on popular support, engulfed most of the country indicates how much more change in policy is needed. The broader implications beyond the confines of Nepal itself, and especially in India, cannot be ascertained.

The reporting on the Maoist Insurgency is extensive and complicated, especially because the situation is changing from day-to-day. Additional scholarly publications include those by Dixit (2002), Thapa (2003), Thapa and Sijapati (2003), and Gellner (2003). As a final point, it is worthwhile to reflect upon the last sentence of Blaikie *et al.*'s book, *Nepal in Crisis* (1980), even though it was written a quarter century ago:

> We see no reason to believe that the peasantry of Nepal will discover a collective political expression of its needs which reaches beyond mere populist rhetoric in time to save millions of people from impoverishment, malnutrition, fruitless migration, and early death.
>
> (Blaikie *et al.* 1980: 284)

The Siachen Glacier: warfare at high altitude

The Siachen Glacier would appear to have had little or no strategic value before the current conflict. The glacier is about 70 km long, extends to altitudes above 6,000 m, and is surrounded by the even higher peaks of the Karakorum. It occupies the centre of a contested triangle of glacierized high mountain terrain lodged between territory claimed by Pakistan, India, and China, hinging on the Karakorum Pass southeast of K-2, the world's second highest mountain. Yet it has been the scene of serious military conflict since 1984. This is reputedly the world's longest lasting armed deployment of the last hundred years.

The dispute arose out of the three wars fought by Pakistan and India in 1948, 1965, and 1971. After the 1948 war Kashmir was divided into Pakistan's Azad Kashmir and India's Jammu and Kashmir. A cease-fire line was laid down in 1949 but surveyed only as far north as a triangulation point known as NJ 9842. From there the divide between territory allocated to Pakistan and India was extended vaguely northward and was simply, yet ambiguously, described to run 'thence north to the glaciers' (Ahmed and Sahni 1998). Following the 1971 war a new 'Line-of-Control' was established in 1972 with only minor changes, but still nothing was done to clarify the boundary beyond point NJ 9842.

It might have been expected that this extremely rugged, high-altitude land of rock faces, avalanches, and glaciers would have remained a forgotten, or at least neutral wilderness. Nevertheless, Pakistan interpreted the vague wording to mean a straight line northeastward from NJ 9842 to the Karakorum Pass while India's reading was a line that followed the watershed along the Saltoro Range north-by-west to the Indira Pass, thus encompassing the entire length of the Siachen Glacier. Before 1949 neither country had any permanent presence in the area.

During the 1970s and early 1980s several international mountaineering expeditions sought climbing permits from the Pakistan Government to enter the area. This was subsequently used by Pakistan in an attempt to enforce its claims. But in 1984 the Indian Army and Indian Air Force occupied a large section of the glacier and interdicted one of the mountaineering expeditions. Pakistan quickly responded by deploying troops. The conflict has escalated from this point and, with several lulls and resurgences, has continued to the present.

Estimates of current troop deployments vary from 3,000 to 10,000 on each side. India holds the high ground with fortified redoubts along the Saltoro ridge and has defeated persistent assaults by Pakistan to dislodge its forces. However, India pays heavily for the tactical advantage of the high ground as it must rely on helicopters to supply its troops, and to bring out its sick and wounded.

It has been estimated that less than 10 per cent of the casualties are due to actual fighting: sickness, altitude, avalanches, rockfall, and extremely low temperatures are the main adversary. Routine artillery exchanges are punctuated periodically by much more serious actions, such as in 1999 when aircraft and surface-to-air missiles were deployed. Yet more typical is the recent *Agence France-Presse* report from Jammu dated 10 March 2003 stating that 25 Indian soldiers and porters were lost in a series of avalanches. Four injured were rescued and 14 bodies were brought out by helicopter before bad weather terminated the operation.

An estimated 1,300 Pakistani soldiers have been killed or have died between 1984 and 1999. India has estimated an expenditure of Rs50 billion and almost 2,000 lives up to 1997. However, estimates of lives lost and of the cost of military operations are always of dubious accuracy. For instance, Ali (2002: 316) states that '[a]bout 97% of the approximately 15,000 casualties have been due to altitude and weather rather than enemy action'.

Over the years the Siachen catastrophe has been the subject of seven major rounds of talks between the combatants, and agreement on a compromise settlement of this absolute stalemate has almost been achieved. Yet suspicion, mistrust, and an obsession for control always seem to thwart the chances for a solution. Suggestions for declaring the area an international park have been on the table for many years. Others have proposed establishment of an international scientific preserve with elements of the Antarctic Treaty as guidelines. One of the resolutions of the Mohonk Conference in 1986 (Resolution 3, *Mountain Research and Development*, 1987, 7 [3]: 185) reads:

> Realizing that nature recognizes no international boundaries and that many of the issues and challenges facing development and conservation cannot be dealt with adequately without co-operation between the countries of the Himalayan region, the Mohonk Mountain Conference strongly urges the governments of the Himalayan region to take steps to establish international parks in border areas (Parks for Peace) to promote peace, friendship, and co-operation in research and management, for the optimal sustainable use of the natural and human resources, and to improve the quality of life of all the peoples of the region.

The most recent recommendation for establishment of a peace park (Ali 2002) prompted a series of letters to the editor of *Mountain Research and Development* (2003, 23 [2]: 208). While there was the expected strong endorsement in principle, Nazir Sabir saw no possibility of significant progress under present circumstances: 'Sadly, I am afraid Siachen would remain a hostage to the Indo-Pak Kashmir policy as has been the case . . . looking at the political will at the top level.'

The pollution of the environment is reaching serious proportions. It is difficult enough to evacuate the sick and injured; all waste is abandoned, adding to water pollution downstream. Ali (2002: 317) states that human waste alone amounts to 1,000 kg per day on the Indian side. The disposal process involves the use of local crevasses for dumping, for instance, an estimated 4,000 metal cans each year.

It could be argued that the Siachen catastrophe represents one of the most futile, even senseless, military engagements of all time. Apart from the large losses of military and support personnel and the high cost of operation, on those occasions when the conflict has expanded, innocent mountain villagers at lower elevations have suffered severely.[1]

Local examples of oppression and discrimination

The widespread discrimination against the mountain minority peoples is not confined to relatively large-scale instances, such as those of the Chittagong Hill Tracts and the Bhutanese refugees, that were considered in the foregoing sections. Observant travellers throughout the Himalayan region will quickly discern a sense of discrimination that may be latent or indirect, but is nevertheless pervasive.

The frequent misdirection of aid agency assistance projects arguably may be considered part of this oppression. Examples abound: the Tehri Dam project; the proposed development of a cascade of dams on the Arun River in eastern Nepal with little concern for the local people and with foreign currency through sale of electricity to India as the prime objective; the construction of the Kaptai Dam in the Chittagong Hill Tracts (discussed above). It appears that many of the large infrastructural undertakings neglect the position of the local mountain minorities.

Micro-hydro development in Himachal Pradesh, India

A recent account of an attempt to encourage the private development of small-scale hydro-electric projects in Himachal Pradesh is provided by Sinclair (2003). The federal and state governments of India and Himachal Pradesh, with support from UNDP, have identified 55 potential micro-hydro sites in the Kulu District; four facilities are under construction, another 12 are in various stages of approval. Two of the four that were nearest completion were surveyed by Sinclair's research team.

The provision of funding by UNDP was based upon a number of assumed benefits. These included: reduction in fossil fuel use and reduction in production of greenhouse gases; protection of mountain forests and biodiversity; enhancement of local economic opportunity; provision of local construction jobs; relief of women's drudgery of collecting fuelwood; and 'capacity-building benefits'.

A survey of people living in the vicinity of the two projects led to a series of conclusions that disagree with benefits proposed by the project proponents. There was no local motivation to substitute electricity for fuelwood as the former would be very expensive while the latter was considered to be free. The time spent in collecting fuelwood was not regarded as relevant (while not stated in Sinclair's report, it has often been reported that mountain women have regarded fuelwood collection as a chance for a social outing). The construction phrase had produced no new local job opportunities as all labour was arranged by the project managers who imported low paid Nepalese workers. There was some concern that the generating plants and associated infrastructure would eliminate a considerable number of trees. The temporary camps for the imported labourers were regarded as a pollution risk. Above all there had been no consultation with the local people.

The overall Kulu project remains far from complete and Sinclair's report calls for a rapid review of the current situation and a more rigorous environmental assessment process. However, he doubts the necessary political will can be marshalled. Although this example barely deserves classification as an instance of conflict, similar projects that have been completed throughout the Himalayan region create tension by disregard of the need for local consultation and involvement. Furthermore, they increase a tendency toward cynicism in relation to outside aid. Another danger that is also widespread is the potential for enhancing a sense of dependency; if 'big brother' enters our territory and establishes major projects, why should we ourselves assist in their maintenance? Further aspects are introduced in relation to the impacts of tourism on Kulu's environment and people (Chapter 7, pp. 165–8).

The 'Love-Suicide Meadow': a tourist attraction, Yunnan, China

On the eastern flank of the Yulong Xue Shan (Jade Dragon Snow Mountains) in Lijiang County, northwestern Yunnan, China, lies a beautiful subalpine meadow. Local Naxi (Dongba) folklore describe this as a sacred place with historical significance. Following the accession of the Qing Emperor Yongzheng in 1723 and the imposition of Han marriage laws and customs on Naxi society (Swope *et al.* 1997), the meadow was selected by young Naxi couples who committed suicide as a form of protest, sometimes in groups of 6 to 12 couples. As tourism developed with the opening of Yunnan to foreign visitors after 1985, the meadow emerged as a tourist attraction (see Chapter 7, pp. 156–65). Access from Lijiang Town by unsurfaced road was originally available to tourists only to within 350 vertical metres of the meadow, the remaining distance requiring a steep walk. Local Yi people were quick to take advantage of the visitors' interest and acquired horses that were gaily ornamented and led by their young women, also in colourful traditional dress (Figure 8.1). A successful business rapidly developed as tourists were willing to pay a fee to be mounted and led to the melodramatically named meadow. The extremely poor Yi villagers neglected their traditional crops to develop this opportunity to 'get-rich-quick'. However, upon realizing the potential,

Figure 8.1 Yi woman with decorated horse waiting to take tourists to the 'Love-Suicide Meadow', Yulong Xue Shan, northwestern Yunnan, China, October 1995.

the government Tourist Office intervened. Investment funds were acquired from Hong Kong and Bangkok and a chair-lift was built, complete with parking lot and surfaced access road. The state policy had become 'development of mass tourism'.

Within three years the investment cost of the chair-lift was recovered and the spontaneous Yi businesses were destroyed. The Yi were left with horses for which they had no use and could not afford to feed. There is much more to this sad story of thwarted entrepreneurship, combined with inappropriate tourist development and environmental damage. The spontaneous Yi development itself was an intrusion into Naxi culture in terms of the meadow's original tradition, but the destruction of the local Yi self-help efforts to rise above their own poverty can only lead to hardship and further tension.

Hill Tribes of northern Thailand and deforestation

The specific issue of deforestation in the mountains of northern Thailand has been explored in Chapter 3 (pp. 60–7). In that context the discourse on opium production, reforestation, and the negative attitudes of mainstream Thai toward swidden agriculture was extensively developed. Here, the central question of the well-being of the Hill Tribe peoples will be examined as a case study relating to conflict and oppression.

First, as stated in Chapter 3, swidden agriculture has been misrepresented, either unwittingly or deliberately, as a justification for adoption of repressive policies. Thus severe measures have been taken to suppress swidden farming (together with opium production) and to convert the hill people to sedentary forms of agriculture. There has been scant acknowledgement of the wide variety of swidden types or, for instance, of how some ethnic groups have practised forms of traditional agriculture that either maintain or even increase the local degree of biodiversity. Regardless, the great increase in population growth among all the different ethnic groups over the last half century has rendered traditional agriculture untenable. Therefore, efforts have been exerted for several decades to introduce a more intensive settled agriculture and the adoption of a variety of cash crops ranging from coffee and vegetables to cut flowers and garden flower seeds. At issue here is the manner in which various policies have been formulated and applied. They have been based on the general assumption that the hill people are the prime cause of environmental damage. The process of environmental degradation has been perceived as over-population leading to deforestation, soil erosion, and downstream siltation and flooding. As argued in Chapter 3, this variant of the Theory of Himalayan Environmental Degradation is insupportable. Furthermore, much of the actual deforestation has been effected by ethnic northern Thai who, as population pressure on land has increased, have augmented their lowland irrigated agriculture by a reckless form of swiddening in the nearby hills. Commercial forest clearance, both legal and illegal, is a further source of environmental degradation.

The impacts of governmental policies for land conservation or development that confront the ethnic minorities have included evictions, forced resettlement, fines, and imprisonment, backed where believed necessary by military force. One of the most serious obstacles that the hill people have faced is denial of citizenship under the existing Thai constitution – in a legal frame, they are not considered as human beings (McKinnon 1983). This lack of legal standing has denied them the right to obtain and own land. It has also allowed unscrupulous Thai to appropriate traditional hill village lands, and even to sell to other Thai. Ganjanapan (1996, 1998) has severely criticized the government's approach, both to its ethnic minorities and to its environmental conservation. The establishment of new national parks and conservation forests has often appeared as a cover to justify eviction of minority people from land they have farmed for generations. It has also been reported that, subsequently, this legalization of land seizure has facilitated timber exploitation by commercial interests.

The assumed environmental damage to the downstream regions that is the argument used for justifying repressive policies that affect the well-being of the ethnic minorities in the Thai mountains, has been challenged by Alford (1992a). He has demonstrated that there is no evidence to support the claims that land cover changes, such as the conversion of hill slopes from forest to Imperata grassland, or clear cutting to extend farming on steep slopes, have had any noticeable effect on downstream river discharge and siltation. A total logging ban introduced in 1998 is a further imposition on the minority hill people. Unfortunately, this type of mistreatment of mountain minorities in northern Thailand has been a widespread phenomenon throughout the greater Himalayan region and can only serve to cause resentment that frequently exacerbates the tendency to increased tension and strife.

The Tajikistan Pamir

The Tajik people were jubilant in 1927 when the Tajik Socialist Soviet Republic was established and its new capital city, Dushanbe, was founded. The Red Army had separated the Tajiks administratively from their traditional enemy, the Uzbeks. Nevertheless, the Tajik SSR evolved as the poorest republic of the Soviet Union, located in an effective cul-de-sac of closed frontiers far removed from the powerful political centre, Moscow (Badenkov 1992).

In the late 1930s the first phase of forced resettlement of the Mountain Tajiks occurred when many were evicted from their villages in the western and northern Pamir and sent to work on the expanding cotton plantations in the southwest. After the Second World War demands for cotton did not slacken, nor did the requirement for cheap labour. Then on 10 July 1949 an earthquake-induced landslide (see Chapter 6 pp. 134–6) produced the tragedy of Khait during which more than 28,000 lives were lost. Not only was knowledge of the disaster withheld from the rest of the world for more than 50 years (Yablokov 2001), but the occasion was used to justify a further wave of forced out-migration in response to the Soviet demand for additional cheap labour for the southwestern cotton plantations; practically all people in the area who survived the disaster were forced out.

The final years of Soviet domination proved an extraordinary time in the Tajik SSR. Water from the Pamir had become the region's primary resource, and the giant Nurek Dam on the Vakhsh River, the world's highest rock-filled dam, was supplying hydro-electricity to a wide area of what was still Soviet Central Asia. A second major dam, Rogoun, was nearing completion upstream of Nurek on the Vakhsh River. By 1990, however, Tajik nationalism, that also nurtured a 'green' party, was beginning to protest the perceived consequences of the Rogoun Dam. If built to its design height, some 40 Tajik villages along the river between the dam and Garm would be submerged. The growing anti-Russian hostility was prompting the emigration of Russian engineers and their families to the Russian heartland. As work on the Rogoun Dam slowed and agreements to renegotiate its ultimate maximum height were under way, the new independent Republic of Tajikistan came into being and a complex civil war broke out, stopping all further progress. There was conflict between a Communist hard-line Stalinist remnant seeking to retain political control, Muslim fundamentalists, neo-liberals, and democrats, and between various local militias. It has been estimated that more than 40,000 were killed, many in a wave of car-bomb assassinations.

As one of Afghanistan's contiguous northern neighbours, Tajikistan is attempting to emerge from this debilitating civil war. A UN negotiated peace was effected in 1997, although several of the involved militia did not participate and the region of the Rogoun Dam remains very dangerous territory. The international border with Afghanistan is also frequently tense (Figure 8.2). In addition, Tajikistan continues to serve as a major conduit for illegal drugs and arms; prospects for stability remain uncertain and a large proportion of the population has no sense of security.

Himalayan dams and their effects on mountain people

How can mountain people be given a fair share of the benefits of large-scale commercial utilization of mountain resources? Development of water resources presents a particularly apt opportunity. In the European Alps, for example, substantial payment is made to mountain farmers from the sale of electricity produced from high dams that hold back reservoirs in mountain valleys. Despite the fact that the periodic negotiations necessitated by changing economic circumstances often entail a degree of controversy, the process of compensation for the mountain people has become the norm. With the recent identification of mountains as the 'water towers' of the world, it would appear axiomatic that some form of compensation to mountain inhabitants would be initiated within the Himalayan region, especially as mountain farming and forest management are often essential elements of upper watershed protection for the water resource itself. Regrettably, the history of water exploitation across the Himalaya and its wider region has provided little evidence of any benefits for the mountain people, except in the special case where micro-hydro facilities have been installed. There

Figure 8.2 Militia with Kalashnikovs along the Tajikistan–Afghanistan border confront a United Nations convoy that was en route to investigate the hazard posed by Lake Sarez in the High Pamir. This turned into a friendly encounter, July 1999.

have been instances of military or administrative repression, displacement of many thousands of people, and total disregard for local needs. Such is the case of the Chittagong Hill Tracts (pp. 175–7), the construction of the Kulekhani Dam and reservoir in Nepal (Bjønness 1983, 1984), the original plan to drown more than 40 villages of the upper Vakhsh valley in Tajikistan (pp. 190–1), and in the planning for the construction of the Tehri Dam.

The Tehri Dam, Uttaranchal, India

The Bhagirathi River is one of the numerous Himalayan head streams of the Ganges. The projected advantages of a large dam some kilometres below the town of Tehri were discussed throughout the 1950s and 1960s. In 1972 the Planning Commission of Uttar Pradesh (of which Uttaranchal was a part until it became an autonomous state in 1999), approved plans for construction of what was intended as one of the world's highest earth- and rock-filled dams. The specifications called for a dam 261 m in height on a base more than 575 m wide with a length of one kilometre. The storage capacity was estimated at 3,500 million m^3 to occupy an area of 34 km^2 above the dam. Twenty-two kilometres downstream, at Koteswar, there would be a concrete dam 103.5 m high with an 86 million m^3 storage capacity. The main functions of the dams would be to produce hydro-electricity to meet peaking requirements

(100,000 MW for phase 1, the same for phase 2, and a further 40,000 MW from the Koteswar Dam downstream), to irrigate 270,000 ha of agricultural land and to stabilize the existing irrigation of 600,000 ha, and to produce 270 million gallons of drinking water per day for Delhi, Arunachal Pradesh, and Uttaranchal. The main reservoir would flood the Old Tehri Town and submerge more than 50 villages.

The Tehri Dam has been the object of considerable controversy since it was first proposed. The most influential and charismatic figure among the opponents of the dam is Shri Sunderlal Bahuguna, Messenger of the famous Chipko (hug-the-trees) Movement and prominent Gandhian environmentalist and supporter of people's rights of access to resources. The dispute has also attracted leading Indian and international seismologists and geophysicists, environmentalists, and many NGOs. Construction has been repeatedly stopped, only to start again amidst protests and riots. A 74-day hunger strike by Sunderlal Bahuguna in 1996 forced the government to set up a new review of the seismic, environmental, and resettlement aspects of the project. The range of issues brought against the dam include:

1   Public safety, because of the attested under-design of the dam in view of projected Great Earthquakes in the vicinity (Chapter 6 pp. 120–4). Assuming that dam opponents are correct and a future earthquake causes the dam to collapse, millions of lives downstream would be endangered and the ensuing flood would result in damage sufficiently severe to place India's entire economy in jeopardy.
2   Such a dam would be a prime target for terrorists or an enemy of the country.
3   Destruction by submergence and partial submergence of more than 50 villages, the Old Tehri Town, and many sacred temples and monuments. More than 100,000 people would be affected, a serious event in light of India's very poor record of providing adequate compensation. By October 2002 provisions had been made to resettle only about half of the fully affected people and very few of those partially affected. The level of compensation has been criticized as totally inadequate.
4   All the benefits of the project would favour downstream interests; mountain villages would not even receive electric lighting.
5   Considerable environmental loss due to submergence and massive manipulation of the Bhagirathi River.

Furthermore, opponents of the dam, including internationally renowned scientists, claim that there has been government intransigence, secret negotiation, breach of commitments, including that of a former prime minister, and corruption. Sunderlal Bahuguna has staged three lengthy fasts, ending each with the public commitment of senior politicians for further testimony and review. Each commitment was subsequently broken, and Bahuguna himself was imprisoned temporarily.

A further pragmatic objection is that the costs will seriously outweigh the benefits. It is projected that the unit cost of electricity will be twice that produced from neighbouring sources. Now, after almost three decades of construction progress and work stoppages, in April 2003 the Supreme Court began review of a petition for stopping work on the dam. On 1 September 2003 a three-judge panel of the Supreme Court dismissed the petition by a two-to-one decision. However, the ruling instructed that the two upper tunnels through the dam should not be closed so that the reservoir could not be filled to its planned level until all resettlement and compensation had been completed. Further protests are pending.

Much has been written about the Tehri Dam controversy, both technical (Chapter 6, pp. 120–4) and socio-economic. In particular, Valdiya (1992, 1993a, 1993b) emphasized that there are viable alternatives to big dams and argued for several smaller run-of-the-river installations, together with a significant modification to downsize the existing Tehri project. He also pointed out that the projected construction would provide no benefit for the people of Garhwal; he foresaw only the misery of enforced evacuation and inadequate restitution.

The Tehri Dam dispute is sufficiently complex and emotionally charged to frustrate any 'outside' balanced judgement. Nevertheless, the honesty of several levels of government officials is clearly questionable. The attitudes of powerful political decision-makers toward the poor mountain people appear to be totally reprehensible. Certainly, the Tehri Dam case, and many others comparable to it, represent a form of oppression of poor mountain people that can only lead to widespread unrest and disaffection.

Conclusion

Only a few examples of the different types of conflicts and oppression within the Himalayan region have been reviewed. This chapter is by no means a catalogue or complete coverage. Regrettably, there are many more, ranging from local injustices to province-wide armed conflict. Some have been very long-lasting, as in the case of the Chittagong Hill Tracts, others are more recent, as in the case of the Bhutanese refugees and the Maoist Insurgency in Nepal. Several appear to be on the point of resolution, such as the Chittagong dispute and the struggle of the Nagas. Others, such as the seemingly pointless conflict on the Siachen Glacier, appear interminable.

The most widely publicized areas of conflict, of course, have been Afghanistan and Kashmir. The defeat of the Soviet Union in Afghanistan and the spread of Taliban fundamentalism led to the infiltration of al-Qaeda and the United States-led war of 2001–2002. The impact of this on Pakistan itself and the brinkmanship along the Kashmir Line-of-Control leaves the entire western Himalayan/Karakorum/Hindu Kush area in a state of sustained crisis. This is made the more serious by the fact that the two chief protagonists of the region are now nuclear powers. The devastating effects of the prevailing lack of security of the mountain people, including the millions of refugees, are hard to imagine (despite extensive television coverage). Moreover, the impacts of the warfare, augmented by prolonged drought and extreme political instability, give rise to extensive environmental damage. The tension between Dharamsala and Lhasa over the progressive Han colonization of Tibet is an additional serious regional problem that has ebbed and flowed for a half century.

It must be concluded that the fleeting notions of Shangri-la, of laughing villagers who pour tea for thirsty Western trekkers, of quiet contemplation amidst majestic mountain landscapes and equally majestic carefully tended emerald green terraces, have indeed been fleeting experiences. Yet the mountain people *are* resilient, gracious, hospitable, beautiful, and they do smile easily. It would appear that it is governments and agencies that provoke conflict, often unwittingly. Obviously, this is too simplistic, yet if the present-day catch phrases, such as, 'participatory development', 'increased local access to resources', and 'sustained mountain development' acquired more political backing, the Himalayan region could well turn away from conflict.

9 Prospects for future development

Assets and obstacles

> Once a threshold is passed, then malnutrition, warfare, and environmental collapse, will feed upon each other. This is supercrisis.
>
> (Ives and Messerli, 1989: 270)

Introduction

Any comprehensive discussion of the assets of the Himalayan region and current obstacles to development would far exceed the scope of this book. The present focus of the work is to trace the theme of Himalayan environmental degradation and thus an overview of the region's future prospects is very relevant. Selective attempts are made to address a small number of pressure points in light of the now firm stipulation that Himalayan environmental and associated socio-economic collapse, as originally popularized, is a myth; in short, the Theory of Himalayan Environmental Degradation no longer should be supported. Yet it is not intended to imply that the region does not face severe environmental and related problems. This was anticipated in the forerunner to this book (Ives and Messerli 1989), although the earlier emphasis was placed on a first assessment of the paradigm of environmental crisis and its assumed causes in light of information available at the time.

With the passage of 15 years since the earlier writing, the World Bank's apocalyptic year, AD 2000, that was predicted to witness elimination of all accessible forests in Nepal, has been safely negotiated. Nevertheless, it is widely presumed that population pressures on natural resources have continued to grow in many parts of the region. At the same time, regional and local accessibility has increased at an accelerating pace, radically changing the situation that prevailed during the decade of the 1980s and earlier. The same period has also seen a major donor policy shift to enhance participation of the local people in decision-making in so far as use of natural resources is concerned. And this is interlocked with identification of the need to achieve better treatment for, and equal participation of, women (Ives 1996; Ortner 1996; UNICEF 1996; Gurung 1999).

One approach to assessing future prospects is to catalogue the available resources of this vast and diverse region and attempt to examine how they are being used or abused in terms of the well-being of the mountain people. Precise population numbers for the Himalayan region are not available, but a reasonable estimate would lie somewhere between 70 and 90 million people. The great majority are poor or extremely poor. Therefore, their dependency on access to, and use of, natural resources is very high. Nevertheless, even at the village level there are great disparities; all levels of any society have elite groups who tend to ensure disproportionate access to resources for their own benefit.

Mountains in general, and the Himalayan region in particular, offer a set of intrinsic and extrinsic economically valuable resources and services. The natural resources include minerals, timber, and water, together with the supply of a wide variety of non-timber forest products, as well as the cultures and human qualities of the large number of ethnic groups. To the extent that agriculture and livestock husbandry produce any surpluses for exchange within the region or for sale on the world market, they also constitute important resources. Also, there are the environmental or ecosystem resources: the protection of watersheds and aesthetic attraction that has led to the extensive development of tourism. Of all these resources, water and its management, including hydro-electricity, irrigation, drinking/ household water, and watershed protection, is probably the most significant. However, while these resources on a regional scale are immense, whether or not they are utilized for improvement in the well-being of the rural poor will depend on who controls them and how such control is exercised.

The approach adopted here to explore these and associated issues cannot be exhaustive. In view of this, two of the major assets of the region, water and aesthetic amenities, are highlighted as macro-phenomena. At the other extreme, the micro-phenomena of the coping strategies of a small village in response to threat to the very survival of its people are examined. While the macro- and micro-phenomena are far apart, in one sense, the degree of pragmatic linkage between them, or lack of linkage, is taken as an indicator of regional stability or instability. The difficulties that lie between the two extremes include: environmental degradation, important, but now seen as much less so than it appeared 20–40 years ago; continued rapid population growth and the apparent inability of regional government, donor agency, or market forces, to prevent such growth from accelerating the decline into abject poverty; maladministration and inappropriate development policy; and political instability, leading to popular unrest, armed insurrection, government-enforced repression, and open warfare. It is proposed to begin this rather experimental essay by examination of a very few village-level investigations as type examples of the reality of livelihood possibilities throughout the entire region. Then the 'big' resources of water and recreational potential will be discussed. But first the general well-being of the mountain people warrants review.

Mountain people of the Himalayan region in the twenty-first century

The topic of the well-being of the mountain people is approached by examination of a series of case studies in Nepal. Bohle and Adhikari (1998) have examined the challenges of food insecurity facing a series of very poor and very isolated villages in Nepal's Middle Mountains to the north of Pokhara. The villages lie at one end of a continuum ranging from extreme poverty, vulnerability, and instability to the other end, one of change and reduced vulnerability, demonstrated by the settlement of Dhulikhel with main road access to Kathmandu. Dhulikhel is characterized by rapid adjustment to the new market conditions and relative affluence. The second study, by Malla and Griffin (1999), is also part of an appraisal of the impacts of variable road accessibility in a continuum from Dhulikhel itself, through two intermediate villages, to the relatively isolated village of Budhakhani on the southern slope of the Mahabharat Lekh. A third investigation, the study of a single village, attempts to understand the effectiveness and working relations of the rapidly growing number of Forest Users Groups, ideally centred on democratic village-level processes (Timsina 2003).

To partially anticipate the conclusion to be drawn from contrasting and comparing the three rural studies, it can be stated that optimism appears warranted for Dhulikhel while black pessimism characterizes the villages of Bohle and Adhikari's field area. The anticipated conclusion is only partial because the fieldwork for each of the studies was completed prior to the outbreak of the Maoist Insurgency. The desperate conditions facing the villagers in Bohle and Adhikari's study, when multiplied by similar conditions in hundreds of other villages, might well have been a presage to insurrection.

Rural livelihoods beneath Annapurna Himal

In the preamble to their study, Bohle and Adhikari (1998) explain that since the early 1990s Nepal has become a net importer of food, although prior to this it had ample production to permit significant exports of food cereals and meat. By the end of the twentieth century nearly half of all Nepal's administrative districts have become deficient in food production. The IDRC Cooperative Research Program (quoted in Bohle and Adhikari) had predicted that by AD 2000 only seven of the country's 75 districts would produce a food surplus, a decline from 40 districts in 1981 when 27 attained self-sufficiency and only eight were deficient. The prediction was substantially correct. The decline in food production has been paralleled by an increase in the proportion of the country's population that falls below the poverty line, reflecting a dangerous level of malnutrition and sickness. Bohle and Adhikari pose a series of questions: What are the most critical regions? Who are the most vulnerable groups? How do these most vulnerable groups cope with the food shortages that they are facing?

In an attempt to identify communities that are characterized by extreme vulnerability, Bohle and Adhikari selected four villages. They are located in Kaski District north of Pokhara in the transition zone between the Middle Mountains and the High Himal on the southern flanks of the Annapurna massif. These four villages are relatively inaccessible to the nearest markets and are located in steep and environmentally fragile localities. In practice this implies comparatively high altitude sites and a day's very steep walk to the nearest road-head for access to market. Furthermore, the connecting trails are impassable for much of the monsoon rainy period. A second group of villages was selected in Nawalparasi District south of Pokhara and located in the transition zone between the Terai and the Siwalik Hills. The research design (Bohle and Adhikari 1998: 322) involved four specific steps to ensure a 'bottom-up' approach along the food chain. It included a food self-sufficiency ranking of all households, supplemented by social and resource mapping. Food self-sufficiency categories were calculated from questionnaires to which 25–35 per cent of all households responded. Next a traffic survey was undertaken from nodal points on the network of mountain paths. Villagers en route to and from market were asked to respond to a brief questionnaire. Finally, at the road-heads leading to the mountain footpaths, where bazaars were set up, approximately half of all the traders were surveyed.

The parameter used for assessing food consumption level was 180 kg of cereals/person/year. This is the WHO-Standard, corresponding to 1,650 cal/person/day, an absolute minimum amount. In comparison, the Nepal National Planning Commission has defined 2,250 cal/person/day as the national poverty line (NPC 1993).

Village food was produced using three distinct agricultural sub-systems. These were: irrigated land (khet) with two to three harvests each year (mainly paddy rice); rain-fed bari that yielded only one harvest each year (mainly maize or millet); and slash-and-burn cropping on the steeper slopes with only one harvest every three years (mainly millet). At the higher locations in Kaski District reliance was predominantly on the bari and slash-and-burn sub-systems.

Surveys in two of the higher altitude villages in Kaski District, Karuwa-Kapuche and Siding, revealed that 41 per cent of all households were not able to produce food adequate to last for more than six months of a year. The lower altitude villages of Nawalparasi District were even more deficient as 68 per cent of all households could not maintain themselves by their own food production for more than six months of a year and only 1 per cent managed a food surplus.

Using the absolute minimum standard for necessary daily food intake (1,650 cal), Bohle and Adhikari demonstrate that the subsistence production of their high-altitude villages ranges from as low as 23 per cent to a high of 77 percent of this minimum. Some households produce less than three months food supply per year. Intra- and inter-village exchange, sale of produce (such as firewood and alcohol), sale of livestock, and handicrafts at the local market or road-heads, are some of the necessary measures to make up for the subsistence food deficit. In addition, remittances from seasonal migrants who manage to find wage employment outside the villages, and income from military service retirees and military remittances are also important.

The Kaski District villages are located close to the upper limits of cereal production with low fertility, although they do have a comparative advantage over the lower elevation villages in terms of relatively easier access to forest and animal grazing. Nevertheless, the extremely low level of overall food self-sufficiency is reflected in a high degree of stunting and chronic illness among the population at large. The lack of educational facilities and the absence of health centres in the villages ensure the persistence of very low levels of literacy and add to the burden of obtaining medical attention. This, together with the lack of well-developed social links to urban centres makes it especially difficult for villagers to find outside employment. Bohle and Adhikari stipulate that, while these poor Nepalese farmers are by no means ignorant 'victims' of their situation but are 'highly active, adaptive and dynamic actors', their coping strategies are hardly adequate to overcome the life risks that they are facing. The authors demonstrate that these villages, despite all the coping strategies that they have adopted remain, on average, able to achieve only about 80 per cent of the required absolute minimum food intake. 'Under these conditions, survival for large and growing parts of the Nepalese rural population has become a permanent livelihood crisis' (Bohle and Adhikari 1998: 332).

There are no currently available data to determine what proportion of the entire Nepalese population is in a situation comparable to that of the surveyed villages on the southern slopes of the Annapurna massif. In addition to rural food vulnerability, Bohle and Adhikari (2002) have demonstrated a comparably alarming situation for slum dwellers in Kathmandu itself. Here for somewhat different reasons food insufficiency, malnutrition, lack of access to education and health care facilities combine in a highly polluted urban slum setting.

The two cases introduced above, relatively inaccessible rural villages and socially inaccessible urban slums, are probably best reviewed as the more extreme situations facing the Nepal Middle Mountains and probably much of the Himalayan region at large. An equally intensive study by Malla and Griffin (2000) is worthy of examination for comparison and contrast with the study by Bohle and Adhikari. The environmental setting is very different, although accessibility is an important component of both studies. Malla and Griffin's work is one of the long series of studies that have been part of the Nepal–Australia Forestry Project, 1976–2000 (see Mahat *et al.* 1987b; Griffin 1989; Jackson *et al.* 1998).

Economic and agricultural change in Kabhre Palanchok District, Nepal

Malla and Griffin (2000) selected for study four settlements extending from a village on the southern flank of the Mahabharat Lekh in a relatively isolated location, through two intermediate villages, to Dhulikhel, a village on the main motor road linking it to Kathmandu. All four are located in Kabhre Palanchok District. These study sites were chosen to provide an accessibility–semi-isolation continuum as a basis for examination of the changes in a section of rural Nepal that result from the recent and increasing expansion of market forces.

Budhakhani, the remotest of the four villages, consists of 436 households (Tamangs 69 per cent; Magars 22 per cent; others, including Brahmins and Chhetris and occupational castes 9 per cent). The village economy is near-subsistence. It occupies 4,036 ha and because of the steepness and ruggedness only about 25 per cent is cultivated, nearly all of which is rain-fed bari. It is located more than a day's walk from the motor road at Dhulikhel.

Chaubas is located to the east of Dhulikhel on the far side of the Sun Kosi River. It contains 419 households (Tamangs 35 per cent; Brahmins/Chhetris 35 per cent; Paharis 25 per cent; and occupational castes 5 per cent). There are 1,035 ha of land, of which 60 per cent is cultivated. Access involves a six-hour steep uphill walk from the motor road at Dolalghat, although in 1995 a fair-weather road was built from across the Sun Kosi River to the village. It is one of the many villages in the district that has received considerable external assistance from the Nepal-Australia Forestry Project (Griffin 1989).

Chhatrebanjh lies 12 km east of Dhulikhel. It contains 580 households (Tamangs 56.6 per cent; Newars 27.7 per cent; Brahmins/Chhetris 12.4 per cent; occupational castes the remainder). It is connected to Dhulikhel by a fair-weather road. There are 850 ha of land, of which 80 per cent is cultivated.

Dhulikhel consists of 1, 411 households (Newars 61 per cent; Brahmins/Chhetris 21.6 per cent; Tamangs 11.5 per cent; occupational castes the remainder). It is a centre for regional government institutions and NGO offices and has excellent access to Kathmandu via a good all-weather road. The settlement includes 1,140 ha of which 71 per cent is cultivated. The road also provides easy access to other urban centres in the Kathmandu Valley, such as Bhaktapur.

The Malla and Griffin study involved extensive data collection in each village, including information that defines household subsistence production, development of industry and trade, and overall economic and agricultural change that has occurred over the last several decades. Some of their main conclusions are given below.

One-hundred-and-two households were surveyed. All possess farmland, although average size of holding is only 1.26 ha and ranges from 0.7 ha in Dhulikhel to 1.7 ha in Budhakhani. If cultivated land only is considered then average holding drops to 0.94 ha with a range from 0.57 ha in Dhulikhel to 1.19 ha in Chaubas. From the household survey, it was calculated that 75 per cent of cultivated land is rain-fed bari, the remainder is khet. Budhakhani possess almost no khet. Most households in Budhakhani and Chaubas grow food for home consumption, while a high proportion of those in Chhatrebanjh and Dhulikhel produce high value grain and other crops for the market.

It is apparent that use of inorganic fertilizer and increased intensity of agricultural activities based on access to improved technology and high-yielding crop species is proportionate to ease of access to markets, with Dhulikhel being in the lead. These developments have resulted in substantial increases in crop yield, especially with maize and paddy, although many farmers report that to maintain the increased yield more inorganic fertilizer is needed each successive year. Thus, there have been some reversals. Nevertheless, some households that were barely able to attain food self-sufficiency 10–15 years ago now have a sufficient level of production to be able to sell grain in the market.

There has also been a shift in the choice of crops from cereals to mustard and rape (for seed oil), potatoes, peas, fruit, and other vegetables, again reflecting access to the market.

Some households use cash acquired from the sale of market crops to purchase rice and maize for home consumption. Furthermore, the expanding market has also produced rapid monetization of other products, such as trees, rice straw, and brooms, that previously had little cash value. It is interesting to reflect on this change from subsistence crops to commercial crops. In the 1980s it was somewhat surprising that the enormous growth in demand for potatoes, vegetables, fruit, meat, and milk prompted by the rapid increase in tourist hotels in Kathmandu was being met by long-distance bulk purchases from sources across the international border in India.[1] It appears that some lag time was required before local farmers and businesses were able to organize themselves to exploit the rapidly growing Kathmandu demand for farm products.

Livestock holdings have also changed substantially. There is an increasing emphasis on production of meat and milk for sale. This in turn has led to fewer farm animals per household, with reduction in the number of cows and increase in buffaloes and goats. Budhakhani alone supplies 800–1,000 goats each year for sale in the Kathmandu market. Ghee (purified butter) is sold in the urban centres, and there are two milk collection centres in Dhulikhel. One household in Chhatrebanjh reported annual earnings of Rs18,000 from milk alone.

The increasing market orientation of farm production has had significant impacts on forest use and off-farm earning activities. The adjustments in domestic animal holdings have led to a great increase in stall feeding (of buffalo) so that the demands on the forests have changed. Emphasis has shifted from fodder and grazing to timber and firewood. In addition, there has been a significant curtailment in production of millet, a low-value crop the cultivation of which is labour intensive. This has encouraged the conversion of bari terraces at the higher levels to commercial tree production, and although this activity is entirely on private land it has reduced pressure on the community forests (Jackson *et al.* 1998; see also Chapter 3).

Another trend has been the proliferation of business enterprises, particularly in Dhulikhel. These range from tea houses and small hotels to furniture making, brick making, carpet weaving, construction, and transport. There has been an attendant growth in the opportunities for employment as artisans, such as carpenters, plumbers, masons, electricians, and drivers. Education levels have risen and most children attend school. Combined with the expansion in the other activities, school attendance has led to an acute shortage in agricultural labour.

Perhaps the most striking development, however, is the increase in commerce and trade, mainly in Dhulikhel and Chhatrebanjh. By the 1990s there were more than 300 commercial enterprises in the study area, of which 65 per cent were in Dhulikhel and 25 per cent in Chhatrebanjh. The driving force is road accessibility. In addition, partial or seasonal migration to urban centres, and even abroad, for wage earning has expanded rapidly.

These inter-linked processes described by Malla and Griffin have greatly affected the local traditional farming systems. The most important aspects are:

- changing crops and cropping patterns;
- changing uses of marginal agricultural lands;
- changing livestock numbers, species composition, and husbandry;
- increasing use of artificial fertilizers;
- an increasing gap between supply and demand for farm labour.

Despite this apparently optimistic representation of recent economic and agricultural change along the Dhulikhel–Budhakhani transect, it must be viewed in terms of changing market conditions in Nepal as a whole and depreciation of the Nepalese currency. Malla and Griffin (2000) calculate that minimum per capita income required to meet the basic needs of the average person in their study area (1990 data) is Rs9.10/day or Rs3,321.50/yr. The average household size is 7.87 persons so that the minimum basic needs for an average household is estimated to be Rs26,140 annually for Dhulikhel, 25,461 for Chhatrebanjh, 25,802 for Chaubas, and 21,824 for Budhakhani. The authors determined that an average of only 46 per cent of the surveyed households exceeded these minima, ranging from 71 per cent in Dhulikhel to 25 per cent in Budhakhani. When subsistence on-farm resources alone are considered, only 15 per cent of the households surveyed exceed the minimum, ranging from 29 per cent in Dhulikhel to none in Budhakhani. The conclusion drawn from these figures is that off-farm income has become of dominant importance for most households and that degree of access to a motor road is an important factor. Nevertheless, poverty still remains a critical problem (while the actual village-level data presented here were accumulated over a decade ago, it is maintained that they are appropriate for demonstrating a trend that has continued to the present).

Although the inhabitants of these settlements close to the Kathmandu Valley show considerable adaptation and overall improved prospects, in terms both of general well-being and the environment, ambiguity remains with respect to differential gains (or losses) among castes and ethnic groups at the village level (see below). In contrast, the households that constitute the villages studied by Bohle and Adhikari (1998) are incomparably more impoverished. Moreover, Malla and Griffin conclude that the political and economic power sources in their study villages are strongly unbalanced in favour of Brahmins, Chhetris, and Newars. For instance, Tamangs, who constitute more than 60 per cent of the total district population, hold only about 6 per cent of the more desirable wage-earning positions, such as government appointments, school teachers, and owners of commercial ventures. This raises a thorny problem that faces Nepal as a whole – perceived widespread ethnic discrimination, a topic that is frequently raised in the context of the current political unrest.

Malla and Griffin point out that the accumulating evidence of widespread change has several important policy implications:

- It is important to realize that rural communities are dynamic, not static.
- The oft-repeated contention at the government level that there is tremendous and growing pressure on Nepal's land resource base in the Middle Mountains is exaggerated, if not totally incorrect. They argue that this prevailing view 'forms one of the underlying assumptions on which the *theory of Himalayan degradation* is based'. They contend that, while land-use problems certainly exist, the political approach to solving them is far too simplistic. The emphasis placed on protection of community forests and reforestation is out-dated and does not reflect the changing value of forests and forest products in the minds of much of the rural population.
- Most agricultural and rural development projects of both government and donor agencies have been oriented toward supporting and maintaining household economy based upon subsistence agriculture. Malla and Griffin's study indicates that rural households actually wish to move away from the subsistence agriculturally-based economy in favour of increased access to off-farm cash income.
- Despite the changes toward off-farm income demonstrated in the study area, Malla and Griffin show that a significant proportion of the rural households have not been able to benefit from this trend. In this respect they point to the rapidly emerging gap between the richer and the poorer households and note that the main factors influencing this are the degree of road access to markets as well as the advantages of political power held by privileged classes.

A village-level study of the effectiveness of Forest User Groups (FUGs)

A third village study was designed by Timsina (2003) to analyse the effectiveness of the relatively new participatory forest policy. It was undertaken specifically to determine ways in which relationships between people have evolved in the context of forest resource management. The concept of participatory forest resource management has developed as part of worldwide donor efforts to improve villager access to local forest resources, especially for women and the poorest of the poor. It was also based on the belief that the quality of forest cover would be maintained and improved to the extent that local users were able to effect management of their own presumed resources. In Nepal, and in the larger Himalayan region, this approach was also intended to enhance participation by members of the 'occupational' (untouchable) castes. It was seen as a means to promote social justice.

In Nepal, at least in numerous government and news media reports, the policy has been rated as a great success in that more than 12,000 Forest User Groups (FUGs) have been established. The Forest User Group policy is a local Nepalese institution that began to expand following introduction of democracy and constitutional monarchy in 1990. Timsina's investigation of its effectiveness centred on the study of a single village in Dolakha District in the Middle Mountains about 120 km east of Kathmandu. Emphasis was placed on the Dhungeshwori community forest that includes three wards of Kavre Village Development Committee (VDC).[2]

The Kavre VDC is dominated by high caste Brahmins and Chhetris. Other groups include Tamangs, Sherpas, Jirels, and occupational castes (untouchables): *Sarki* (shoe makers), *Kami* (blacksmiths), and *Damai* (tailors). Subsistence agriculture is the main occupation of the village although productivity is low and most households face food shortages and must participate in off-farm activities, including wage labour for the few rich farmers who own khet land. A significant number are daily workers in Charikot and Kathmandu during the farm off-seasons. Several obtain jobs on the trekking circuit since Lukla, the start of the trekking route into the Khumbu, is in easy walking distance. Some have permanent off-farm employment in the army and police department.

Timsina classified the forest user households into three broad categories according to relative wealth. He then interviewed 54 households comprising '6 rich, 23 medium, and 25 poor'. During this process he lived for two months in the study area and conducted interviews in the homes and in the fields. He also participated in small group discussions and attended village general assemblies and committee meetings. The results demonstrated, for this particular FUG, both positive and negative aspects.

The Dhungeshwori forest had been completely denuded before it was allocated by the government to the Village Development Committee as a community forest. The newly established FUG organized tree planting and imposed rigid restrictions on access for several years. Regulations for management were enacted and all users, including women and lower castes, were involved in the decision-making process. A true forest was established and it had begun to fulfill the resource needs of the local people to some degree. Landslides and soil erosion were completely controlled. It follows that, from an environmental point of view, the FUG must be accorded credit for complete success.

The negative aspects all seem to derive from the functioning of the FUG itself, in that the poor and the women remained marginalized. The majority of the committee members who controlled the decisions in the general assembly were the rich and elite of the village. To illustrate this statement, Timsina (2003: 240) describes his observations during one of the FUG committee meetings that he attended. On this occasion opinions were requested on a new operational plan for the forest. Of the 175 people present only five responded, all from the elite group. The meeting decided to levy heavy fines for people who infringed on the forest protection rules. It also adopted an increase in membership fees aimed at enlarging the FUG fund to provide for a new school and a road. Cash fines were proposed for those households who failed to provide volunteer labour for forestry and other development activities. When the agenda was put forward for adoption by the few powerful members, the rest remained silent.

The irony of this process can be explained by the following comments: (1) the poor are the ones who are forced to infringe on forest rules as they cannot afford to buy firewood, grazing materials, or litter – this illegal use of the community forest is an extreme activity on which their survival depends; (2) the poor have no cash to pay for increase in membership fees; (3) the poor cannot benefit from a school since all household members, including children, must work for their very survival; and (4) many cannot even afford the time to attend the sometimes lengthy committee meetings. Timsina noted that the FUG records showed that the majority of the fines for illegal forest use were paid by the poorest households, thus marginalizing them still further.

Timsina concludes by faulting the system whereby the central government has made no provision for external neutral facilitators who would enable the marginalized households to present their viewpoints and expect fair treatment. This raises mixed reactions to the whole concept of participatory forest policy and social justice where there is no process for the democratization of local institutions and power structures.

Commentary

The three sets of village studies presented above provide direct and detailed insights into problems and opportunities facing only a minute fraction of the rural people of Nepal. Nevertheless, it is believed that they are also informative of the type of situations that are widespread throughout the Himalayan region. No general conclusions will be given beyond emphasizing their specific relevance to Nepal itself. Perhaps the single most important point is Bohle and Adhikari's statement that Nepal has become a country unable to produce enough food to feed its own people despite the fact that it was able to export food products as recently as the late 1980s. Another important point, highlighted by Malla and Griffin, is that the central government agencies and, to some extent, donor agencies, have tended to base development policy on a static approach to a perceived problem. There has been little response to the significant change in the attitudes of many rural communities to their valuation of accessible forests and, in particular, to their view that the forests are providers of timber for sale rather than sources of fodder and grazing by domestic animals. Similarly, the tendency remains on the part of the authorities to attempt to improve the means of subsistence agriculture when much of the rural community is moving toward a preference for off-farm employment for cash which induces major changes in agricultural practice. There are several other well recognized trends that run parallel with the foregoing initiatives on the part of mountain people:

- Increasing urbanization: Goldstein *et al.* (1983) had indicated 20 years ago that Nepal was already showing signs of changing from a 'mountain-rural to a plains-urban society'. This trend has continued to accelerate, but urban growth has occurred both on the plains (Terai) and in the Middle Mountains;
- Increased participation in tourism as a means of obtaining off-farm cash income;

- Cash cropping on the Terai for sale in the markets of the rapidly expanding urban centres of Biratnagar, Birtemode, Nepalganjh, and others;
- Transport of tree products, especially lumber and firewood, from the Terai for sale in the Middle Mountain growth centres, such as Dhulikhel;
- Increasing out-migration, seasonal and longer-term, for wage earnings in Nepalese urban centres, in India, and further afield.

Similar growth patterns and economic changes are occurring in many other sectors of the broader Himalayan region beyond the limits of Nepal. One of the best-known is the long-standing and extensive development of horticulture (especially apple orchards) in Himachal Pradesh. Rapid expansion in domestic tourism throughout Himachal Pradesh has accompanied this growth, a process that is also typical of the former British Raj hill stations of Darjeeling, Nainital, Simla, Mussoorie, and others. This in turn has nurtured cash cropping for expanding local markets, in addition to the extensive activities associated with tea plantations, and the production of fresh fruit, cardamom, potatoes, vegetables, and a wide variety of medicinal plants. Of primary influence here has been the rapid increase in accessibility between many rural mountain areas and large population centres of the lowlands. Simultaneously, population centres in the Middle Mountains throughout the region from Kashmir to Assam, in Yunnan, and in northern Pakistan have also expanded. Thus the lives of many farming families have been changed although, as in Dhulikhel, this trend has produced a large gap in relative affluence between the richer and poorer sections of the communities.

These developments have been reinforced with the concomitant spread of modern communication systems – radio, telephone, television, and now, the Internet and all facets of information technology. This is having a remarkable impact on rural thinking and is influencing the formation of opinion in areas where two decades ago, or less, poverty and lack of awareness of events outside the village prevailed. Perhaps the most extreme case is the impact of the recent decision by the Government of Bhutan to overturn the hitherto tight control over access to television and the Internet. A recent article (*The Economist*, 23 August 2003) indicates the rapidity with which quite serious negative impacts are occurring, such as crime, hitherto claimed as virtually non-existent in Bhutan before the relaxation of control. The article notes a surprising upsurge in drug addiction, theft, and violence, especially among young people.

Major change is undoubtedly occurring throughout the greater Himalayan region. It seems that one of the most potent factors driving change stems from the impacts of policies to encourage mass tourism, for instance, as in Yunnan and northern Thailand, discussed in detail in Chapter 7, but in general, in scattered localities through the entire region. Another driving force for change is the continuing expansion of the road network.

Major resources of the Himalayan region

Two of the major resources, or attributes, are introduced here: water and aesthetic amenities, or recreational attraction. Himalayan water, especially for production of hydro-electricity and provision of irrigation of large areas of agricultural land, represents one of the largest sources of actual and potential renewable energy in the world. Tourism has long been identified as the world's largest and fastest growing industry and has been cited as the most important source of foreign exchange for countries, such as Nepal, and for specific regions, such as Yunnan, northern Thailand, Himachal Pradesh, Sikkim and the Darjeeling area, and northern Pakistan. As will be related below, tourism is acutely vulnerable to political instability. This is exemplified by the situations that have persisted for decades in Kashmir, that beset northern Pakistan after 11 September 2001, and are currently unfolding in Nepal with the disruptions caused by the Maoist Insurgency.

Each of these major resources has been discussed in some detail in preceding chapters: water, especially development of hydro-electricity facilities, was brought into focus in Chapters 6 and 8. The whole of Chapter 7 was devoted to tourism. Each is reintroduced here and their related critical importance for the future development of the entire region is discussed.

Himalayan water resources

The International Year of Mountains (2002) clearly identified the inter-related importance of mountains and water as vital to world society at large. Mountains as the 'water towers of the world' received further emphasis during 2003, the United Nations 'Year of Fresh Water'. The combined water resources of the greater Himalayan region are immense, almost beyond comprehension. Consequently, the need for appropriate and efficient management cannot be exaggerated. Not only the economic stability of several of the individual countries of the region, but the very well-being of many hundred million people, depend on rapid progress toward appropriate management. Yet, as Libiszewski and Bächler (1997) have pointed out, with increasing real and perceived water shortages facing world society as we enter the twenty-first century, controlled access to water has become one of the major causes of political confrontation and armed conflict.

There have been some very successful water developments in the Himalayan region. The accords between India and Pakistan over management of the major headstreams of the River Indus – the Jhelum, Chenab, Beas, and Sutlej – have worked smoothly and persistently despite the long period of tension across the 'Line-of-Control' in Kashmir. Similarly productive has been the co-operation between China and Nepal involving valuable joint surveys of the growing number of dangerous glacier lakes, especially in the Arun River basin, about 90 per cent of which lies in Tibet, China (see Chapter 6).

Nevertheless, these examples are off-set by decades of governmental in-fighting without resolution between India and Bangladesh over management of the Ganges and Brahmaputra. A similar disappointment marks Indo-Nepalese water negotiations. Additionally, Indian aid to Bhutan has been characterized by India's emphasis on development of hydro-electricity projects on Bhutan's mountain rivers aimed at sale of electricity to India at preferential rates (Dhakal 1987).

Under the title 'Water Projects in Trouble: What Lessons?', Ramaswamy Iyer, a former Secretary, Water Resources, Government of India, and member of two major review committees (Sardar Sarovar Project – 1993–5 and Tehri Hydro-Electric Project – 1996–7), has attempted a balanced review of the problems facing mega-water projects in the Himalayan region (Iyer 1998). He deplores the prevailing tendency of proponents and opponents of major water projects to almost invariably assume antagonistic positions. One of the results is that the ensuing conflicts have been part of the cause of delays in final decision-making that have extended over decades. This has been extremely damaging and costly, whether or not a project is completed or rejected. He makes a series of recommendations:

1   Mega-projects should be treated as projects of last resort to be undertaken only after all other possibilities have been evaluated.
2   Integrated and holistic planning for an entire watershed should be a prerequisite, including social and environmental aspects.

3   Construction should not begin until after completion of the evaluation process.

4   Large water-related projects should never be allowed to pre-empt a disproportionate share of the available resources.

5   Ensure and institutionalize the fullest collaboration with the people to be affected and the NGOs representing them from the beginning, and make all information freely available.

6   Establish credible grievance-redress machinery and build safeguards against harassment of people and against corruption.

7   Maintain constant review and institute corrective actions and remedial measures as soon as difficulties manifest themselves.

Iyer (1998) concludes it is often the state itself, through failure to consult, secrecy, and unimaginative implementation, that forces the people and NGOs into a process of confrontation. That such remarks by an experienced professional and former senior government bureaucrat are assumed necessary, indicates the magnitude of the water resource development issue facing the region.

Subba (2001, 2003) and Gyawali (2003) have deplored the serious mismanagement of Nepal's enormous hydro-electric and irrigation potential. Subba (2001: 220) comments that 'the solutions to the region's problems are not necessarily stored in giant reservoirs in the mountains'. Subsequently he stipulates that 'conventional ideas . . . need to undergo a major transformation'. Those developments that have been undertaken and retained within total control of the central government produce electricity at consumer prices three to four times those of India. At least within the last few years there has been some relaxation of Nepalese government control and a number of small hydel projects have been developed and are being operated privately. The government's apparent preference for vast projects as prestige acquisitions has squandered large sums of money and has met with a number of seriously negative results. The long-drawn-out Arun III controversy is one example, yet efforts are still being made to reactivate this particular water-development game with possible financial support from the Asian Development Bank. The Kulekhani hydro-electricity project (see Chapters 4 and 6) is a disaster of a different kind, at least due in part to inadequate preconstruction watershed survey, in part to the severe misfortune of a phenomenal rainstorm.

The Rogoun Dam in Tajikistan and the Tehri Dam in India provide examples of central government determination to proceed with development regardless of the serious hardship faced by many thousands of local people. Work on the Rogoun Dam was suspended due to civil war, that on the Tehri Dam is being pushed ahead despite vigorous local and international protest (see Chapter 8, pp. 192–4). The logging ban of 1998 that affects almost half the area of China, under the guise of protecting the huge investment in the Three Gorges project and downstream security, represents an extreme case in terms of scale. At the other extreme – the construction of many small run-of the-river hydro projects in Himachal Pradesh, development also seems to proceed regardless of either environmental concerns or the welfare of the local people (see Chapter 8, pp. 187–8).

The social, economic, and political aspects of the Arun III project are introduced as an example of the kind of difficulties that formed the basis of Iyer's (1998) admonitions. This is because, with the introduction of democratic government in Nepal in 1991 and the termination of Arun III in 1995, it has been possible to obtain many of the background documents for analysis and publication (Gyawali 2003).

Arun III: autopsy of a misconceived hydro-electric project

The following account is based upon Gyawali (2003: 68–86). Reference should also be made to the description of the mountain hazards in Chapter 6 (pp. 126–33) that were part of the Arun III discourse.

The River Arun is a tributary of the Sapta Kosi which, in turn, joins the Ganges in India. It rises on the Tibetan Plateau and flows southward through the Nepal Himalaya as one of the few antecedent rivers (Chapter 2). It was first identified for its large hydro-electricity potential in a river basin study financed by the Japanese International Cooperation Agency (JICA) between 1983 and 1985. Also in 1985 the electricity department of the Nepal Ministry of Water Resources and the Nepal Electricity Corporation were merged at the insistence of the World Bank and the Asian Development Bank. The new unit became the Nepal Electricity Authority (NEA).

According to Gyawali (2003: 69), Arun III was the culmination of 'monism from the large multilateral donors side, as opposed to institutional pluralism, a reality on the recipient's side [Nepal], given its wide-ranging social diversity'. Gyawali goes on to imply that these initial institutional adjustments were a reflection of the philosophy of 'efficiency of size' rather than that of 'local institutional needs' and raises the question of 'compulsions to market certain types of equipment such as turbines, generators and services such as consultancies'. During the early negotiations other possible contending hydro-electricity projects, such as the 225 MW Sapta Gandaki project, at a much more accessible site and at about half the cost of Arun III, were eliminated by the simple expedience of arbitrarily raising their assumed costs despite absence of supporting data.

In 1987 the Government of Nepal requested the World Bank to become the lead donor agency to ensure initiation of Arun III, the first time in Nepal's development history the Government had passed its authority to an external institution to negotiate on its behalf with other potential donors. Germany agreed to contribute DM 260 million for the feasibility and detailed engineering study, thereby ensuring that a German consultancy firm received the main design contract. Several competitors, including JICA, and many alternative and possibly more efficient projects, were eliminated by what, in effect, became a monopolistic institution.

From its earliest stages, Arun III had many technical, economic, and social anomalies that were not fully investigated. At an estimated cost of US$1.1 billion, Arun III became the largest investment ever contemplated in Nepal. It was, therefore, an investment with the largest risks, and the loans were negotiated in such a way that any losses would be carried by Nepal and not by the donors. Of the total budget, Nepal faced the prospect of raising US$300–500 million internally, the remainder being supplied through foreign aid in the form of grants, soft loans, and technical assistance. Legitimate national institutions, such as the Water and Energy Commission and the National Planning Commission, were not given the opportunity to assess the risk of such a large project on such a small economy. Even the World Bank neglected to undertake a thorough macro-economic impact assessment prior to 1993.

Several studies were conducted by the World Bank to justify the decision to proceed with the project. One study was undertaken under contract to the World Bank by Électricité de France (EdF). This brought to light subsequently that conditions had been imposed on the study, one being that none of the hydro-electricity projects identified as possible alternatives could be commissioned before 1998/99 (that is, before the planned completion of Arun III). Another condition required EdF to apply arbitrary cost-increase coefficients of up to

30 per cent to the alternative projects with no obvious rationale other than to render them less competitive.

The need to construct a high grade 118 km access road with side feeders through difficult mountain terrain was ignored in the cost and time estimates. Power for the actual construction process would depend on helicopter support. Idle Nepali personnel, including about 500 engineers and 10,000 other employees, at best were given only liaison duties with the expatriate consultants.

Above all, the estimated cost of power production per kW in 1989 US dollars would be among the highest in the world. Gyawali (2003: 73) characterizes the process as leading into an 'iron triangle' of vested interests that are given a free hand to trap Nepal into a 'no-options scenario and then escalate project costs'. The 'iron triangle' consists of three apexes: low-paid Nepali bureaucrats in powerful positions who control large sums of money; donor agencies where promotion depends on the number of loans skillfully negotiated rather than on projects successfully implemented; and middlemen, the commission agents who bring together international industries and consultancies with local decision-makers.

In the early stages of opposition to Arun III, NGOs and concerned individuals were branded as unpatriotic. It was only when the Nepali activists were able to establish effective liaisons with international groups, and together initiated a sustained campaign from 1993 onward, that there were real prospects for eliminating an expensive and dangerous undertaking. Gyawali implies that, the potential for environmental and social damage notwithstanding, it was on James Wolfensohn's realization, as newly appointed President of the World Bank, that the project would likely be an economic disaster, that its cancellation was assured.

Furthermore, to present Arun III as a potential financial and environmental disaster avoided, it has been calculated that, following collapse of the project, five much smaller hydroelectricity projects have been completed in half the time and they are producing more power at half the cost. Environmental disturbance has been comparatively slight and independent Nepalese enterprises have evolved (Dipak Gyawali pers. comm. 22 November 2003).

Although the derailing of Arun III may appear as a victory for common sense, the victory is by no means complete. Other projects remain under negotiation. The Mahakali River negotiations between India and Nepal seem to have no end in sight. In this instance, Thapa (2003) insists that the problem appears to be that India wants to acquire hydro-electric power whereas Nepal needs its irrigated agricultural land extended and the two parties appear to be talking past each other. It is also argued that the Kathmandu Valley, with a population exceeding one million and continuing to grow rapidly, is becoming seriously short of water. Hence the Melamchi Water Supply Project, whereby a 26-kilometre tunnel is being planned to transfer water from the headstreams of the Indrawati River into the valley. The calculations for the amount of water needed in the valley compared with that available from normal precipitation are claimed to be faulty and, given the encouragement of village-level corporations, a large increase in water storage capacity could be implemented at very low cost. This also would provide wage employment for a large number of poor people and encourage local economic development. Furthermore, a 26-kilometre tunnel through a geologically imprecisely mapped area subject to high seismic activity, in the best case, would risk losing much of its water en route to the valley, and in the worst case, losing all of it. Gyawali (pers. comm. 22 November 2003) raised the question, for how long can Nepal afford to receive this kind of 'aid'?

In an overview, 'Water, Power and People: A South Asian Manifesto on the Politics and Knowledge of Water', three authors, Imtiaz Ahmed (Bangladesh), Ajaya Dixit (Nepal), and Ashis Nandy (India) make the point: 'The last fifty years of water management in South Asia has been a story of an unfolding disaster' (Ahmed *et al.* 1999: 121). At least, a growing community of outspoken and highly educated voices are beginning to effect change. They should lead a new generation of 'South Asians away from the stereotypes and clichés of the past and give them more confidence in envisioning the future relations among water, power and people in a less encumbered fashion' (Ahmed *et al.* 1999: 121).

The temptation here is to suggest that, based on current showings, the long-term prospects for appropriate water management and concomitant fair treatment of mountain minority peoples are dismal. Possibly, as the overall situation deteriorates, the relevant governments will be induced to apply a higher level of wisdom to ensure more appropriate development and management of this major resource.

Aesthetic amenities

Reference is made to the general treatment of tourism in Chapter 7. Here the intention is to emphasize that, as with the failure to realize the potential for extensive benefits offered by the water resources of the Himalayan region, the opportunities presented by a rapid growth in tourism, both domestic and international, are also being largely squandered in the context of ensuring benefits for the rural poor. Exceptions have been pointed out in Chapter 7 – where a local tourist destination is under some degree of control by the local people, the positive tend to outweigh the negative aspects. The Sherpas of the Khumbu are cited as an example. This is the regional example *par excellence* where an internationally admired homogeneous, and hence socially and traditionally well-knit, community, because of its degree of control and 'pre-development' experience, has been able to advance in terms of affluence, health, education, and in spiritual ways. The explanation for the Sherpas' success lies with their ability to compete effectively with outside interests. There have been negative aspects, inevitable given the sheer numbers of visitors in proportion to the local inhabitants and the rapid rate of change, but these have been reasonably well accommodated.

The rural people of other regions have not been so fortunate. In places such as north-western Yunnan or the Kulu Valley of Himachal Pradesh, outside interests seem to have profited rather than local people. This is largely determined by the manner in which mass tourism, requiring large monetary investments, takes over control of local development. While an entire region that has experienced development of tourism may have benefited overall from an economic point of view, increasing local disparities, together with environmental pressures, tend toward negative outcomes.

Perhaps a good example that lies between the extremes of Yunnan and Kulu, for instance, where the majority of the rural people have received little benefit, and the Khumbu Sherpas, is that of the Annapurna Conservation Area (see Chapter 7). This is an instance where central government control has been avoided (excepting the newer addition of Upper Mustang) and the successes of the conservation and development policies of the King Mahendra Trust have led to a wide range of schemes for improvements at the village level across the entire conservation area.

Nevertheless, the single most serious impediment to tourism as a universal panacea is the impact of sudden political instability and violence. While the 11 September 2001 terrorist attacks on the United States and their aftermath have affected tourism worldwide, the effects have been especially severe in areas such as northern Pakistan where local people had invested scarce resources in small hotels, and where tourism virtually ceased in 2002. In addition to the relatively affluent entrepreneurs, the poorer people who had come to depend on employment as porters and guides, or on the sundry associated jobs, have been left without support in a country that has not recognized, or been able to achieve, social security legislation. Here the hardship is severe. The Maoist Insurgency is another example of induced political instability that has had a dramatic effect even on the successful tourism development in the Khumbu. More significantly, the economy and political stability of Nepal as a whole is put at risk. Again, the most seriously impacted are the local poor who have developed a degree of dependency on wages earned as porters and other menial adjuncts to tourism development. Kashmir was formerly a major attraction for both international and domestic visitors, although tourism has been virtually eliminated for several decades because of the long-continued violence. Himachal Pradesh has certainly benefited from the collapse of tourism in Kashmir although, as mentioned above, its Kulu Valley has proved a source primarily for outside commercial profit.

In conclusion, it is emphasized that the enormous potential of the Himalayan amenity resource has been realized in a highly asymmetric form. Certainly, tourism has been extremely important as a source of foreign currency and pockets of affluence have evolved throughout the region. Yet the original optimism that it would prove a major factor in equitable development has not been realized. Furthermore, the last decade, and especially the post-11 September period, has demonstrated the extreme vulnerability of tourism, particularly in remote mountain tourist destinations where the local inhabitants had become highly dependent upon it.

10 What are the facts?

Misleading perceptions, misconceptions, and distortions

It is more from carelessness about truth than from intentional lying that there is so much falsehood in the world.

(Samuel Johnson 1778)

Introduction

The aim of this chapter is to address the confusion brought about by a combination of lack of academic rigour in the early stages of the propagation of the myth of Himalayan environmental degradation, aid agency and news media carelessness, and the unsubstantiated basis for some of the policies of regional governments. Thompson and Warburton's (1985a) response to the leading phrase of the chapter's title (*What are the facts?*) has been introduced earlier. Nevertheless, even with the great increase in research across many disciplines and inter-disciplines since 1989, their own provocative response to the question remains relevant: 'what would you like the facts to be?'

Central government agencies in India, Thailand, China, and Nepal, for instance, certainly appear to want the 'facts' to support their policies that are frequently based upon the assumption that 'ignorant' mountain minority farmers are devastating the forests and so causing serious downstream environmental and socio-economic damage. The Government of Bhutan largely fabricates its perception of 'truth'. And there has been a continual flow of news media and environmentalist publication to the effect that death and destruction on a large scale are imminent, whether the result of unwise resource extraction by mountain people or due to global forces, such as climate warming. Is it all part of a game? If so, it is a very serious and dangerous game.

This chapter will examine the larger issues of how Himalayan perceptions have arisen, how many have been misleading, misconceptions, even seemingly deliberate distortions. In contrast, many commentaries and recommendations have been eminently reasonable and have contributed to the eventual inclusion of Chapter 13 in Agenda 21 following the 1992 Rio de Janeiro *Earth Summit*. This in turn was the vital turning point that led to the United Nations designation of 2002 as the *International Year of Mountains*. It is appropriate, therefore, to go back to the origins of the Theory of Himalayan Environmental Degradation and to work forward from there.

The Theory of Himalayan Environmental Degradation

Although this topic has been discussed extensively in Chapter 1, it is reintroduced here to set the stage for examination of the way in which perceptions of Himalayan development and environmental stability are being distorted. Other scholars, or developmental practitioners, or environmentalists may select alternate starting points. However, the GTZ-UNESCO conference of December 1974 in Munich can be regarded as the initiation of a worldwide discourse on environmental problems of the Himalaya. The formal topic in Munich was *The Development of Mountain Environment* and it brought together a diverse group of particip-ants – diverse nationalities, disciplines, and professions. The impacts of the 1972 Stockholm Conference on the Human Environment had only recently been felt. Similarly, the recent winding down of the International Biological Programme (IBP) had begun to influence the formulation of the UNESCO MAB–Programme, Project–6, and had demonstrated the applicability of computer modelling.

The Munich participants were presented with a series of well-intentioned, if disturbing, scenarios. Many were based on apparently first-hand experience in the Himalaya, others were derived from experience elsewhere, and yet others depended upon rational thought arising from formal conventional education, or a combination of the above. The participants were alerted informally to GTZ's plans for providing funds to establish an international mountain research and development institution; that it would probably be headquartered in Tehran because the Shah of Iran had indicated that he would provide many more millions of dollars.

A feeling of dire emergency was generated at the conference, together with a sense of opportunity. Something must be done to save the world's mountains; mountain regions in the developing countries were most seriously at risk; and the Himalaya warranted special attention. A Munich Manifesto was deliberated and unanimously approved. There were suggestions that a 'Club of Munich' should be formed to imitate environmentally the Club of Rome, proceedings were published, and press releases were initiated. Nevertheless, the proceedings (Müller-Hohenstein 1974) were eminently constructive and constrained. A request was made for accentuated mountain research linked to development policy and the creation of a scholarly publication outlet for the results of such research. The need for informing United Nations agencies, national governments, and world opinion at large was underlined. Frank Davidson (in Müller-Hohenstein 1974: 186) urged establishment of an independent mountain research institution with appropriate links to United Nations agencies and universities and, taking cause from the widely recognized contributions of *Oceanography*, recommended consideration for establishment of *mountainology* (to become *montology* – Oxford English Dictionary 2002 edition). Very little of the informal discussions about an environmental crisis in the Himalaya appeared in the proceedings. The closest, yet oblique, reference appears in the summary report of the proceedings:

> But these mountain regions are seriously and increasingly affected by processes of deforestation, soil erosion, improper land use, and poor water management. Overuse of mountain environments has a widening impact on the plains with downstream floods, the siltation of dams and harbours and on the damage of crops and of homesteads.
>
> (Müller-Hohenstein 1974: 5)

Thirty years later, following a considerable increase in mountain research (both academic and applied) and much wider recognition of the importance of mountain regions, the general statements emanating from the Munich Conference read as eminently rational. But in terms of the last three decades of melodramatic recounting by the news media of Himalayan deforestation causing catastrophic flooding in Gangetic India and Bangladesh, Eric Eckholm's statement in the Munich proceedings is revealing:

If deforestation in Nepal and Kashmir threatens the survival of three-quarters of a billion people in South Asia, and indirectly will affect the political and economic well-being of people in Tokyo, New York, and Munich, then these facts should be in the newspapers every week in all of these countries. But I read several newspapers every day, and have followed the accounts of many major devastating floods over the last few years, and I have discovered that *the news accounts never mention deforestation as a cause of the flooding.* The collective knowledge of the minds in this room, if distilled in the proper form, would horrify and astound millions of people and hopefully goad them into the needed actions. The question is: How will we help them find out before it's too late.

(my emphasis)
(Müller-Hohenstein 1974: 131)

Thus it is reasonable to conclude that the Munich Conference of 1974 served indirectly, rather than directly, as the flashpoint for propagating widespread acceptance of the notion of imminent environmental catastrophe in the Himalayan region.[1] The innumerable litera-ture references to Eckholm's paper in *Science* (1975) and to his book (1976) show how the assumptions, portrayed with such skill and intellectual appeal in these two publications, dominated mountain environment and development thought over the next 15 years; and the catastrophe discourse has remained highly influential in many areas of government and institutional decision-making to the present time.

Despite earlier cautious reaction to the deforestation/landslide /downstream flooding scenario (Ives 1970) I recall being swept up by the sense of urgency in Munich. Nevertheless, the seeds were sown for eventual publication of the journal *Mountain Research and Development* in 1981, and for the establishment in 1982 of the International Centre for Integrated Mountain Development (ICIMOD) in Kathmandu.

Following the Munich Conference, however, it appeared that writers, academics, agency personnel, and politicians were seeking to out-perform each other by moving progressively through repetition to hyperbole. No new 'facts' were needed, only the repetition and enlarge-ment of existing 'facts'. Thompson *et al.* (1986) argued that these 'facts' were precisely what the agency personnel required in seeking to enlarge their development budgets and to expand and prolong their presence in Nepal, long regarded as one of the most attractive locales for appointment of expatriate bureaucrats by donor agencies.

The now notorious World Bank (1979) prediction of total loss of accessible forest cover in Nepal by 2000 was very powerful. The 'State of India's Environment: A Citizen's Report' (1982) spoke with great authority in similar terms, as did the World Resources Institute (1985) and the Asian Development Bank (1982). Likewise, internationally respected foresters and environmentalists raised the spectre of Khumbu forest devastation, perceived as a necessary part of the struggle for establishment of the Sagarmatha National Park. It should be noted, however, that the Khumbu was a special case, for it was there that the imminent disaster scenario had unfolded early and independently of the Munich Conference and only later merged with the general demand for mountain forest protection as a prime approach to averting environmental disaster. All of the foregoing were powerful institutional forces that drove the complex of assumptions for which the shorthand term *Theory of Himalayan Environmental Degradation* was coined.

During the first 10–15 years following the Munich Conference the majority of academic publications concerned with the Himalayan region, or parts of it, both echoed and replen-ished the news media campaign and the *myth* of Himalayan environmental degradation became firmly embedded in world opinion. However, after about 1983, first a trickle, and then a flow of academic publications began to discredit the *myth* although, for the most part, the news media continued on course, as did many of the vested interests of the region. The process of Himalayan environmental discourse and its split into two opposing streams will be illustrated by a selection of short quotations, citations, and comments.

Academic and research publications

There were innumerable references in scholarly and research publications that advanced and reinforced the Theory:

Eckholm (1975 *Science*, 189: 764–70: referred to above)
Rieger (*in* Lall and Moddie 1981: 351–76, and earlier unpublished consultancy reports)
    These papers provide a parallel discourse to Eckholm (1976) except that Rieger (1981), in particular, develops a series of computer simulations demonstrating relations between population growth, deforestation, soil erosion, and downstream impacts. However, Rieger's approach does foresee a much longer time interval for total elimination of all Himalayan forests.

Ives and Messerli (1981: 229–30 – based on an initial reconnaissance for fieldwork in the Kakani area, Nepal):

> Loss of soil and loss of agricultural land through gullying and landsliding are occurring more rapidly than the local people with their existing resources can replenish. This is true without considering the deterioration to be anticipated by projecting the current rate of population growth into the future.

To be somewhat redeemed by the following:

> It is also believed that involvement of the local people in every planning stage and incorporation of their experience will prove critical.

Karan and Iijima (1985: 81):

> One-fourth of the forests of the country has been cut in the past decade. If this trend persists, the remaining forest area may be denuded in another twelve to twenty years.

Karan and Iijima (1985: 84):

> The Kulu Valley, formerly a picturesque scene of deodar trees, some forty-five meters high . . . is now almost barren.

    This statement should be compared with the relevant section in Chapter 3 (p. 69) that emphasizes the excellent degree of preservation of the Kulu Valley forests.

Myers (1986)
    This paper is also a parallel statement to those of Eckholm (1975) and Rieger (1981).

Literature on deforestation in the Khumbu Himal, Nepal

Blower (1972, cited in Mishra 1973: 2):

> . . . depleting forests of the Khumbu . . . since destruction would result in disastrous erosion leading to enormous economic and aesthetic loss to the country.

Lucas *et al.* (1974) wrote that the members of the New Zealand mission:

> . . . saw too much evidence of incipient erosion to feel other than a sense of deep concern for the future.

Fürer-Haimendorf (1975: 97–8):

> Forests in the vicinity of the villages have already been seriously depleted, and particularly near Namche Bazar whole hillsides which were densely forested in 1957 are now bare of tree growth and the villagers have further and further to go to collect dry firewood.

Speechly (1976: 2):

> . . . forest areas in the proposed Sagarmatha National Park are, as a result of a combination of influences, in a depleted state, such that if present pressure of use is continued, severe environmental damage will result.

Hinrichsen *et al.* (1983: 204):

> . . . more deforestation [has occurred in the Khumbu] during the past two decades than during the preceding 200 years.

In contradiction to the above, Charles Houston (1982, 1987), as a member of the 1950 Mount Everest reconnaissance from the south, had revisited the Khumbu in 1981. He wrote that, with the exception of a thicket of dwarf juniper at Pheriche there was:

> as much or more forest cover than there was in 1950 and I have the pictures to prove it.

International agencies

World Bank (1979):

> Nepal has lost half of its forest cover within a thirty-year period (1950–80) and by AD 2000 no accessible forests will remain.

Asian Development Bank (1982):

> . . . distinct danger that all accessible forests, especially in the Hills, will be eliminated within less than 20 years.

> (ADB 1982, Vol. 1, p. 12)

On page 63 of ADB Volume 2, the alarm is somewhat heightened by the prediction of forest elimination within 14 years.

World Resources Institute (1985):

> . . . a few million subsistence hill farmers are undermining the life support of several hundred million people in the plains.

United Nations Environment Programme was reported to have commented on the seriousness of the threat of deforestation in *The Bangladesh Observer*, Dhaka, 2 June 1990 under the headline *Deforestation in the Himalaya Aggravating Floods*. The article was reporting on an address to the National Seminar on Environment and Development by Dr Mustafa K. Tolba, Executive Director of UNEP, organized by the Environment and Forestry Ministry, UNDP and UNEP. It quoted Dr Tolba as stating that:

> . . . the chronic deforestation in the Himalayan watersheds was already complicating and compounding seasonal floods in Bangladesh.

And added the comment that 700,000 people died in Bangladesh in 1970 because of flooding.

News media reportage

Sterling (1976 *Atlantic Monthly*, 238 [4]: 14–25 – one of the earliest and most melodramatic reports).

Between 1976 and 1986 most of the world's newspapers were predicting imminent disaster in the Himalaya and on the plains of the Ganges and Brahmaputra. The coverage ranged from *The Times*, London, to almost every local newspaper in the Western world, and in India, Nepal, Pakistan, and China. The coverage extended to leading periodic magazines, such as *Newsweek* and *Atlantic Monthly* and the conservationist literature. Television coverage was also extensive worldwide. Examples are restricted to the more recent period following 1986.

Farzend Ahmed in *India Today*, under the title *Bihar Floods: Looking Northwards*, 15 October 1987:

> Each time north Bihar is devastated by floods, the state Government performs two rituals. It holds neighbouring Nepal responsible and promises to implement a master plan for flood control . . . Nepal is invariably held guilty because most of the rivers . . . originate there before flowing into the Ganga. The Bihar Government maintains that Nepal's non-cooperation lies at the root of the annual cycle of human misery . . . This time the chorus of accusation reached fever pitch when Prime Minister Rajiv Gandhi . . . demanded to know what preventative measures had been taken . . . Predictably, the [response] referred to the hill kingdom's lack of cooperation. The Nepal-bashers also scored a major victory at the Second National Water Resources Council meeting in New Delhi last fortnight. State Irrigation and Power Minister Ramashray Assad Singh managed to have the national water policy draft amended to say that the solution to Bihar's flood problems lay beyond its borders.

Begley *et al.* 1987 in *Newsweek*, under the title *Trashing the Himalayas – that once fertile region could become a new desert*:

> Dense alpine forests once covered the lower slopes of Mount Everest, and the Khumbu Valley below the mountain used to blush dark green from its carpet of junipers. But that was the Everest of 1953, when Sir Edmund Hillary and Tenzing Norgay became the first men to conquer the highest peak on earth. Today the forest at Everest's base is 75 percent destroyed, replaced by a jumble of rocks interspersed with lonesome trees. All the Khumbu's junipers have fallen to axes . . . The degradation of the Himalayas is not confined to the tall peaks. In Pakistan, India, Nepal and Tibet, deforestation has eroded fertile top-soil from the hills, triggering landslides and clogging rivers and reservoirs with so much silt that they overflow when they reach the plains of the Ganges . . . At the rate trees are being felled for fuel and cropland, the Himalayas will be bald in 25 years . . . Although a significant fraction of the erosion stems from nature . . . most of the damage is man-made.

New York Times: 9 September 1988:

> United Nations expert Tom Enhault, director of projects in Bangladesh – asserted that the environmental havoc wreaked by the destruction of the Nepalese forests have done the most damage [referring to the flooding of 1987 and 1988] . . . he also blamed over-grazing.

Sunday Star–Bulletin: Honolulu, 11 September 1988:

> *Bangladesh flood disaster blamed on deforestation*
>
> Flooding on a massive scale may soon become the norm . . . remarkable collapse of the Himalayan ecosystem.
>
> A. Atiq Rahman, director Institute of Advanced Studies, Dhaka, stated 'the main environmental problem is the widespread and growing deforestation of the Indian and Nepalese mountains.' . . . B. M. Abbas, Bangladesh's leading authority on water control and for many years Minister of Water Resources said 'For so many years I have told people that trends in the mountains would destroy us.' Hassan Saeed stated that there had been 1,451 deaths and that 700,000 flood refugees had been forced to find shelter in Dhaka.

Dawn: Sunday magazine, Islamabad, 4 October 1992:

> Minister for Environment and Urban Affairs, Anwar Saifullah Khan said 'the destructive power of the floods has increased manifold as a result of deforestation which has been continuing unabated in the Northern Areas of the country.'

Sacramento Bee: Sunday 1 August 1993:

> Bangladesh has renewed demands that India and Nepal agree to control the powerful rivers that flow through their countries. Officials in Bangladesh say the flooding has killed at least 150 people and displaced 7 million people.

World Tibet Network News: Beijing, 28 August 1998, under the headline: *Asian Disasters Blamed Partly on Shrinking Forests: Deforestation Leads to Floods*:

> Floods kill more than 2,000 people along China's Yangtze River and 370 others along the Ganges and Jamuna in Bangladesh . . . Rain across the region has been much heavier than normal this year, but World Watch Institute President, Lester Brown, said recently that a 'human hand lurks behind the floods. That hand often wields the ax or chainsaw, denuding the highlands that feed Asia's great river systems and sending greater volumes of water and silt to compound the catastrophes downstream. The forests that once absorbed and held huge quantities of monsoon rainfall, which could then percolate slowly into the ground are now largely gone. The result is much greater runoff into the rivers.

Apart from the interspersed explanatory remarks, no further comment will be added to the quotations introduced above, with a single exception. This is because the statement in the Islamabad magazine *Dawn* (4 October 1992), attributed to Environment and Urban Affairs Minister Anwar Saifullah Khan, is especially out-of-step with reality. The cause of the devastating floods, to which the Minister refers, has been assessed by Hewitt (1993 – see Chapter 6 pp. 139–40). Hewitt was present in northern Pakistan during the event and was able to obtain many observations on the extent of the damage and subsequently to analyse the records of relevant climate stations throughout the region. The cause was, without doubt, unusual and excessive rainfall. Furthermore, as reported in Chapter 3 (p. 74), even prior to the opening of the Karakorum Highway and the accelerated and illegal logging, total forest cover of northern Pakistan was such a minute percentage of total land area that, even if complete removal of trees had been accomplished by 1992, impact on flood magnitude would have been imperceptible.

The political implications of the official statements, however, warrant careful attention. For instance, Bihar and New Delhi authorities and politicians blame Nepal for downstream disasters due to assumed mountain deforestation; Bangladeshi authorities blame India and Nepal; the Chinese government blames the irresponsible and illegal logging by minority peoples in the upper watersheds of the Yangtze. Herein lies part of a possible explanation for institutional adherence to the Theory up to the present. This will be discussed further below.

How was the academic tide turned?

As indicated above (pp. 7–9) academics undertaking research in the Himalayan region began to reverse the tide of support for the concept of an environmental super-crisis in the early to mid-1980s. Increasingly since 1989 the Theory of Himalayan Environmental Degradation has come to be regarded as an insupportable myth and today, while some confusion and misunderstanding remains, there is little support within academia for the totality of the notion of Himalayan environmental collapse in the form in which it originated in the 1970s. So, how was the tide turned?

A large part of the explanation is that several research groups and individuals began detailed studies about the same time (late 1970s to 1980) and became aware of each other's work. The 'coming together' was greatly facilitated by emergence of the quarterly journal *Mountain Research and Development*, that in turn led to organization of the Mohonk Conference on the Himalaya–Ganges Problem in May 1986. From that point most of the linkages in the eight-point scenario that was constructed to illustrate the Theory came under increasingly critical investigation. Very little rigorous environmental research had been carried out in the Himalayan region prior to about 1980. The foregoing account of the alarmist discourse in both the academic and popular literature was based upon supposition and emotion that entered policy formulation. It also entered the environmental and development politics of the region and, in turn, encouraged even greater commitment to the 'cause' of addressing Himalayan environmental degradation. Examination of many of the reports prepared for aid agencies and local governments was particularly revealing – successive consultants simply reproduced the conclusions of their predecessors. There were exceptions, although the 'white noise' was almost overwhelming.

For the UNU research team, the tide turned on entering Balami/Chhetri/Tamang villages with Nepalese students and Western university fieldworkers. Johnson, Olsen, and Manandhar (1982) quickly learned how well the villagers understood landslide mechanics and witnessed their ability to manage, even to propagate landslides themselves. The research team was able to analyse the complexities of the environmental–socio-economic situation; year-round research with the subsistence agricultural systems helped to explode the myth, and it became apparent that it had been based upon reports of 'experts', prepared in Kathmandu's best hotels, heavily dependent on earlier reports by other 'experts' also based on Kathmandu hotels but preferably not during the summer monsoon, the peak season for landslides, leeches, and maximum discomfort for field travel.

By 1983 the research progress of the UNU team was sufficiently advanced for a public review of early results to be organized in Kathmandu. This, together with the regular publications scheduled through *Mountain Research and Development*, became one element in the turning of the tide. There were others equally effective. Most important were the Nepal–Australian Forestry Project and the involvement of the East–West Center, Honolulu. The Australian foresters and their Nepalese graduate students appreciated the 'truth' from living and working with the indigenous mountain farmers (Bajracharya 1983; Mahat *et al.* 1986a, 1986b, 1987). Hamilton's basic forest ecology led him to attack the notions that forests act as a sponge for excessive rainfall and that 'deforestation' is necessarily bad. He argued that the very term 'deforestation' had been abused to the point of it being reduced to the level of emotion; finally, there was his 'rain on the plain' motif (Chapter 5 p. 104). Intellectually, one of the most satisfying contributions was Thompson and Warburton's (1985a) adaptation of Fürer-Haimdendorf's (1975) 'careful cultivators and adventurous traders' phrase leading to 'uncertainty on a Himalayan scale'. All of these separate strands came together as the 'Mohonk Process' (Thompson 1995; Forsyth 1996).

The answer to the question 'how was the academic tide turned?' is that the very melodrama seems to have aided in prompting the first phase of rigorous research in the Himalaya by scholars who had no restricting vested interests.

But are these the 'facts' and what are the next steps? It appears that as specific myths are identified and explained, modified, or demolished, or used to good effect (Thompson 1995), new ones spring up to take their place.

Some current myths on a Himalayan scale

A series of examples, or case studies, are introduced to illustrate the problem of misrepresentation. It is unlikely that proof can be obtained to demonstrate a causal relationship between popular reporting and policy formulation, or the reverse. It is also difficult to determine how particular exaggerations are manufactured because the news media as the channel of communication between the field research and sometimes casual observation, and popular presentation is rarely a direct line. Nevertheless, the following examples are offered because the degree of misinformation appears to be both extensive, widespread, and continuing. They are introduced, not so much because of their inherent importance, but as examples that could be multiplied many times over. They could be dismissed as part of a phenomenon that pervades all spheres of world society. Reporting on global warming, the world economy, international terrorism, or almost any disaster has become comparable to the campaign speeches politicians tend to make at election time. It has also been understood for several decades now that 'green' movements have felt compelled to exaggerate in order to compete for attention with the possible bias of well-financed campaigns of big business and industry. Regardless, the examples of 'latter-day myths' are set forth because their pervasiveness tends to clutter the sustainable development landscape and perpetuate the Himalayan scale of uncertainty.

The cause of flooding in Bangladesh

The infamous 1979 World Bank prediction of a nearly treeless Nepal within 20 years has been referred to in different contexts and its re-introduction here may be criticized as out-dated and over-used. Yet it forms a good starting point as a frequently argued explanation for flooding in Bangladesh. Two decades later, in response to the severe flooding of 1998, the *Basler Zeitung*, among numerous major newspapers, published the following on 15 September 1998:

> The severe floods in eastern India and Bangladesh are not the result of a natural disaster, but of ruthless exploitation of the forests which has been practiced over many centuries in the Himalayas.

The Canadian Broadcasting Corporation (CBC-TV) produced a documentary for its Newsworld programme on 21 March 2000. The topic was the cyclone of the previous September that caused extensive damage and loss of life in Orissa, India. Amidst dramatic film footage, the commentator warned the viewers that:

> . . . conditions will deteriorate further in the future because the sea level is rising as a result of deforestation in the Himalayas.

Following the Bangladesh flooding of 1998, the news media were awash with hyperbole. Yet the following quotation from the *Bangladesh Daily Star* (repeated from Chapter 5 p. 103) should provoke a reflective pause:

> Have no fear, the children are enjoying diving in the River Jumuna

The melodrama is surely recognizable as such, yet the fact remains that the governments of India, China, and Thailand have all legislated logging bans on their upper mountain watersheds. Their prime justification is that large-scale commercial logging, as well as that of the mountain minority peoples, is causing extensive environmental, economic, and social losses downstream. The linkage with the Three Gorges Dam in China is a prime cause–effect assumption. The danger herein is that, even if the logging bans can be enforced, assumed or actual deforestation in the upper watersheds has not been shown scientifically to propagate downstream devastation. Although some kind of control over logging is certainly needed, in the form of consistently applied forest laws and effective forest management, the Government's policy represents an example of trying to solve a problem by confronting an assumed yet unproven cause. It is certain, however, that upstream losses are occurring. However, much of the loss is in the form of considerable hardship placed on the shoulders of the very poor people whose livelihoods depend on the forests and a shift in forest pressure onto the village community forests and the reserves of neighbouring countries.

The above commentary is not intended to denigrate the importance of mountain forests since they are vital to the survival of viable mountain agriculture and also have an important aesthetic value. In addition, given good management practices, they are vital for their commercial products.

Meltdown in the Himalaya

'Meltdown!' is the title of one of four papers published by the *New Scientist* as part of its celebration of the International Year of Mountains – 2002 (Pearce 2002: 44–8). The core of this presentation is an explanation for the undoubted increase in flash flooding that is occurring when glacier lakes in the Himalaya (and elsewhere) break through their end moraine dams to produce destructive mudflows/debris flows/floods for many kilometres downstream. These glacial lake outburst floods (jökulhlaup, or GLOFS) have been discussed at length in Chapter 6 (pp. 126–33).

There is no question that they represent a serious threat. Nevertheless, Pearce (2002) quotes John Reynolds, an experienced geotechnical consultant, as predicting that:

> . . . the 21st century could see hundreds of millions dead and tens of billions of dollars in damage . . .

from the outbreak of glacier lakes worldwide, but principally in the Himalaya and Andes. There is also the prediction that the downstream extent of such outburst floods could extend for hundreds of kilometres, cross the borders of Nepal and Bhutan, and cause extensive damage to the large Indian cities on the Ganges flood plain.

There is a factual base for Pearce's reported predictions. As noted in Chapter 6, two recent surveys have identified the initiation and growth of about 2,700 such lakes in Bhutan alone (Mool *et al.* 2002a) and about 2,300 in Nepal (Mool *et al.* 2002b), of which 44 have been designated as dangerous, although a majority were little more than tiny ponds. Outburst floods that have occurred have barely penetrated more than 50–75 kilometres downstream (the unusual event of the Indus, also related in Chapter 6, cannot be placed in the same category as the Nepalese and Bhutanese Himalayan glacier lakes). There is no intention here, however, to deny that GLOFS are dangerous, nor to imply that serious efforts to mitigate their potential effects are not needed. But it would remain an understatement to suggest that Pearce's reporting represents an exaggeration.

Rolwaling, Nepal, and the threat from Tsho Rolpa glacial lake

The history of formation and the mechanics of development of potentially dangerous glacial lakes, including Tsho Rolpa, have been described in some detail in Chapter 6. Here emphasis is placed, rather, on socio-economic and psychological consequences that arose in 1997 from reactions to a report that Tsho Rolpa was on the brink of a catastrophic outbreak.

The discussion is taken from a published blow-by-blow account by Gyawali and Dixit (1997) and personal comments (Gyawali 22 November 2003).

Concerns for the safety of the inhabitants of Rolwaling valley were expressed by the lake survey team in 1996, and the Government of Nepal requested a more detail examination of the end moraine that forms the dam for the expanding lake. This was undertaken in May 1997 by Reynolds Geo-Science Ltd., in collaboration with the Nepal Department of Hydrology and Meteorology (DHM), funded by the British Government. Following the field survey, a seminar was held in Kathmandu to facilitate public and government review. The report presented by the consultants was cited as eminently cautious and responsible. However, following the seminar, oral presentation to the news media appears to have created the impression that a catastrophic flood was about to be released momentarily. This produced panic among the public and government departments and among many of the inhabitants living below the lake all the way down to the frontier with India. The panic prompted the local Member of Parliament to demand immediate government action. It was considered that such a flood would directly affect 4,000 people in 600 households of 18 villages. The warden of the Kosi Tappu Wildlife Preserve in the Terai reported that 175 km² of the preserve would be destroyed with the loss of 200 wild buffaloes, 400 species of birds, as well as crocodiles, deer, wild boar, snakes, dolphins, and other precious animals. It was also contended that the flood would wash away the Kosi Barrage threatening enormous losses in Bihar, India.

The Royal Nepal Airline Corporation (RNAC) suspended flights to the lower area, villagers were evacuated, and workers at the Khimti hydro-electricity project, as well as 90 per cent of the people of Kirnetar, began to evacuate. Police and army posts were set up in the Rolwaling valley.

Many more details of the panic are provided by Gyawali and Dixit (1997) who also estimated considerable personal loss on the part of many people who were induced to leave their homes. Yet the villagers living near Tsho Rolpa, who had observed the seasonal fluctuations in the lake level for years refused to move, 'asking the police not to speak nonsense' (Gyawali and Dixit 1997: 24).

RNAC resumed its regular flights on 13 July 1997. By the end of July the flood level of the Tama Kosi, into which the Rolwaling drainage empties, had fallen to almost winter flow conditions and the people who had fled their homes began to return. The results of the affair in Kathmandu included widespread journalistic charges that the rumour of a possible Tsho Rolpa outburst flood had aided expatriate consultants and Department of Hydrology and Meteorology officials to prepare an outrageously expensive proposal for artificial lowering of the lake level for their own financial benefit (Gyawali and Dixit 1997: 33).

The discussion illustrates the severe problem of how authorities should react to potentially lethal mountain hazards that are notoriously difficult to predict with any precision. It underlies the need, not only for extensive survey and monitoring of hazardous mountain phenomena, but also for the establishment of a responsible review and reporting mechanism. In the Tsho Rolpa case by far the most serious losses were caused by the panic reaction to what appears to have been a rumour. Glacial lake outburst floods do occur, as the carefully surveyed case of Dig Tsho of August 1985 illustrated. Following that event, the Government remained lethargic for nearly a decade; by 1997 it appears that the reaction had moved to the opposite extreme – one of panic.

On a related theme *The Times* of London (21 July 2003), reporting on an international meeting held at the University of Birmingham, noted that 'Himalayan glaciers could vanish within 40 years because of global warming . . . 500 million people in countries like India could also be at increased risk of drought and starvation.' Syed Hasnain is quoted as affirming that 'the glaciers of the region [Central Indian Himalaya] could be gone by 2035'.

According to Barry (1992: 45) the average temperature decrease with height (environmental lapse rate) is about 6° C/km in the free atmosphere. The dry adiabatic lapse rate (DALR) is 9.8° C/km. If it is assumed that the equilibrium line altitude (comparable with the 'snow line') in the Central Himalaya is about 5,000 m asl and it will need to rise to about 7,000 m if all the glaciers are to be eliminated, then the mean temperature increase needed to effect this change would be about 12–18°C. Given that degree of global warming, summers in Calcutta would be a little uncomfortable.

The Khumbu and Sagarmatha National Park

As indicated earlier, myths tend to be self-perpetuating. In practice their longevity is often encouraged by vested interests of one form or another. Sagarmatha National Park is perhaps the most likely location in the entire greater Himalayan region for such perpetuation. Conflicting reports and stories here began with Byers's disagreement with the claims for extensive deforestation by Fürer-Haimendorf and the New Zealand foresters as part of the campaign to ensure the gazetting of the world's highest national park (Byers 1986, 1987b, 1997: Ives and Messerli 1989: 59–65).

Byers's most recent work indicates the persistence of healthy forests throughout the Sagarmatha National Park area and little change since the 1950s, very long-term indigenous landscape modification, and significant disturbance of the subalpine juniper belt along the approaches to the Mount Everest base camp (see Chapter 3 p. 49). The successful reforestation in the vicinity of the park headquarters was certainly an improvement in the park-like landscape although it risks distracting attention from the serious damage in the upper treeline belt. It was with considerable interest, therefore, that Paul Deegan (February 2003) requested review of a manuscript dealing with the dangerous loss of forest cover in the Himalaya, especially in Sagarmatha National Park. The manuscript was sent for critical comment to Alton Byers and Stan Stevens, active current researchers in the Khumbu. The result was a much more balanced account that was submitted by Deegan to *Geographical*, the London-based monthly magazine. Press deadlines did not permit the author, let alone the informal reviewers, to read the final edited version. The ensuing article was published in the March 2003 issue under an editorially imposed title: *Appetite for Destruction*. Essential passages accredited to Byers in the original submission had been eliminated and the tone of the conclusions substantially altered. Upon publication, Deegan protested and alerted his informal reviewers to his disappointment. The editor promised to redress the situation by inclusion of the following statement in the June issue of the magazine.

> Correction: During the editing process, text was removed from Paul Deegan's article on forest-related issues in Nepal . . . that highlighted the difference between healthy forest cover below the treeline in Nepal's Sagarmatha National Park and the clearing that is taking place in the alpine zone. Extensive research by Dr. Alton Byers has shown that not only did the forest cover below the forest treeline remain constant between mid-1950s and the 1980s, but it has increased over the past 20 years.

This article, however, brings another aspect into focus that does involve unfortunate environmental destruction. Fear of Maoist Insurgency activity had prompted Nepalese military personnel to eliminate '[t]housands of young trees around the park headquarters . . . to give army personnel clear fields of fire in the event of a rebel attack' (Deegan 2003: 34). Seth Sicroff, who was chairing a conference at Namche, was asked to check the details directly and replied: 'Mendelphu Hill (site of SNP headquarters) has been trashed . . . trees cut, foxholes and trenches dug, barbed wire everywhere' (Seth Sicroff pers. comm. 20 May 2003). Nevertheless, Deegan's article does serve to identify illegal tree felling south of the park boundary by local mountain people, as well as provide a firmly documented example from the park itself (Mendelphu Hill) as an act sponsored by government authorities, regardless of whether or not such an act is justifiable in light of the insurgency threat.

The foregoing discussion requires some qualification. Cutting of trees south of the national park boundary in Pharak has been observed for several decades. This has been reported by Stevens (1993, 2003), Ortner (2000), and others (Figure 10.1). During the 1979 UNU reconnaissance, considerable numbers of porters carrying heavy timbers toward Namche were noted. Similarly, firewood was being carried, not only to the Mount Everest View luxury hotel above Namche, but also for use in the new trekking lodges that were springing up throughout the area (Figure 10.2). Additionally, illegal cutting was occurring within the park, especially for firewood and construction timber. Nevertheless, this cutting, while likely to be damaging in the long term if continued unabated, had not been sufficient to cause even local *deforestation* (i.e. clear cutting), nor to affect the area's hydrological regime and cause accelerated soil erosion. In relation to the pre-1950 landscape changes it was insignificant.

Figure 10.1 The forests of Pharak, Solu Khumbu, immediately south of Sagarmatha (Mount Everest) National Park have been extensively logged. The timber is used for lodge construction inside the park. Nevertheless, the result is open woodland, not complete deforestation. Photograph, 1994, by Stan Stevens, first published as Plate 8 in *Geographical Journal*, 169(3): 273.

Figure 10.2 Sherpani carrying firewood for use in the Mount Everest View Hotel above Namche. The source of the firewood is the forests within the national park boundary, April 1979.

Lake Sarez, Pamir Mountains: prediction of a flood of 'biblical proportions'

Lake Sarez began to accumulate behind a massive earthquake-induced landslide in 1911. By 1998 the upper course of the Murghab River had formed a lake 62 km long and with a volume of about half that of Lake Geneva (see Chapter 6 p. 134). Soviet scientists had been monitoring the lake for several decades but with the collapse of the Soviet Union observations had ceased. Understandably, the government of the newly independent Tajikistan began to express its concern about the possibility of the dam collapsing, leading to catastrophic drainage of the lake. Since the lake surface stands at 3,200 m asl and the landslide dam is more than 500 m high, it was eminently reasonable to examine the prospects for a 'worst-case scenario' evaluation. Based on research by staff of the United States Geological Survey on landslides, mudflows, and the dangers of landslide dams (Schuster 1995) the United States Army Corps of Engineers produced a computer simulation. This predicted that if total failure of the dam were to occur (by any measure, a worst case) then the impacts would be profound. According to the computer simulation, any total lake outburst would produce a very high speed (100s km/hr) mudflow, varying with the topography of the valley below and the availability of loose slope material, and would eventually extend over 2,000 kilometres to the Aral Sea. Five million lives would be at risk in four different Central Asian countries, together with untold destruction of property. Nevertheless, it is emphasized that this was a computer simulated model of the *worst case scenario* of the type that is frequently set up in such circumstances to provide a basis for field test and not a vehicle for public alarm.

At the urgent request of the Government of Tajikistan, the UN International Strategy for Disaster Reduction (ISDR), based in Geneva, and the World Bank formed a team of experts to investigate the actual nature of the Lake Sarez hazard. With close support, including scientific and military personnel, from the Government of Tajikistan, the team of geophysicists, engineers, geologists, and geographers examined all aspects of the hazard during June 1999 (Alford and Schuster 2000; Alford *et al.* 2000). In brief, the unanimous conclusion was that the worst-case scenario was such a remote possibility that it could be discounted. Nevertheless, because the mountain slopes above the lake were highly unstable, and also subject to frequent earthquakes, there were inherent secondary hazards. The most likely event, although there was insufficient data available for real-time prediction, would be a large rockfall/landslide hitting the lake surface and generating a seiche wave to over-top the dam. This, in turn, would splash down the steep outer slope of the dam into the Bartang Gorge and imperil the 32 villages that are strung along the floor of the gorge for more than 120 kilometres as far as the confluence with the Pianj River. In view of this, recommendations were made for the installation of fully automatic lake-level monitoring, slope stability monitoring, and advanced warning systems. In addition, a series of 'safe havens' were proposed, to be located above estimated flood levels and stocked with food and supplies for use in an emergency. Installation is proceeding at time of this writing (August 2003).

So far only verifiable facts have been introduced. However, knowledge of the perceived hazard constituted by Lake Sarez was sufficiently widely known that the UN/World Bank team of experts organized a press conference on their return to Geneva. More than 20 eminent news media were represented. Pains were taken to diffuse the relevance of the worst-case scenario; in fact all team members who made presentations emphasized that discussion of such a disaster could be dismissed as wild speculation, if not irresponsible. The facts, as reiterated above, were set forward together with a plea for consideration of the Mountain Tajiks living in the Bartang Gorge who already had to contend with a great range of 'normal' natural hazards and, in any event, needed food relief support from the Aga Khan Rural Support Programme to survive there.

It was unfortunate, therefore, that two inflammatory reports appeared (Pearce, *New Scientist*, 19 June 1999; Burke, *The Observer*, 20 June 1999) prior to the Geneva press conference. Each article cited as its main source Scott Weber of the 'UN Department for Humanitarian Affairs' and 'who organized the expedition' [to survey the degree of hazard posed by Lake Sarez]. Some of the more inflammatory phrases include: 'Scott Weber said . . . they [the research team] had found an enormous disaster waiting to happen.'; 'Five million people could die.'; 'When the natural dam which holds back the water breaks – which experts say could be at any moment – a wave as high as a tower block will blast a trail of destruction a thousand miles through the deserts and plains once crossed by the fabled Silk Road and now covered in farms, fields and cities.'; 'we don't know when it could go, but it could go at any time'. Many details were added to include information on the high seismicity of the region, the recent civil war in Tajikistan, and problems of establishing an early warning system. In general, all the news media who were represented at the Geneva press conference reflected the calm assessment of the Lake Sarez team. To underline the exaggerated nature of the reports published by *The Observer* and *The New Scientist* the response obtained from an interview (aided by local interpretation) with an elderly widow is reproduced. Her home is located close to the junction of the Bartang and Pianj rivers. When asked to what extent she feared the possibility of a flood from Lake Sarez, she replied:

My parents were living in this house when the 1911 earthquake and landslide occurred and I was born here in 1932. Neither they nor I worried about Lake Sarez. I intend to stay here until I die. If Allah decides that the dam will burst, so be it; but I don't think he will.

After the mission report was presented, the Government of Tajikistan accepted the recommendations and plans went ahead for design and installation of the monitoring and warning systems. All seemed calm. Then, in early April 2003, an alarm was sounded on a Russian website (www.strog.ru):

> In Central Asia an accident on a planetary scale is expected. . . . Today, Uzbek scientists have deciphered space images from the Japanese film-making system *Aster* using the satellite *Terra*. They discovered that Lake Sarez has over-topped the dam that is now being destroyed as if cut by a giant circular saw.

The ensuing prediction referred to a 100 metre-high mudflow destroying cities for 2,000 kilometres downstream to the Aral Sea with 600,000 to five million lives lost (translated loosely from the Russian by the United States Embassy in Dushanbe). Tense reaction reverberated throughout Central Asia and all the way to Washington DC, as well as to members of the 1999 evaluation team. Sober, authoritative responses calmed the possibility of panic, although the post-1999 Lake Sarez Risk Mitigation Project planned to send a reconnaissance mission to the lake. No recent information has appeared and the very absence of news certifies that there has been no flood 'of biblical proportions' with the loss of millions of lives, and that the April alarm was false.

There is need here for a pause to reflect on the possible events that might have occurred had the 1999 evaluation mission to Lake Sarez not made a responsible assessment. One of the serious risks envisaged at that time was the prospect of governmental over-reaction to the hazard that could prompt a forced, and unnecessary, evacuation of the 32 small villages along the Bartang Gorge together with all the hardship that would entail, even to the collapse of an important, if poverty-stricken mountain culture (Alford and Schuster 2000: 83–90).

A final anecdote from 1988

This series of anecdotes and commentaries intended to illuminate the regrettable misunderstandings created by the manner in which the Himalayan–Ganges Problem has been reported is brought full circle by returning to the coverage of the serious 1987 and 1988 floods in Bangladesh. Piers Blaikie (pers. comm. 24 June 2003) recalled his interview with the BBC in preparation for the Nine O'clock News programme. When he expressed his conviction that the Theory of Himalayan Environmental Degradation had no factual basis, this caused the interviewer's face to fall. She responded, 'Oh, but I have already had all the upstream/downstream diagrams prepared.' Thus, when the actual news was broadcast the accompanying cartoons showed hectares of trees felled and rising flood waters. All mention of Blaikie's explanation of the socio-economic management of the floods and the lack of any relationship between deforestation in the Himalaya and flooding downstream had been eliminated. He relates that the TV image of his face was seen to jump a little where the section of the film track that explained his opposition to the Theory had been edited out.

Conclusion

The aim of this chapter has been to highlight the misrepresentation and exaggeration that have been perpetrated for decades and are still being generated today. It is firmly believed that such misrepresentation inhibits urgently required definition of some of the many problems that do beset the region. The single biggest obstruction that dominated the development of thought during the 1970s and 1980s was the widespread assumption that linked increase in mountain rural populations with massive deforestation, soil erosion, and damaging downstream consequences. Some of the real underlying problems that have persisted for decades have been exacerbated by lack of adequate attention or by attempts to solve perceived problems that did not exist, or were of less importance. Although Thompson *et al.* (1986) expressed doubt that the 'uncertainty' could be dispelled and thus should be accepted as part of the Himalayan scene, it is believed that an attempt should be made to reduce the level of uncertainty as far as possible. Hence the need to ask how the misunderstandings arose and why they have been carried into the present century when, at the same time, the academic perceptions have changed significantly.

This discussion is not intended to minimize the profound complexity of the greater Himalayan region and of its many problems. It would be a disservice to imply that deforestation is not occurring in some specific areas, or that soil depletion and landsliding are unimportant. But these considerations should not be exaggerated and generalized to characterize the entire region, nor should they be articulated to a single simplistic and unsubstantiated cause. This only serves to deflect attention from the extent of poverty, mistreatment of poor minority peoples, and the cruel and self-destructive violent conflicts that are engulfing large parts of the region and so may forestall any attempt at resolution. Nor is it the intention to blame the news media for a large share of the misinformation. Although many elements of the news media are certainly culpable, it is bilateral aid agencies, United Nations institutions, governments, NGOs, and non-rigorous scholars that frequently have failed to show real determination to separate cause and effect, whether intentionally or not.

11 Redefining the dilemma

Is there a way out?

A problem ignored is a crisis invited.
(Henry Kissinger)

The last 15 years have witnessed the rapid and accelerating penetration of the entire Himalayan region by modern technology, including television, electronic mail, and the Internet. This has been reinforced by the equally rapid expansion of surface and air communications and almost unrestricted access by trekkers and tourists. Many previously isolated villages with little knowledge of day-to-day events in the outside world have acquired instant access to happenings in London, Washington, Tokyo, and Beijing, as well as in their own capital cities. Furthermore, an increasing number of mountain dwellers are travelling abroad in search of wage employment or higher education. Many return home, adding to the flood of new ideas and knowledge and bringing money with them. Others join growing ethnic communities in the industrialized countries and send home remittances and opinions and initiate a two-way exchange. In parts of the region, terrorism and warfare have produced an influx of foreign military personnel and journalists that has opened the door to the outside world still further. The result has been the generation of change on an unprecedented scale; it overwhelms the earlier inroads made by foreign aid and tourism that began to expand during the third quarter of the twentieth century.

Such a deluge of information and new experience must be seen as a catalyst for further acceleration of the process of change, whether or not this be inherently advantageous. As Frederick Starr (Chapter 8) remarked, the increase in contact has brought HIV to hitherto remote villages but not the medical assistance needed for its relief.

In light of these circumstances, any attempt to forecast how the region will respond to these influences, even in the near-future, would be foolhardy in the extreme, yet the process of change must be regarded as a primary force within the Himalayan region.

During the 1970s and 1980s most of the national governments, donor agencies, and many scholars appeared to hold a collective conviction that environmental degradation constituted the foremost threat to stability. The simplistic view was that rapid growth of rural populations in the mountains was inducing widespread deforestation that led to soil erosion and land-sliding and to the acceleration of downstream flooding and siltation. It was a development paradigm that confused cause and effect, resulted in the misdirection of large financial resources, and sidelined some of the real needs of the people. Attempts were made to solve a problem that did not exist, or at least, one that had been exaggerated beyond measure. Thus, 'develop-ment' was distorted and the identification and prioritization of circumstances that demanded attention was delayed. Malla and Griffin (2000) argue, for instance, that the impact of the environmental degradation theory has negatively influenced forestry policy in Nepal to the turn of the century. This they characterize as a fixation with trying to ensure protection of community forests rather than assisting the rural poor to use forest and tree products to their advantage as the changing market situation allows.

Foreign aid has been injected into parts of the Himalayan region for a half century. Regional governments have promoted 'development' of their own mountain regions at considerable expense, either independently or in conjunction with multinational, bilateral, and non-governmental organizations. There has also been a significant input of volunteer assistance from charitable foundations, ranging from Sir Edmund Hillary's Himalayan Trust to the World Conservation Union, the World Wildlife Fund, and the Aga Khan Foundation. To this is added a large body of independent scholarly and applied research that has attempted to analyse problems and evaluate development successes and failures. Yet, despite notable exceptions, looking back one senses failure, or at least lack of significant success. If 'development' induced by international and bilateral aid is taken to mean the consistent improvement in the well-being of the peoples of the region, then failure, or the lack of success, are unavoidable epithets. It was some decades ago that *The Times* of London published a frequently quoted definition of foreign aid that can be paraphrased as 'a system whereby the poor people of the rich countries transfer money to the rich people of the poor countries'. As cynical as this criticism may appear, unfortunately it contains an element of reality.

There is also the enigma of defining 'development', more currently 'sustainable development'. In many instances the 'donor' has attempted to impose a fabricated development project on a country or region that is unprepared or unable to absorb it. The major infrastructure projects, such as the large hydro-electricity facilities that the 'target' community has difficulty in maintaining once outside support is withdrawn, are an example. Worse, if financing such a project requires the recipient to take on a large loan, the local economy may be impaired and the profits will flow to the international banking community and expatriate consultants. Dipak Gyawali (2003) has illustrated this part of the persisting dilemma in his autopsy of the infamous Arun III hydro-electricity project (Chapter 10); his analysis is highly relevant in the present context.

The Tehri Dam affair, discussed at some length in Chapters 6 and 8, has become a *cause célèbre* with underpinnings of misinformation, mistrust, secrecy, and questionable disbursement of funds; above all it is a project that aims to develop an important mountain resource for the benefit of lowland vested interests at the expense of the mountain people and the mountain environment.

Likewise, the Rogoun Dam on the Vakhsh River in Tajikistan was initiated by the former Soviet Union for the benefit of the central authority with the planned inundation of about 40 Tajik villages. Although construction was truncated by the Tajik civil war, the project exemplifies development of an important resource with total disregard of the local population. Similarly the 'gift' to Bangladesh from the World Bank and USAID of the Kaptai Dam in the Chittagong Hill Tracts precipitated disaffection, unrest, and eventually 20 years of conflict, extensive loss of life, as well as environmental damage (see Chapter 8).

It appears that senior bureaucrats of the central governments of the 'target' countries encourage mega-projects as prestige undertakings and, not infrequently, donor agencies proffer significant personal advantages to key local personnel to ensure that the wheels of governmental approval turn smoothly. Furthermore, the donor countries influence preparation of the contracts so that their home industrial corporations supply the expensive equipment, the 'necessary' highly paid consultants, executive staff, and technicians; qualified local professionals are often excluded.

In contrast, successful small voluntary and carefully focused projects are those that first ask the local people what they most urgently need. Frequently the local response centres on education and health. Two well respected organizations exemplify this approach. The Aga Khan Rural Support Programme in northern Pakistan has been able to encourage the establishment and improvement of local schools, hospitals, and first aid posts and to support modest but vital initiatives, such as establishing additional irrigated fields in small villages. Yet even with these successes there may be unintended negative impacts on the position of village women, as reported by Felmy (1996) and Azar-Hewitt (1999). Much of the success of the Himalayan Trust in the Khumbu again lies in hospitals, first aid facilities, and schools. It seems that small projects that facilitate the participation of the local people hold a strong potential for success. The very many improvements on this scale serve a form of counter-balance to the frequently inappropriate large-scale interventions. Nevertheless, the large-scale infrastructural programme that has been exceptionally beneficial is the expanding road network; it has been a major factor in the introduction of change throughout the region.

The title of this book, *Himalayan Perceptions*, is intended to raise questions. From the initiation of the 'Mohonk Process' it has been apparent that widely differing perceptions about the region have co-existed. When uncertainty overwhelms the approach to problem definition (Thompson and Warburton 1985), how can there be agreement on feasible and appropriate measures? Twenty years ago Michael Thompson and David Griffin each identified several elements of the dilemma. In particular, Thompson stressed the need for a total change in attitudes toward foreign 'aid' – from the perception that it was 'charity' to an understanding that it should be restructured as one-half of 'gift exchange' because the poor mountain dwellers, through their extensive environmental knowledge and traditions, have much to offer the people and institutions of the industrialized countries.

Griffin (1987, 1989) noted two important considerations. One is that those who come from afar, intent to improve the livelihood of mountain peoples, should enter an area with a 'clean slate' and first seek to understand the functioning of local institutions. Only then should they work with the local communities to identify projects that might be amenable to co-operative solution. Griffin's second point was the need to help ensure that the relevant departments of the local and national government were staffed with appropriately trained and educated professionals.

It is extremely risky for anyone to attempt to define what is going wrong in the Himalayan region, let alone to suggest how to take corrective action. Nevertheless, one of the conundrums that has faced mountain regions in general has been widespread lack of understanding and appreciation of the importance of mountains and of their complex relationships with the lowlands. Much progress has been made over the last 15 years, culminating in the greatly increased awareness promoted by the International Year of Mountains (2002). The winds of change seem to have blown into the corridors of many of the big agencies. James Wolfensohn's cancellation of the Arun III project, shortly after becoming President of the World Bank, is a good example. But more change is needed, and quickly, if Dipak Gyawali's (2003) prediction is correct and the end of foreign aid is fast approaching. The first essential requirement is to summarize the status of knowledge resulting from the change in perceptions since 1989; the second is to identify the assets and challenges of the region in such a way that all the major decision-makers, including national governments, reach a measure of agreement.

Surely the old paradigm of poverty and rapid mountain population growth leading to deforestation and environmental collapse has been laid to rest. In this case the basis for the imposition of large-scale logging bans, in contrast to effective forest management, needs to be fully explored. The overriding issue, while seemingly related to population pressure and environmental degradation, is the apparent inability to alleviate poverty, in some instances because of outright disregard of the mountain minority peoples, in others because of discrimination and oppression. This has certainly resulted in the spread of disenchantment and unrest, sometimes leading to armed insurrection.

There is also the need to recognize the dangers of the various forms of conflict across the region. In Kashmir, the state of armed neutrality, cross-border terrorism, occasional exchange of artillery fire, and continual active warfare on the Siachen Glacier between the two nuclear powers of the region, has not only brought about immense direct losses on both sides, but has inhibited the regional co-operation so urgently needed. The conflict also has consumed large amounts of money and resources that otherwise could have been applied constructively.[1]

In addition to the major conflicts, numerous relatively minor instances of deliberate or unintentional discrimination against ethnic and religious minorities characterize much of the region. At the very local level, as exemplified by the research of Hoon and Chatravarty-Kaul (see Chapter 8), Bakrwals, Gaddis, and Bhotiya transhumance peoples are marginalized quite unnecessarily, just as local initiatives for self-improvement by mountain minorities in northwestern Yunnan are undercut by ill-conceived government-sponsored development.

At the country level there is the unconscionable treatment of the Lhotsampa refugees by the Government of Bhutan and the insurgency spreading across Nepal, at least in part because of government ineptitude, political in-fighting, and corruption. These create dangerous flash-points that could engulf a much larger region.

It would appear that many of the real problems have been ignored for decades. One suggestion is that, instead of mega-projects, such as the Arun III or the Chisapani hydro-electricity projects, an equal level of funding should be invested in assisting Nepal to establish a social security system, or clean drinking water and adequate sewage disposal systems that are still lacking in many of the towns and cities, despite the vast amounts spent on development projects. Adequate provision for the health and welfare of older people might begin to alleviate the continuing trend of rapid population growth.

The growing poverty and lack of self-sufficiency in food, as well as many of the local crises, have evolved to the point that they present a danger to the entire region. There are no obvious solutions. Nevertheless, the thoughts that follow might prove useful for any region-wide attempt to grapple with the prospects for sustainable mountain development.

The unprecedented change that is engulfing the region, with both positive and negative aspects, needs to be carefully analysed. In particular, the elements within the process of change, such as the way many people are adapting to market opportunities advantageously, must be identified and aided by the development of more constructive government policies.

Undoubtedly, environmental degradation is characteristic of much of the greater Himalayan region. Yet there are many areas where environmental gains have been achieved. What is now needed is a concerted effort to effect a precise assessment of the environmental status, region by region. An important step has been taken by Zurick and Karan (1999). Their book *Himalaya: Life on the Edge of the World* represents an important first attempt to compile and analyse a large databank, collected on an area-by-area, valley-by-valley basis, and to provide an assessment of the Himalaya proper in a disaggregated format. This should greatly help to open the curtains of generalization that have led to so much confusion and misunderstanding in the past. The approach needs to be pursued further.

Access to resources must be fair and equitable. The worldwide intercourse prompted by the Mountain Forum as part of the International Year of Mountains (2002) has provided many suggestions, some of which are especially applicable to the Himalayan region. Above all, ways must be found to ensure that the considerable resources of the mountains are not developed mainly for the vested interests on the neighbouring lowlands or on the other side of the globe. A multinational and interdisciplinary effort should be undertaken to achieve consensus in identifying and prioritizing the many specific options available and to address the difficulties that face the Himalayan region at large.

One proposal is that a 'Himalayan Summit' be organized with participation from all regional sectors, from governments to universities, to citizens' committees and NGOs. It should also include representation from relevant institutions and individuals from outside the region. Such a summit, to reduce the possibilities of political tensions, could be modelled on the Rio de Janeiro Earth Summit (UNCED), although on a correspondingly smaller scale. A respected umbrella organization would be needed.

In conclusion, I believe that there may be a way out of the encroaching crisis, although it will be long and arduous; sustained stability and prosperity for all may be a utopian dream. Yet, for millennia the Himalaya have been a source of inspiration and they are part of the world's priceless natural and cultural heritage. The security of a very large proportion of humankind may be determined on how the resources of the Himalaya are managed.

Epilogue

The need for an 'epilogue' was anticipated during final preparation of the manuscript for this book because of the rapidly changing political scene in many sections of the Himalayan region during 2003. This remark is especially relevant to the situation in Kashmir, including the Siachen Glacier, northern Pakistan, Nepal, Bhutan, and the Indian province of Nagaland. The dispute raging over completion of the Tehri Dam (Chapter 8, pp. 192–4) was also in a state of flux in November 2003, and the stability of Afghanistan remains uncertain. Nevertheless, recent information and statements of opinion on only a selection of these areas have been accumulating since November 2003 when the manuscript was submitted to my Routledge editor. These have been acquired mainly by e-mail correspondence with friends and colleagues within the region, and also by scanning of the news media. What is presented below, therefore, is not based on carefully undertaken research; it lies in the grey area between documented scholarship and journalism. It should be read as such, particularly in light of my criticism of the news media in sections of the main body of the book. Nevertheless, this epilogue is appended because I believe that, although much is unsubstantiated, the view-points expressed may assist the reader to obtain additional insights into the complexity of problems facing the Himalayan region. Special acknowledgement is made to the following friends and colleagues for their assistance, although I accept responsibility for the overall presentation:

Dr Dinesh N.S. Dhakal, Lhotsampa refugee camp leader,
Dr Dipak Gyawali, former Minister of Water Resources, H.M.G. Nepal,
Mr Agha Iqrar Haroon, President, Eco-tourism Society of Pakistan,
Mr Luingam Luithui, Human Rights Activist on behalf of the Naga People,
Dr Seth Sicroff, Ithaca, New York State.

Drs Dhakal and Gyawali are also alumni of the Mohonk Conference of May, 1986.

Siachen glacier and Kashmir

Reference was made to the November 2003 Indo-Pakistan cease-fire in Kashmir in an end note to Chapter 8 (p.235). The cease-fire was met with widespread international enthusiasm and a resurgence of interest in prospects for resolving the Siachen Glacier impasse by creation of a 'peace park'. Despite the initial burst of optimism and the fortunate maintenance of the cease-fire, no further clear-cut progress appears to have been made. Regardless of the occasional terrorist atrocity in Kashmir, however, planning for provincial elections remains on track.

Northern Pakistan and tourism

The collapse of tourism in the Pakistan Karakorum and Himalaya that occurred in 2002 has persisted, despite personal reports on the comparative safety for travellers. The Pakistan Ministry of Tourism halved the summit fee for all the high mountaineering peaks in an effort to induce a continued growth in mountaineering visits, although this seems to have had little impact. The 50-year anniversary celebrations to commemorate the first ascent of K-2 have been twice postponed.

Nagalim

Mr Luingam Luithui, Nagalim activist, after attending consultations in New Delhi in February 2003 (Chapter 8, p. 178), returned to Ottawa and provided a detailed account of the history and post-1947 independence struggle of his people. In particular, he emphasized that the basis of Nagalim aspirations for independence was predicated on peaceful negotiations by this predominantly Christian nation. These hopes had been subverted by the Indian military aggression that had enforced an armed struggle for ethnic survival. The cease-fire has now been in effect for more than six years although negotiations have not yet produced a resolution acceptable to both sides.

Bhutan and the Lhotsampa refugees

The fifteenth round of bi-lateral talks between Nepal and Bhutan was held in Thimphu 23–25 October 2003. It was agreed on this occasion that all willing refugees who were declared eligible would be repatriated, the first batch to be allowed into Bhutan on 15 February 2004. The Joint Verification Team visited the Khundunabari camp in late-December 2003 to explain the results of the verification process and the conditions of repatriation. The leader of the Bhutan delegation, Dr Sonam Tshering, briefed the refugees harshly, using words highly offensive to Hindus. In particular, he stipulated that although citizenship rights would be restored to those confirmed in category 1 (see Chapter 8, pp. 178–83) they would not be allowed to reclaim their old homes or land. Refugees in category 2, while permitted to return, would have to wait from two to twenty years before full citizenship rights were restored; until such time they would be confined to prison camps and only one member of each family would be allowed to work as a construction labourer to support his family. Refugees in category 3, determined to have no claims to Bhutanese citizenship, would be excluded from further consideration; no appeal process would be allowed. Those in category 4, classed as criminals, would be escorted back to Bhutan under police escort and subjected to the Bhutan process of law. No further discussion or appeal would be tolerated.

Dr Tshering's provocative statements instigated an episode of stone throwing by some of the refugees. Several Bhutanese and Nepalese members of the verification team were injured. The Bhutanese delegation departed for Thimphu the following day and did not inform the Nepalese authorities of their departure.

Since December 2003 the negotiations have remained stalled. Small groups in the refugee camps have been calling for an armed struggle against Bhutan and it is feared that the refugee leaders who continue to urge adherence to a peaceful process will soon find themselves in the minority. UNHCR has alerted the Nepalese government to help control the attempts to provoke military action by the refugees.

The Sikkimese news media reported the Bhutanese Foreign Minister as stating that the Lhotsampa still residing in Bhutan represented 30.8 per cent of the country's total population. If the refugees were to be added to this the combined groups would exceed 46 per cent of the population. Such a situation would not be tolerated by the Government of Bhutan.

Dr Dhakal believes that, unless the international community, including India, can be persuaded to share the burden of the refugees, the chances of finding a durable solution in the near future, is slight.

Maoist insurrection and the current situation in Nepal

As one of the few countries of Asia never directly colonized, Nepal presents an interesting case of a poor country coming to terms with modernization and development. This process has not been easy or painless as the current turmoil indicates. The root of the problem is considered by many to lie, not only in the violent Maoist Insurgency, but also in the failure of the current framework of governance and model of development. Indeed, it has been said in discussions leading to the meeting of Nepal's *Development Forum*, that Nepal is NOT a failed state: rather it is country where inappropriate development paradigms and projects based on them have failed. This reflects on the failure of mainstream political parties in Nepal – at the helm of affairs since the restoration of multi-party democracy in 1990 – to articulate an alternative grassroots vision of state policy and development. This has been compounded by the massacre of many members of the Royal Family.

Maoist Insurgency

The joblessness of Nepali youth and rural poverty have been indicated by many as the primary cause of the Maoist Insurgency. The heartland of the Maoist revolt in mid-west and Western Nepal is one of the country's poorest regions with very high permanent and seasonal out-migration. Since the 1970s, the area's traditional livelihoods, such as Gurkha service and animal husbandry, have been declining. Sale of hashish had been the major source of income for many families and provided what they perceived to be a good standard of living. In 1976 the government declared the sale of hashish illegal. This was the first interaction for most of the people with the hitherto remote Kathmandu government, and it reduced that relative prosperity to abject poverty. Imposition of the ban was perceived as arbitrary and no effort was made to replace the loss of income, or even to provide a road that would facilitate the marketing of alternative crops. The action generated enduring bitterness. The seeds had been sown, not because of some deep-rooted ideology, but as a result of poverty and indifference on the part of the government.

In reality, a low-level insurgency in Nepal has existed for several decades but it burst into the open only with the declaration of the 'people's war' on 13 February 1996 by the Communist Party (Maoist), or CPN(M), the most radical offshoot of the left-wing spectrum in Nepali politics. Desultory action ended when the 'Maoists', as they are universally termed, unilaterally abrogated talks with the government and launched a nationwide general offensive in November 2001. The United States, the United Kingdom, India, and China have played an important role in assisting the Nepalese government to counter this violence by ruthless radical actors whose pronouncements often seem to change by the expediency of the moment.

Maoists became a serious force within five years of their declaration of the 'People's War' because successive Nepalese governments took no action to deal with them, and the Indian government seemed to allow them to operate almost openly on Indian soil.

Nepal and India share an open common border of about 1,600 kilometres. The Indian attitude toward the Maoists was seen in Nepal as ambivalent. Nevertheless, it appears to have hardened recently as they have been found to be co-operating with similar Maoist groups in India (PWG, Naxalite, Bodo, Kamtapura). Indeed, it was the attempted assassination of an Indian Chief Minister, Chandra Babu Naidu, that prompted the Indian government to take more serious action as, presumably, the Maoists and their activities came to be seen as opposed to the larger Indian interests. Some top Nepalese Maoist leaders, for instance, have been arrested in India. Recently, there has also been a measure of co-ordination amongst United States, Indian, and Chinese responses to the insurgency problem.

Impotence of Political Parties

In mid-2002, the tenure of Nepal's elected officials was about to lapse. The conflict with the Maoists had reached its highest intensity and a State of Emergency had been declared. Clearly, new elections were not feasible in many places across the country. Nonetheless, in May 2002, the Nepalese Congress Party's Prime Minister, Mr Sher Bahadur Deuba, dissolved the national Parliament in anticipation of elections. Seven weeks later, in July, with the conflict still raging, he also dissolved all local elected bodies, leaving not a single elected representative in office throughout the country. It immediately became apparent, predictably, that elections could not be held.

The dissolution of Parliament in mid-2002, together with that of all the local elected officials that link rural villages with the government, was a colossal blunder. It provided the Maoists with a prize that they had sought from the outset: to empty the rural areas of local elected leaders who opposed them and to deflate the government's presence at a moment when it was most critically needed.

An indication, in turn, of Maoist political weakness, is that they were unable to take full advantage of the leadership vacuum that dissolution had created. They were not able to capitalize on this mistake because they are not generally perceived to offer a better alternative to the autocratic feudal system which the people believed the post-1990 democratic system was replacing. Rather, the insurgency's own autocratic, single-party approach and its wide-spread use of intimidation are more comparable with, or in some cases worse than, the feudal system under which these areas had laboured for centuries.

The most abused elements in the present situation appear to be the political parties. Indicted by the King, aid donors, foreign governments, diplomats, the common people, and academicians for their irresponsible behaviour, frequent changes in government, bad gover-nance, and corruption, the political leaders and their parties have been called upon to accept responsibility for the present difficulties. Unfortunately, the various governments of Nepal *had* failed to deal with the changes in the political and socio-economic climate in the face of weakening traditional institutions. They sought to cultivate their own narrow political gains, sacrificing ideals and principles. The outcome has been discontent in society, Maoist violence, and non-violent protests from others.

The stalemate continues. Even though the political parties united in criticizing the King's action, they were not able to produce any consensual strategy. The Nepal Congress (establishment) wanted the King to recall the Parliament; the main opposition Communist Party of Nepal – United Marxist/Leninist (CPN-UML) – wanted a government of national consensus; and former Prime Minister Deuba's Nepal Congress (rebels) described the King's move as unconstitutional and clamoured for re-instating Mr Deuba as Prime Minister with his old cabinet. Most parties think that the political environment is not conducive to holding elections by April 2005, as proposed by the King, yet they have not suggested an alternative. Changing Prime Ministers in an interim government is not the solution. Furthermore, it has been argued that the main reason that parties do not want to decide in favour of elections has little to do with the Maoist threat: if elections can be held in Kashmir and in Sri Lanka, they can also be held in Nepal. The main reason seems to be the political vacuum in the villages where the Nepal Congress Party and the Communist UML do not have much of a presence. This is mainly because the corruption in which they engaged during their tenure in office (12 governments in 12 years) has made it impossible for their cadres to face the people.

In order to counter the monarchy and in support of democracy, most of the political parties have made a concerted effort to establish a united front. However, they do not have any common strategy to continue with their agitation. The political leaders distrust each other and carry personal vendettas. The parties are still based on personality rather than on principles or ideology and are bereft of democratic culture. Hence, the people have no motivation to trust and support them.

At present there is virtually no mass support for the movement started by the political parties. The people, for the most part, have refused to join in the demonstrations, in contrast to 1990. Support is largely restricted to the party cadres, affiliated organizations, such as student unions and politically aligned professional unions (for example, teachers' unions), and hired demonstrators. The last group command a price of Rs.125 to 300 per day, more for vehicle tyre-burners if they bring their own tyres. There are a few genuine demonstrators, but hardly countrywide. In other words it is NOT another Bolshevik October Revolution. The people have chosen to stay aloof, in part, because of the presence of well-known corrupt faces in the street demonstrations that inspire a degree of disgust.

The future

It does not appear that the Maoists have the political or military competence to prevail. This is particularly striking because they were virtually unchallenged militarily for the first six years of the insurgency, and since mid-2002 the entire democratically-elected political apparatus of the country has been dismantled creating a destabilizing political vacuum. Despite the Army's increasing effectiveness, the conflict will not likely be resolved by military means alone. A political track, including the re-empowerment of local elected officials and the seizing of any opportunity for serious negotiations, would likely be crucial elements.

Neither India nor, ironically, China would likely approve of a Maoist state in their neighbourhood. It would be impossible for Nepal to go against the wishes of these two countries and if they are opposed to a Maoist republic it will not come about.

The bargaining between the King and the political parties may continue for some weeks but it must be finalized in the near future. There is extensive international and internal pressure on both the King and political parties to resolve their differences and go on to solve the Maoist problem. After all, the differences between the King and the political parties are minor in nature: who should be the Prime Minister; when should new elections be held; and how involved should the King be in governing the country? Both sides have publicly and repeatedly expressed their support for multi-party constitutional monarchy.

A widely appreciated commentary by Ajaya Dixit at the civil society consultative meeting before the aid donors' Nepal Development Forum 2004 was published by Yugasambad Weekly, Kathmandu, on 4 May 2004. The following is a paraphrase of parts of that account:

Contrary to popular press perception, it was not the King who dissolved the Parliament: it was the Nepal Congress Prime Minister who feared that his rival, Mr Koirala, was going to defeat his government. Thus he hastily recommended the dissolution of parliament. The King, according to the strictures of the Westminster model constitution, was obliged to announce the dissolution and to recommend holding elections within six months. After six months, when the then Prime Minister admitted he could not hold the elections and requested the King to extend his term for another 14 months, a constitutional crisis was precipitated. The King was obliged, on 4 October 2002, to invoke Article 127 of the constitution that empowers him to take measures to resolve such a deadlock.

Nepal's army, if suppression of the Khampa uprising in the 1960s is excepted, had not been in action in Nepal since the Anglo-Nepal war of 1815–1816. The Maoist Insurgency has transformed a largely ceremonial army into a fighting force and, perhaps ominously but realistically, into a political force for the future, since it has now doubled its strength and is the most organized structure in the country.

The Chinese government, when referring to the Nepalese Maoists, consistently has refused to use the term 'Maoist', preferring, instead, to call them 'anti-government forces', or even bandits.

There have also been many news media reports of atrocities and violence. Amnesty International has pointed to the deteriorating human rights situation at the 60th session of the UN Commission on Human Rights held in Geneva in mid-March, 2004. In Nepal itself, the Bhote Koshi hydro-electricity facility was temporarily shut down in October 2003, apparently because of Maoists threats. The Royal Nepal Army had been guarding the plant for the previous two years, although security was poor at the dam site three kilometres distant. Public television in Canada and the USA, for instance, occasionally shows scenes of mass protests in Kathmandu, accompanied with burning car tyres and apparently brutal police and army responses. However, there are also many sources that insist the seriousness of the situation, while real, has been greatly exaggerated.[1]

Conclusion

The final word must be that the overall situation across the entire Himalayan region is fluid and essentially seriously troubled. At least the writing of this book has been an attempt to identify some of the major problems facing the region and to place other incorrectly perceived or exaggerated problems in better perspective.

Ottawa, 7 May 2004

Notes

1 The myth of Himalayan environmental degradation

1 The term was popularized by the famous English author and Nobel Laureate Rudyard Kipling.
2 Since 1990 over 12,000 Forest User Groups (FUGS) have been established in Nepal (Piers Blaikie pers. comm. 22 July 2003).

2 The Himalayan region: an overview

1 The Soviet names for the most prominent high mountains will most likely be changed in the near future. For instance, Peak Communism is now listed as Pik Imeni Ismail Samani (7,495 m) in the new edition of *The Times Atlas*, September 2003.

3 Status of forests in the Himalayan region

1 There were other bilateral forestry projects in Nepal that experienced a measure of success, although the Nepal–Australia Forestry Project is widely regarded as by far the most successful.

4 Geomorphology of agricultural landscapes

1 All data from the Kakani area relate to 1979–1983.

5 Flooding in Bangladesh: causes and perceptions of causes

1 The thickness of 'perhaps over 12 km' (Curry and Moore 1971) is probably an excessive estimate.

6 Mountain hazards

1 The Richter scale (a measure of surface-wave energy) is accurate below M=8; above that Mw, a measure of total energy, is preferred – Mercalli Intensity. This extends from M=3 (barely felt by people) to M=12 (total destruction). M=7.5 is the equivalent of a 30 megaton bomb (Roger Bilham pers. comm. 24 April 2003).
2 A 15 kw hydro-electricity plant began operating in early November 2003. It only serves the three-man monitoring station established to observe lake fluctuations and does not provide electricity for the Sherpa villages located more than 5 km downstream (Seth Sicroff pers. comm. 26 January 2004).

7 Tourism and its impacts

1 'Buddhist Pass' – Nangpa means 'inside-person' and is a term generally used by Tibetans to describe themselves.
2 This has recently been modified so that solo climbers need only pay US$25,000, while a party of four are charged US$56,000. The fee remains at US$70,000 for a party of seven. In addition, in its attempt to further encourage tourism, the Government of Nepal has 'opened' 100 new peaks to trekkers and climbers (American Alpine Club, April 2002).

3 The presence of military personnel must be qualified as a mixed blessing. The local community certainly found their presence reassuring throughout the first phase of the Maoist Insurgency, although they have also been responsible for a significant amount of environmental destruction (see Chapter 8).

4 'Tiger Leaping Gorge' is incorrectly translated from 'Hu Tiao Xia' which means, literally 'Tiger Leap Gorge'. This is not a gorge that leaps over tigers. Despite this purist protest the popularly-used term has been adopted here.

8 Conflict, tension, and the oppression of mountain peoples

1 It is to be hoped that the November 2003 formal Indo-Pakistan cease-fire in Kashmir will provide the long-desired opportunity for resolution of the Siachen Glacier crisis, if not for a permanent peace treaty between the two nuclear powers.

9 Prospect for future developments: assets and obstacles

1 Personal observation.

2 A VDC is the smallest political and administrative unit in Nepal, usually divided into nine wards. Each committee is governed by a council with elected members from each ward, together with a chair and vice-chair.

10 What are the facts? Misleading perceptions, misconceptions, and distortions

1 While there had been earlier warnings of perceived environmental degradation in Nepal (Kaith 1960), they had not entered the mainstream discourse. In addition, alarm had been expressed concerning the Himalaya and other Asian mountain areas within India, China, and Thailand.

11 Redefining the dilemma: is there a way out?

1 The November 2003 cease-fire between India and Pakistan has been referred to above.

Epilogue

1 Immediately after shipping the corrected page proofs to London, I received an email from Kathmandu explaining that Prime Minister S.B. Thapa had resigned on 7 May after eleven months in office. It is to be hoped that negotiations between the official political parties and the King can now go ahead as the first essential step in resolving Nepal's lengthy political impasse.

References

Adhikari, J. (1996) *The Beginnings of Agrarian Change: A Case Study in Central Nepal*, Kathmandu: TM Publications.

Ahmad, M. (ed.) (1989) *Floods in Bangladesh*, Dhaka: Community Development Library.

Ahmed, I., Dixit, A. and Nandy, A. (1999) 'Water, power and people: a South Asian manifesto on the politics and knowledge of water', *Water Nepal*, 7 (1): 113–21.

Ahmed, S. and Sahni, V. (1998) 'Frozen frontline', *Himal*, 11/12 December: 13–21.

Alford, D. (1992a) 'Streamflow and sediment transport from the mountain watersheds of the Chao Phraya basin, northern Thailand: a reconnaissance study', *Mountain Research and Development*, 12(3): 257–68.

—— (1992b) *Hydrological Aspects of the Himalayan Region*, Kathmandu, ICIMOD, Occasional Paper No. 18.

Alford, D. and Schuster, R.L. (eds) (2000) *Usoi Landslide Dam and Lake Sarez: An assessment of hazard and risk in the Pamir Mountains, Tajikistan*, New York and Geneva: United Nations.

Alford, D., Cunha, S.F. and Ives, J.D. (2000) 'Lake Sarez, Pamir Mountains, Tajikistan: mountain hazards and development assistance', *Mountain Research and Development*, 20(1): 20–3.

Ali, A. (2002) 'A Siachen Peace Park: The solution to a half-century of international conflict?', *Mountain Research and Development*, 22(4): 316–19.

Allan, N.J.R. (1987) 'Impact of Afghan refugees on the vegetation resources of Pakistan's Hindu Kush', *Mountain Research and Development*, 7(3): 200–4.

Aris, M. (1979) *Bhutan: The Early History of a Himalayan Kingdom*, Warminster, England: Aris and Phillips.

Asian Development Bank (1982) *Nepal Agricultural Sector Strategy Study*, 2 vols, Kathmandu: Asian Development Bank.

Awasthi, K.D., Sitaula, B.K., Singh, B.R. and Bajracharaya, R.M. (2002) 'Land-use change in two Nepalese watersheds: GIS and geomorphometric analysis', *Land Degradation and Development*, 13(6): 495–513.

Azhar-Hewitt, F. (1998) 'All paths lead to the spring: conviviality, the code of honour, and capitalism in a Karakorum village, Pakistan', *Mountain Research and Development*, 18(3): 265–72.

—— (1999) 'Women of the high pastures and the global economy: reflections on the impacts of modernization in the Hushe Valley of the Karakorum, northern Pakistan', *Mountain Research and Development*, 19(2): 141–51.

Badenkov, Y.P. (1992) 'Mountains of the former Soviet Union: value, diversity, uncertainty', in P.D. Stone (ed.), *The State of the World's Mountains: A Global Report*, London and New Jersey: Zed Books, pp. 257–99.

—— (1997) 'Violent conflicts in the mountains of the former Soviet Union', in B. Messerli and J.D. Ives (eds) *Mountains of the World: A Global Priority*, London and New York: Parthenon, pp. 113–14.

—— (1998) 'Mountain Tajikistan: a model of conflictory development', in I. Stellrecht (ed.) *Karakorum – Hindukush – Himalaya: Dynamics of Change*, Köln: Rüdiger Köppe Verlag, Part II: 187–206.

Bahadur, J. (2003) *Indian Himalayas: An integrated view*, New Delhi, C–24 Qutab Institutional Area: Virgin Prasar.

Bajracharya, D. (1983) 'Fuel, food, or forest? Dilemmas in a Nepali village', *World Development*, 11(12): 1057–74.

Bandyopadhyay, J., Jayal, N.D., Schoettli, U. and Singh, C. (1985) *India's Environment, Crises and Responses*, Dehra Dun, U.P., India: Natraj Publishers.

Bansal, R.C. and Mathur, H.N. (1976) 'Landslides – the nightmare of hill roads', *Soil Conservancy Digest*, 4(1): 76–7.

Banskota, M., Papola, T.S. and Richter, J. (eds) (2000) *Growth, Poverty Alleviation and Sustainable Resource Management in the Mountain Areas of South Asia*, Feldafing, Gemany: Deutsche Stiftung für internationale Entwicklung.

Barry, R.G. (1992) *Mountain Weather and Climate*, London and New York: Routledge.

Barsch, D. and Caine, N. (1984) 'The nature of mountain geomorphology', *Mountain Research and Development*, 4(4): 287–98.

Bartarya, S.K. and Valdiya, K.S. (1989) 'Landslides and erosion in the catchment of the Gaula River, Kumaun Lesser Himalaya, India', *Mountain Research und Development*, 9(4): 405–19.

Begley, S., Moreau, R. and Mazumdar, S. (1987) *Newsweek*, 9 November, 1987.

Benn, D.I. and Evans, D.J.A. (1998) *Glaciers and Glaciation*, London and New York: Arnold/Wiley.

Berkes, F., Davidson-Hunt, I. and Davidson-Hunt, K. (1998) 'Diversity of common property resource use and diversity of social interests in the western Indian Himalaya', *Mountain Research and Development*, 18(1): 19–33.

Bhutan, National Planning Commission (1992) *Bhutan: Towards Sustainable Development in a Unique Environment*, Official country report presented to the United Nations Conference on Environment and Development, Rio de Janeiro, Thimphu: National Environmental Secretariat.

Bilham, R. (2003) 'Earthquakes in India and the Himalaya: tectonics, geodesy and history', *Annals of Geophysics*, Special issue: proceedings of a workshop – Investigating Past Records of Earthquakes (Erice, Sicily, July 2002), in press.

Bishop. B.C. (1978) 'The changing geoecology of Karnali Zone, western Nepal Himalaya: a case of stress', *Arctic and Alpine Research*, 10(2): 531–43.

—— (1990) *Karnali Under Stress: Livelihood Strategies and Seasonal Rhythms in a Changing Nepal Himalaya*, Chicago: University of Chicago Press.

Bjønness, I.-M. (1980) 'Animal husbandry and grazing, a conservation and management problem in Sagarmatha (Mt Everest) National Park, Nepal', *Norsk Geografisk Tidsskrift*, 34: 59–76.

—— (1983) 'Kulekhani Hydro-Electric Project: Research working paper No. 3: Socio-economic analysis of the effects from the Kulekhani Hydro-Electric Project, Nepal', Oslo: University of Oslo, Department of Geography.

—— (1984) 'Kulekhani Hydro-Electric Project: Research working paper No.4: Strategies for survival – subsistence agriculture versus work at the Kulekhani Hydro-Electric Project, Nepal', Oslo: University of Oslo, Department of Geography.

Blaikie, P. (1995) 'Environmental management in the farming community' in R.A.M. Gardner and A. Jenkins (eds), Land Use, Soil Conservation and Water Resource Management in the Nepal Middle Hills,' unpublished report, UK Overseas Development Administration, 7.1–7.14.

—— (2000) 'Development post-, anti- and populist: A critical review', *Environment and Planning*, A 32: 1033–50.

Blaikie, P. and Coppard, D. (1998) 'Environmental change and livelihood diversification in Nepal: where is the problem?, *Himalayan Research Bulletin*, 18 (2): 28–39.

Blaikie, P.M. and Sadeque, S.Z. (2000) *Policy in High Places: Environment and Development in the Himalayan Region*, Kathamndu, Nepal: ICIMOD.

Blaikie, P.M. and Muldavin, J. (2004) 'Upstream, downstream, China and India: the politics of the environment in the Himalayan region', *Annals Assoc. Amer. Geogr.*, 94(4), in press.

Blaikie, P.M., Cameron, J. and Seddon, D. (1980) *Nepal in Crisis: Growth and Stagnation at the Periphery*, Oxford: Clarendon Press.

Blower, J.H. (1972) 'Establishment of Khumbu National Park: outline project proposal, memorandum', Department of National Parks and Wildlife Conservation, Kathmandu.

Bohle, H.-G. and Adhikari, J. (1998) 'Rural livelihoods at risk: how Nepalese farmers cope with food insecurity', *Mountain Research and Development*, 18(4): 321–32.

Bolt, B.A. (1993) 'The estimation of seismic risk for large structures in regions like the Himalaya', in V.K. Gaur (ed.) *Earthquake Hazard and Large Dams in the Himalaya*, Indian National Trust for Art and Cultural Heritage, New Delhi: Indraprastha Press.

Brammer, H. (1982) 'Agriculture and food production in polder areas', paper presented at International Polders Conference, Lelystadt, Netherlands.

Breu, H. and Hurni, H. (2003) *The Tajik Pamirs: Challenges of Sustainable Development in an Isolated Mountain Region*, Berne: Centre for Development and Environment (CDE), University of Berne, Switzerland.

Brichieri-Colombi, S. and Bradnock, R.W. (2003) 'Geopolitics, water and development in South Asia: cooperative development in the Ganges – Brahmaputra delta', *Geographical Journal*, 169(1): 43–64.

Brower, B. (1990) 'Range conservation and Sherpa livestock management in Khumbu, Nepal', *Mountain Research and Development*, 10(1): 34–42.

—— (1991) 'Crisis and conservation in Sagarmatha National Park, Nepal', *Society and Natural Resources*, 4: 151–63.

—— (1996) 'Geography and history in the Solukhumbu landscape', *Mountain Research and Development*, 16(3): 249–56.

Brower, B. and Dennis, A. (1998) 'Grazing the forest, shaping the landscape? Continuing the debate about forest dynamics in Sagarmatha National Park', in K.S. Zimmerer and K.R. Young (eds) *Nature's Geography*, Madison, WI: University of Wisconsin Press, pp. 184–208.

Brown, A. (1985) A study of soil erosion on the Kosi Hills of Eastern Nepal, unpub. BSc thesis, King's College, London.

Bruijnzeel, L.A. and Bremmer, C.N. (1989) *Highland-Lowland Interactions in the Ganges–Brahmaputra River Basin: a review of published literature*, Occnl. Paper No. 11, Kathmandu: ICIMOD.

Brunsden, D. and Thornes, J.B. (1979) 'Landscape sensitivity and change', *Trans. Institute of British Geographers*, 4: 463–84.

Brunsden, D., Jones, D.K., Martin, R.P. and Doornkamp, J.C. (1981) 'The geomorphological character of part of the Low Himalaya of eastern Nepal', *Zeitschrift für Geomorphologie*, N.F., Suppl.-Bd. 37: 25–72.

Buchroithner, M.F., Jentsch, G. and Wanivenhaus, B. (1982) 'Monitoring of recent geological events in the Khumbu area (Himalaya, Nepal) by digital processing of Landsat MSS data', *Rock Mechanics*, 15: 181–97.

Burke, J. (1999) 'Biblical flood poised to drown a nation', London: *The Observer*, 20 June 1999, p. 25.

Byers, A.C. (1986) 'A geomorphic study of man-induced soil erosion in the Sagarmatha (Mount Everest) National Park, Khumbu, Nepal', *Mountain Research and Development*, 6(1): 83–7.

—— (1987a) 'Landscape change and man-accelerated soil loss: the case of the Sagarmatha (Mount Everest) National Park, Khumbu, Nepal', *Mountain Research and Development*, 7(3): 209–16.

—— (1987b) 'An Assessment of landscape change in the Khumbu region of Nepal using repeat photography', *Mountain Research and Development*, 7(1): 77–81.

—— (1987c) 'Geoecological study of landscape change and man-accelerated soil loss: the case of Sagarmatha (Mt. Everest) National Park, Khumbu, Nepal', unpublished doctoral dissertation, University of Colorado.

—— (1996) 'Historical and contemporary human disturbance in the upper Barun Valley, Makalu-Barun National Park and Conservation Area, East Nepal', *Mountain Research and Development*, 16(3): 235–47.

—— (1997) 'Landscape change in Sagarmatha (Mt. Everest) National Park, Khumbu, Nepal', *Himalayan Research Bulletin*, XVII(2): 31–41.

—— (2002) 'Historical and contemporary landscape change in the Sagarmatha (Mt. Everest) National Park, Khumbu, Nepal', unpub. MS.

Caine, N. and Mool, P.K. (1981) 'Channel geometry and flow estimates for two small mountain streams in the Middle Hills, Nepal', *Mountain Research and Development*, 1(3–4): 231–43.

—— (1982) 'Landslides in the Kolpu Khola drainage, Middle Mountains, Nepal', *Mountain Research and Development*, 2(2): 157–73.

Carpenter, C. and Zomer, R. (1996) 'Forest ecology of the Makalu-Barun National Park and Conservation Area, Nepal', *Mountain Research and Development*, 16(2): 135–48.

Carson, B. (1985) *Erosion and Sedimentation Processes in the Nepalese Himalaya*, Occnl. Paper No. 1, Kathmandu: ICIMOD.

Carter, A.S. and Gilmour, D.A. (1989) 'Increase in tree cover on private farm land in Central Nepal', *Mountain Research and Development*, 9(4): 381–91.

Casimir, M.J. and Rao, A. (1985) 'The pastoral economy of the nomadic Bakrwal of Jammu and Kashmir', *Mountain Research and Development*, 5(3): 221–32.

Cenderelli, D.A. (2000) 'Floods from natural and artificial dam failures', in E.E. Wohl (ed.) *Inland Flood Hazards: Human, Riparian and Aquatic Communities*, Cambridge, UK: Cambridge University Press, pp. 73–103.

Centre for Science and Environment (1982) *The State of India's Environment, 1982: A Citizen's Report*, New Delhi: Centre for Science and Environment.

Centre for Science and Environment (1985) *The State of India's Environment, 1984–85: The Second Citizen's Report*, New Delhi: Centre for Science and Environment.

Chakravarty-Kaul, M. (1998) 'Transhumance and customary pastoral rights in Himachal Pradesh: claiming the high pastures for Gaddis', *Mountain Research and Development*, 18(1): 5–17.

Champion, H.G. and Seth, S.K. (1968) *A Revised Survey of the Forest Types of India*, Delhi: Government of India, Manager of Publications.

Chapman, E.C. and Sabhasri, S. (eds) (1983) 'Natural resource development and environmental stability in the highlands of northern Thailand', *Mountain Research and Development*, 3(4): 309–431.

Chaudhary, R.P., Subedi, B.P., Vetaas, O.R. and Aase, T.H. (eds) (2002) *Vegetation and Society: Their interaction in the Himalayas*, Kathmandu and Bergen: Tribhuvan University, Nepal, and University of Bergen, Norway.

Clemens, J. and Nüsser, M. (1997) 'Resource management in Rupal Valley, northern Pakistan: The utilization of forests and pastures in the Nanga Parbat area', in I. Stellrecht and M. Winiger (eds), *Perspectives on History and Change in the Karakorum, Hindukush, and Himalaya*, Köln: Rüdiger Köppe Verlag, pp. 235–63.

Coward, W. (2001) 'The Kulu Valleys in motion', *Himalayan Research Bulletin*, XXI (2): 5–14.

Cunha, S. (1994) 'The applicability of the biosphere reserve model to the Pamir Mountains, Tajikistan', unpublished doctoral dissertation, University of California, Davis.

—— (1995) 'Hunting of rare and endangered fauna in the mountains of post-Soviet Central Asia', Proceedings of the Eighth International Snow Leopard Symposium, Islamabad, Pakistan.

Curry, J.R. and Moore, D.G. (1971) 'Growth of the Bengal deep-sea fan and denudation in the Himalayas', *Geological Society of America Bulletin*, 82: 563–72.

Deegan, P. (2003) 'Appetite for destruction', *Geographical*, March, 2003: 32–6.

DFRS (1999a) *Forest Resources of Nepal (1987–88)*, Department of Forest Research and Survey, Ministry of Forest and Soil Conservation, HMG Nepal Forest Resources Information Project, Government of Finland, Kathmandu.

—— (1999b) *Forest and Shrub Cover of Nepal (1989–96)*, Department of Forest Research and Survey, Ministry of Forest and Soil Conservation, HMG Nepal, Forest Resources Information Project, Government of Finland, Kathmandu.

Dhakal, D.N.S. (1987) 'Twenty-five years of development in Bhutan', *Mountain Research and Development*, 7(3): 219–21.

—— (2000a) 'The unresolved problem', *Bhutanese Refugees Against Violence (BRAVVE)*, 1(1): 3–5, Birtamode, Nepal.

—— (ed.) (2000b) *Bhutan Political Problem: Opinions, Viewpoints and Critical Analysis*, Annual Publication of the Bhutan National Democratic Party (BNDP), Siliguri, North Bengal: Jatiya Bhandar Press.

Dhakal, D.N.S. and Strawn, C. (1994) *Bhutan: A Movement in Exile*, New Delhi: Nirala Publ., S.S. Enterprises, Darya Ganj.

Dhital, M.R., Khanal, N. and Thapa, K.B. (1993) 'The role of extreme weather events, mass movements, and land use changes in increasing natural hazards', Proceedings of a workshop on Causes of the Recent Damage Incurred in South-Central Nepal, July 19–21. Kathmandu: ICIMOD.

Dixit, K.M. (2002) 'Insurgents and innocents: the Nepali army's battle with the Maobaadi', *Himal South Asian*, Kathmandu, June 2002.

Dobremez, J.F. (1986) *Les Collines du Nepal Central: Ecosystemes Structures Sociales et Systemes Agraires*, 2 Vols., Paris: Institut National de la Recherche Agronomic.

Donovan, D.G. (1981) 'Fuelwood: how much do we need?' DGD Newsletter 14, Hanover, New Hampshire: Institute of World Affairs.

Dow, V., Kienholz, H., Plam. M. and Ives, J.D. (1981) 'Mountain Hazards Mapping: the development of a prototype mountain hazards map, Monarch Lake Quadrangle, Colorado, U.S.A.', *Mountain Research and Development*, 1(1): 55–64.

Duffield, C., Gardner, J.S., Berkes, F. and Singh, R.B. (1998) 'Local knowledge in the assessment of resource sustainability: case studies in Himachal Pradesh, India', *Mountain Research and Development*, 18(1): 35–49.

Eckholm, E. (1975) 'The deterioration of mountain environments', *Science*, 189: 764–70.

—— (1976) *Losing Ground*, Worldwatch Institute, New York: W.W. Norton and Co.

Exo, S. (1990) ' Local resource management in Nepal: limitations and prospects', *Mountain Research and Development*, 10(1): 16–22.

Felmy, S. (1996) *The Voice of the Nightingale: a personal account of the Wahki Valley culture in the Hunza*, Karachi: Oxford University Press.

Finn, W.D.L. (1993) 'Seismic design considerations for dams in Himalaya with references to Tehri Dam', in V.K. Gaur (ed.) *Earthquake Hazard and Large Dams in the Himalaya*, Indian National Trust for Art and Cultural Heritage, New Delhi: Indraprastha Press, pp. 116–36.

Finsterwalder, R. (1960) 'German expeditions to Batura Mustagh and Rakaposhi', *Journal of Glaciology*, 3: 787.

Fisher, J.F. (1990) *Sherpas: Reflections on Change in the Nepal Himalaya*, Berkeley: University of California Press.

Flohn, H. (1968) 'Contributions to a meteorology of the Tibetan Highlands', *Atmos. Sci. Paper* No. 130, Fort Collins: Colorado State University.

Forsyth, T. (2003) *Critical Political Ecology: The politics of environmental science*, London and New York: Routledge.

Forsyth, T.J. (1994) 'The use of Cesium-137 measurements of soil erosion and farmers' perceptions to indicate land degradation amongst shifting cultivators in northern Thailand', *Mountain Research and Development*, 14(3): 229–44.

—— (1996) 'Science, myth, and knowledge: testing Himalayan environmental degradation in northern Thailand', *Geoforum*, 27(3): 375–92.

—— (1998) 'Mountain myths revisited: integrating natural and social environmental science', *Mountain Research and Development*, 18(2): 107–16.

Fort, M. (1987) 'Geomorphic and hazards mapping in the dry, continental Himalaya: 1 : 50,000 maps of the Mustang District, Nepal', *Mountain Research and Development*, 7(3): 222–38.

—— (2000) 'Glaciers and mass wasting processes: their influence on the shaping of the Kali Gandaki valley (Higher Himalaya of Nepal)', *Quaternary International*, 65–66: 101–19.

Fort, M.B. and Freytet, P. (1982) 'The quaternary sedimentary evolution of the intra-montane basin of Pokhara in relation to the Himalaya Midlands and their hinterland (West Central Nepal)', in A.K. Sinha (ed.) *Contemporary Geoscientific Researches in Himalaya*, Vol. 2: 91–6, Dehra Dun, India.

Fort, M. and Peulvast, J-P. (1995) 'Catastrophic mass-movements and morphogenesis in the peri-Tibetan ranges: examples from West Kunlun, East Pamir and Ladakh', in O. Slaymaker (ed.) *Steepland Geomorphology*, New York: John Wiley & Sons, pp. 172–98.

Fox, J. (1993) 'Forest resources in a Nepali village in 1980 and 1990: the positive influence of population growth', *Mountain Research and Development*, 13(1): 89–98.

Fox, J., Krummel, J., Yarnararn, S., Ekasingh, M. and Podger, N. (1995) 'Land and landscape dynamics in northern Thailand: assessing change in three upland watersheds', *Ambio*, 24(6): 328–34.

Frankel, M. and Roberts, P. (1995) 'A sound like thunder', *Newsweek*, 27 November 1995, pp. 17–18.

Froehlich, W. and Starkel, L. (1987) 'Normal and extreme monsoon rains – their role in the shaping of the Darjeeling Himalaya', *Studia Geomorphologica Carpatho-Balcanica*, 21: 129–60.

Funnell, D. and Parish, R. (2001) *Mountain Environments and Communities*, London and New York: Routledge.

Fürer-Haimendorf, C. Von (1975) *Himalayan Traders: Life in Highland Nepal*, London: John Murray.

Fushimi, H., Ikigami, K. Higuchi, K. and Shankar, K. (1985) 'Nepal case study: catastrophic floods', in G.J. Young (ed.) *Techniques for Prediction of Runoff from Glacierized Areas*, IAHS, Publ. No. 149, pp. 125–30.

Galay, V. (1986) 'Glacier lake outburst flood (jökulhlaup) on the Bhote/Dudh Kosi – August 4, 1985', internal report, Water and Energy Commission, Kathmandu.

Ganjanapan, A (1996) 'Will community forest law strengthen community forestry in Thailand?', in B. Rerkasem (ed.) *Montane Mainland Southeast Asia in Transition*, Proceedings of a symposium, Chiang Mai University, Thailand.

—— (1998) 'The politics of conservation and the complexity of local control of forests in the Northern Thai highlands,' *Mountain Research and Development*, 18(1): 71–82.

Gansser, A. (1966) 'Geological research in the Bhutan Himalaya', *Mountain World 1964/65*, Zurich: Swiss Foundation for Mountain Research.

—— (1970) 'Lunana, the peaks, glaciers and lakes of northern Bhutan', *Mountain World 1968/89*, Swiss Foundation for Alpine Research, pp. 117–31.

Gardner, J.S. (2003) 'Natural hazards risks in the Kullu District, Himachal Pradesh, India', *Geographical Review*, 92(2): 282–306.

Gardner, R.A.M. and Jenkins, A. (1995) 'Land use, soil conservation and water resource management in the Nepal Middle Hills', unpublished report, UK Overseas Development Administration, 11.1–11.60.

Gardner, R.A.M. and Gerrard, A.J. (2001) 'Soil loss on non-cultivated land in the Middle Hills of Nepal', *Physical Geography*, 22(5): 376–93.

—— (2002) 'Relationships between runoff and land degradation on non-cultivated land in the Middle Hills of Nepal', *International Journal of Sustainable Development of World Ecology*, 9: 59–73.

—— (2003) 'Runoff and soil erosion on cultivated rainfed terraces in the Middle Hills of Nepal', *Applied Geography*, in press.

Gaur, V.K. (ed) (1993) *Earthquake Hazards and Large Dams in the Himalaya*, Indian National Trust for Art and Cultural Heritage, New Delhi: Indraprasthra Press.

Gautam, C.M. and Watanabe, T. (2004) 'Reliability of land use/cover assessment for montane Nepal: a case study in the Kangchenjunga Conservation area', *Mountain Research and Development*, 24(1): 35–43.

Gautam, K.H. (1991) Indigenous forest management systems in the hills of Nepal, unpub. M.Sc. thesis, Australian National University, Canberra.

Gellner, D. (ed.) (2003) *Resistance and the State: Nepalese Experiences*, New Delhi: Social Science Press.

Gerrard, A.J. (2002) ' Landsliding, runoff and soil loss in the Likhu Khola drainage basin, Middle Hills, Nepal: Case study. Section E: Bishkek Global Mountain Summit 2002. Mountain Forum E Consultations', Bishkek – A.J.W.Gerrard@bham.ac.uk

Gerrard, J. and Gardner, R. (1999) 'Landsliding in the Likhu Khola drainage basin, Middle Hills of Nepal', *Physical Geography*, 20: 240–55.

—— (2000a) 'The nature and management implications of landsliding on irrigated terraces in the Middle Hills of Nepal', *International Journal of Sustainable Development and World Ecology*, 7: 229–36.

—— (2000b) 'The role of landsliding in shaping the landscape of the Middle Hills, Nepal', *Zeitschrift für Geomorphologie*, Suppl.-Bd. 122: 47–62.

—— (2000c) 'Relationships between rainfall and landsliding in the Middle Hills, Nepal, *Norsk Geografisk Tidsskrift*, 54: 74–81.

—— (2002) 'Relationships between landsliding and land use in the Likhu Khola drainage basin, Middle Hills, Nepal', *Mountain Research and Development*, 22(1): 48–55.

Gersony, R. (2003) Sowing the Wind . . . History and dynamics of the Maoist revolt in Nepal's Rapti Hills, unpub. consultant's report to Mercy Corps International (prepared for USAID), Washington, D.C.

Gilmour, D.A. (1986) 'Reforestation and afforestation of open land – a Nepal perspective', in A.J. Pearce and L.S. Hamilton (eds), *Land Use, Watersheds, and Planning in the Asia-Pacific Region*, FAO and RAPA Report: Bangkok, pp. 157–69.

—— (1988) 'Not seeing the trees for the forest: a re-appraisal of the deforestation crisis in two hill districts of Nepal', *Mountain Research and Development*, 8(4): 343–50.

—— (1991) 'Trends in forest resources and management in the Middle Mountains of Nepal', in P. Shah, H. Schreier, S. Brown and K. Riley (eds) *Soil Fertility and Erosion Issues in the Middle Mountains of Nepal*, Vancouver: Department of Soil Science, University of British Columbia.

Gilmour, D.A. and Fisher, R.J. (1991) *Villagers, Forests and Foresters*, Kathmandu: Sahayogi Press.

Gilmour, D.A. and Nurse, M.C. (1991) 'Farmer initiatives in increasing tree cover in central Nepal', *Mountain Research and Development*, 11(4): 329–37.

Gilmour, D.A., Bonell, M. and Cassells, D.S. (1987) 'The effects of forestation on soil hydraulic properties in the Middle Hills of Nepal: a preliminary assessment', *Mountain Research and Development*, 7(3): 239–49.

Goldstein, M.C., Ross, J.L. and Schuler, S. (1983) 'From a mountain-rural to a plains-urban society: implications of the 1981 Nepal Census', *Mountain Research and Development*, 3(1): 61–4.

Grabs, W.E. and Hanisch, J. (1993) 'Objectives and preventative methods for glacial lake outburst floods', *Snow and Glacier Hydrology, IAHS Publications*, 218: 241–52.

Griffin, D.M. (1987) 'Implementation failure caused by institutional problems', *Mountain Research and Development*, 7(3): 250–3.

—— (1989) *Innocents Abroad in the Forests of Nepal: An account of Australian Aid to Nepalese Forestry*, Canberra: ANUTECH, Pty. Ltd.

Griffin, D.M., Shepherd, K.R. and Mahat, T.B.S. (1988) 'Human impacts on some forests of the Middle Hills of Nepal. Part 5: comparisons, concepts, and some policy implications', *Mountain Research and Development*, 8(1): 43–52.

Grötzbach, E. (1990) 'Man and environment in the West Himalaya and the Karakoram (Pakistan)', *Universitas*, 32: 17–26.

Grove, J.M. (1988) *The Little Ice Age*, London and New York: Methuen.

Guangwei, C. (ed.) (1999) *Biodiversity in the Eastern Himalayas: Conservation through dialogue*, Kathmandu: ICIMOD.

Guha, R. (1989) *The Unquiet Woods: Ecological Change and Peasant Resistance in the Himalaya*, Oxford: Oxford University Press.

Gurung, D. (1996) Ecotourism development in the Kangchenjunga region, WWF Nepal Program Report Series 27, Kathmandu: World Wildlife Fund.

Gurung, J.D. (ed.) (1999) *Searching for Women's Voices in the Hindu Kush–Himalayas*, Kathmandu: ICIMOD.

Gurung, S.M. (1988) 'Beyond the myth of eco-crisis in Nepal; local response to pressure on land in the Middle Hills', unpublished doctoral thesis, University of Hawaii, Honolulu.

Guthman, J. (1997) 'Representing crisis: the theory of Himalayan environmental degradation and the project of development in post-Rana Nepal', *Development and Change*, 28: 45–69.

Gyawali, D. (2001) *Water in Nepal*, Kathmandu: Himal Books.

—— (2003) 'Yam between Bhot and Muglan: Nepal's search for security in development', in K.M. Dixit and S. Ramachandaran (eds) *State of Nepal*, Kathmandu: Himal Books, pp. 212–34.

Gyawali, D. and Dixit, A. (1997) ' How distant is Nepali science from Nepali society? Lessons from the 1997 Tsho Rolpa GLOF panic', *Water Nepal*, 5(2): 5–43.

Hagan, T. (1969) 'Report on the geological survey of Nepal', *Denkscheiften der Schweizerischen Naturforschenden Gesellschaft*, 86(1): 1–185.

Haigh, M.J. (1982a) 'Road development and rural stresses in the Indian Himalaya', *Nordia*, 16: 135–40.

—— (1982b) 'A comparison of sediment accumulations beneath forested and deforested micro-catchments, Garhwal Himalaya', *Himalayan Research and Development*, 1(11): 118–20.

—— (1984) 'Landslide prediction and highway maintenance in the Lesser Himalaya, India', *Zeitschrift für Geomorphologie*, Suppl-Bd. 51: 17–37.

Haigh, M.J. (1991) 'Reclaiming forest lands in the Himalaya: notes from Pakistan', in G.S. Rajwar (ed.) *Recent Researches in Ecology, Environment, and Pollution*, Vol. 6, New Delhi: Today and Tomorrow's Printers and Publishers, pp. 199–210.

Hamilton, L.S. (ed.) (1983) *Forest and Watershed Development and Conservation in Asia and the Pacific*, Boulder, CO: Westview Press.

—— (1987) 'What are the impacts of Himalayan deforestation on the Ganges–Brahmaputra lowlands and delta? assumptions and facts', *Mountain Research and Development*, 7(3): 256–63.

Hammond, J. E. (1988) 'Glacial lakes in the Khumbu region, Nepal: an assessment of the hazards', unpublished Master's thesis, University of Colorado, Boulder.

Handel-Mazzetti, H. (1921) 'Übersicht über die wichtigsten Vegetationsstufen und -formationen von Yunnan und SW-Setschuan', *Bot. Jahrbuch für Systematik*, 56: 578–97.

Hatley, T. and Thompson, M. (1985) 'Rare animals, poor people, and big agencies: a perspective on biological conservation and rural development in the Himalaya', *Mountain Research and Development*, 5(4): 365–77.

He, Y. and Yang, F. (1998) *A Village-Level Survey of Environment and Development in the Lijiang Jade Dragon Mountain Region*, Kunming, Yunnan, P.R.C.: Yunnan People's Publishing House (in Chinese).

Herzog, M. (1952) *Annapurna: The First 8,000 Metre Peak*, Oxford: Alden Press.

Heuberger, H., Masch, L., Preuss, E. and Schocker, A. (1984) 'Quaternary landslides and rock fusion in Central Nepal and the Tyrolean Alps', *Mountain Research and Development*, 4(4): 345–62.

Hewitt, K. (1964) 'A Karakoram ice dam', *Indus*, Journ. Water and Power Development Authority, Pakistan, 5: 18–30.

—— (1982) 'Natural dams and outburst floods of the Karakoram Himalaya', in J.W. Glen (ed.), *Hydrological Aspects of Alpine and High Mountain Areas*, IAHS, Publ. No. 138, pp. 259–69.

—— (1985) 'Snow and ice hydrology in remote high mountain areas; the Himalayan sources of the River Indus', Snow and Ice Hydrology Project, Working Paper No.1, Waterloo, Ontario, Wilfrid Laurier University.

—— (1988) 'Catastrophic landslide deposits in the Karakoram Himalaya', *Science*, 242: 64–7.

—— (1993) 'Torrential rains in the Central Karakorum, 9–10 September, 1992: geomorphic impact and implications for climatic change', *Mountain Research and Development*, 13(4): 371–5.

—— (1997) 'Risks and disasters in mountain lands', in B. Messerli and J.D. Ives (eds) *Mountains of the World: a global priority*, London and New York: Parthenon, pp. 371–408.

Hinch, T. and Butler, R. (1996) 'Indigenous tourism: a common ground for discussion', in R. Butler and T. Hinch (eds) *Tourism and Indigenous Peoples*, London: International Thomson Business Press, pp. 3–19.

Hinrichsen, D., Lucas, P.H.C., Coburn, B. and Upreti, B.N. (1983) 'Saving Sagarmatha', *Ambio*, 11(5): 274–81.

Hirschboeck, K.K., Ely, L.L. and Maddox, R.A. (2000) 'Hydroclimatology of meteorological floods', in E.E. Wohl (ed.) *Inland Flood Hazards: Human, Riparian and Aquatic Communities*, Cambridge, UK: Cambridge University Press.

HMG/FINIDA (1995) *Woody Vegetation Cover of the Eastern Development Region*, 63, Forest Survey Division, Forest Research and Survey Centre, Kathmandu.

—— (1996) *Woody Vegetation Cover of the Central Development Region*, 67, Forest Survey Division, Forest Research and Survey Centre, Kathmandu.

Hofer, T. (1993) 'Himalayan deforestation, changing river discharge, and increasing floods: myth or reality?', *Mountain Research and Development*, 13(3): 213–33.

—— (1998) Floods in Bangladesh: A Highland-lowland Interaction?, *Geographica Bernensia*, G-48, University of Berne, Switzerland.

Hofer, T. and Messerli, B. (1997) 'Floods in Bangladesh – process understanding and development strategies', synthesis paper prepared for Swiss Agency for Development and Cooperation, Institute of Geography, University of Berne, Switzerland.

—— (2002) Floods in Bangladesh: history, dynamics and rethinking the role of the Himalayas, unpublished MS, Berne, Switzerland.

Hoon, V. (1996) *Living on the Move: Bhotiyas and the Kumaun Himalaya*, New Delhi, London, and Thousand Oaks, California: Sage Publishers.

Houston, C.S. (1982) 'Return to Everest: a sentimental journey', *Summit*, 28: 14–17.

—— (1987) 'Deforestation in Solu Khumbu', *Mountain Research and Development*, 7(1): 76.

Hughes, R., Adnan, S. and Dalal-Clayton, B. (1994) 'Floodplains or flood plans? A review of approaches to water management in Bangladesh', IED, London, RAS, Dhaka.

Hurni, H. (1982) 'Soil erosion in Huai Thung Choa – northern Thailand: concerns and constraints', *Mountain Research and Development*, 2(2): 141–56.

Hurni, H. and Nuntapong, S. (1983) 'Agroforestry improvements for shifting cultivation systems: soil conservation research in northern Thailand', *Mountain Research and Development*, 3(4): 338–45.

Imhof, E. (1974) *Die Grossen Kalten Berge von Szetschuan*, Zurich: Orell Füssli Verlag.

IUOTO (1963) *Conference on International Travel and Tourism*, Geneva: United Nations.

Ives, J.D. (1970) 'Himalayan Highway', *Canadian Geographical Journal*, 80(1): 26–31.

—— (1981) 'Applied mountain geoecology: can the scientist assist in the preservation of the mountains?' in J.S. Lall and A.D. Moddie (eds), *The Himalaya: Aspects of Change*, New Delhi: Oxford University Press, pp. 377–402.

—— (1983) Preface in E.C. Chapman and S. Sabhasri (eds), 'Natural resource development and environmental stability in the highlands of northern Thailand', Special Issue, *Mountain Research and Development*, 3(4): 309–11.

—— (1984) 'Does deforestation cause soil erosion?' IUCN *Bulletin*, 2.

—— (1985) 'Yulongxue Shan, Northwest Yunnan, People's Republic of China: A geoecological expedition', *Mountain Research and Development*, 5(4): 382–85.

—— (1986) *Glacial Lake Outburst Floods and Risk Engineering in the Himalaya*, Occasional Paper No. 5, Kathmandu: ICIMOD.

—— (1987a) 'The Theory of Himalayan Environmental Degradation: its validity and application challenged by recent research', *Mountain Research and Development*, 7(3): 189–99.

—— (1987b) 'Repeat photography of debris flows and agricultural terraces in the Middle Mountains, Nepal', *Mountain Research and Development*, 7(1): 82–6.

—— (1989a) 'Deforestation in the Himalayas: the cause of increased flooding in Bangladesh and northern India?', *Land Use Policy*, 1989: 187–93.

—— (1989b) 'Mountain environments', in G.B. Marini-Bettòlo (ed.) *A Modern Approach to the Protection of the Environment*, Vaticana, Pontificia Academia, Scientiarum, Scripta Varia 75, pp. 289–353.

—— (1991) 'Floods in Bangladesh: who is to blame?', *New Scientist*, 13 April, 130(1764): 34–7.

—— (1996) *Children, Women and Poverty in Mountain Ecosystems*, Primary Environmental Care (PEC) Discussion Papers, No 2, New York: UNICEF.

—— (2002) 'Bhutanese refugees in Nepal', *Mountain Research and Development*, 22(4): 411–14.

Ives, J.D. and He, Y. (1996) 'Environmental and cultural change in the Yulong Xue Shan, Lijiang District, NW Yunnan, China', in B. Rerkasem (ed.), *Montane Mainland Southeast Asia in Transition*, Chiang Mai University: Thailand, pp. 1–16.

Ives, J.D. and Ives, P. (eds) (1987) 'The Himalaya-Ganges Problem: proceedings of the Mohonk Mountain Conference', *Mountain Research and Development*, 7(3): 181–344.

Ives, J.D. and Messerli, B. (1981) 'Mountain hazards mapping in Nepal: introduction to an applied mountain research project', *Mountain Research and Development*, 1(3–4): 223–30.

—— (1989) *The Himalayan Dilemma: Reconciling Development and Conservation*, London and New York: Routledge.

Ives, J.D., Messerli, B. and Rhoades, R.E. (1997) 'Agenda for sustainable mountain development', in B. Messerli and J.D. Ives (eds) *Mountains of the World: A Global Priority*, London and New York: Parthenon, pp. 455–66.

Ives, J.D., Messerli, B. and Thompson, M. (1987) 'Research strategy for the Himalayan region', *Mountain Research and Development*, 7(3): 332–44.

Ives, J.D., Sabhasri, S. and Voraurai, P. (eds) (1980) *Conservation and Development in Northern Thailand*, NRTS-3/UNUP-77, Tokyo: United Nations University Press.

Iwata, S., Sharma, T. and Yamanaka, H. (1984) 'A preliminary report of Central Nepal and Himalayan uplift', *Journal Nepalese Geological Society*, 4: 141–9.

Iyengar, R.N. (1993) 'Dynamic analysis and seismic specification of Tehri Dam', in V.K. Gaur (ed.) *Earthquake Hazard and Large Dams in the Himalaya*, Indian National Trust for Art and Cultural Heritage, New Delhi: Indraprastha Press, pp. 137–52.

Iyer, R.R. (1998) 'Water projects in trouble: what lessons?', *Water Nepal*, 6(1): 5–11.

Jackson, W.J., Tamrakar, R.M., Hunt, S. and Shepherd, K.R. (1998) 'Land-use changes in two Middle Hills districts of Nepal', *Mountain Research and Development*, 18(3): 193–212.

Jayal, N.K. (1993) 'Preface', in V.K. Gaur (ed.) *Earthquake Hazard and Large Dams in the Himalaya*, Indian National Trust for Art and Cultural Heritage, New Delhi: Indraprastha Press.

Jodha, N.S. (1997) 'Mountain agriculture', in B. Messerli and J.D. Ives (eds) *Mountains of the World: a global priority*, London and New York: Parthenon, pp. 313–35.

Johnson, K., Olson, E.A., and Manandhar, S. (1982) 'Experimental knowledge and response to natural hazards in mountainous Nepal', *Mountain Research and Development*, 2(2): 175–88.

Joshi, B.C. (1987) 'Geo-environmental studies in parts of Ramganga catchment, Kumaon Himalayas', unpublished doctoral dissertation, University of Roorkee, Roorkee, India.

Joshi, D.P. (1982) 'The climate of Namche Bazar: a bioclimatic analysis', *Mountain Research and Development*, 2(4): 399–403.

Joshi, S.C. (ed.) (1986) *Nepal Himalaya: Geoecological Perspectives*, Nainital, UP: Himalayan Research Group.

Kaith, D.C. (1960) 'Forest practices in control of avalanches, floods, and soil erosion in the Himalayan front', Vol. III, Fifth World Forestry Congress Proceedings: Seattle.

Karan, P.P. (1967) *Bhutan: A Physical and Cultural Geography*, Lexington: University of Kentucky Press.

—— (1977) 'Cultural geography of the Himalaya', in R.C. Eidt, K.N. Singh and R.B.P. Singh, *Man, Culture and Settlement*, research publication No. 17, Varanasi: National Geographical Society of India, pp. 24–30.

—— (1990) *Bhutan: Environment, Culture and Development Strategy*, New Delhi: Intellectual Publishing House.

Karan, P.P. and Iijima, S. (1985) 'Environmental stress in the Himalaya', *Geographical Review*, 75(1): 71–92.

Kean, D. (1997) 'Grandmother's footsteps, an interview', *Geographical*, LXIX (12): 30–3.

Kienholz, H., Hafner, H., Schneider, G. and Tamrakar, R. (1983) 'Mountain hazards mapping in Nepal's Middle Mountains with maps of land use and geomorphic damages (Kathmandu-Kakani area)', *Mountain Research and Development*, 3(3): 195–220.

Kienholz, H., Schneider, G., Bichsel, M., Grunder, M. and Mool, P. (1984) 'Mountain hazards mapping project, Nepal: base map, and map of mountain hazards and slope stability, Kathmandu-Kakani area', *Mountain Research and Development*, 4(3): 247–66.

Kreutzmann, H. (1991) 'The Karakorum Highway: the impact of road construction on mountain societies', *Modern Asian Studies*, 25(4): 711–36.

Kreutzmann, H. (1993) 'Challenge and response in the Karakorum: socioeconomic transformation of the Hunza, Northern Areas, Pakistan', *Mountain Research and Development*, 13(1): 19–39.

—— (1994) 'Habitat conditions and settlement processes in the Hindukush-Karakorum', *Petermanns Geographisches Mitteilungen*, 6(94): 337–56.

—— (1995) 'Globalization, spatial integration, and sustainable development in northern Pakistan', *Mountain Research and Development*, 15(3): 213–27.

—— (1998) 'Trans-montane exchange patterns prior to the Karakorum Highway', in I. Stellrecht (ed.) *Karakorum – Hindukush – Himalaya: Dynamics of Change*, Köln: Rüdiger Köppe Verlag, Part II, pp. 21–43.

—— (2000) 'Improving accessibility for mountain development. Role of transport networks and urban settlements', in M. Banskota, T.S. Popola and J. Richter (eds) *Growth, Poverty Alleviation and Sustainable Resource Management in the Mountain Areas of South Asia*, Nepal Zentralstelle für Ernährung und Landwirtschaft (Flood and Agricultural Development Centre): Feldafing, Germany, pp. 485–513.

—— (2003) 'Ethnic minorities and marginality in the Pamirian Knot: survival of Wakhi and Kirghiz in a harsh environment and global contexts', *Geographical Journal*, 169(3): 215–35.

Kumer, D. (1995) *Tourism Potential Study of Districts Lahaul and Spiti*, Shimla, India: Institute of Vocational Studies, Himachal University.

Kundstadter, P., Chapman, E.C. and Sabhasri, S. (eds) (1978) *Farmers in the Forest*, Honolulu: University Press of Hawaii.

Laban, P. (1979) 'Field measurements on erosion and sedimentation in Nepal', Department of Soil Conservation and Watershed Management, Kathmandu (1WM/SP/05).

Lall, J.S. and Moddie, A.D. (eds) (1981) *The Himalaya: aspects of change*, New Delhi: Oxford University Press.

Lambert, C. (2003) 'At the crossroads: a survey of Central Asia', *The Economist*, 26 July 2003.

Lhamu, C., Rhodes, J.J. and Rai, D.B. (2000) 'Integrating economy and environment: the development experience of Bhutan', in M. Banskota, T.S. Papola and J. Richter (eds), *Growth, Poverty Alleviation and Sustainable Resource Management in the Mountain Areas of South Asia*, Feldafing, Germany: Deutsche Stiftung für internationale Entwicklung, pp. 137–70.

Li, T. (1994) 'Landslide disasters and human responses in China', *Mountain Research and Development*, 14(4): 341–46.

Li, Z., Chen, J., Hu, F. and Wang, M. (1982) *Geologic Structure of Gongga Mountainous Region, Expedition in the Gongga Mountain*, Chengdu Institute of Geography, Beijing: Academia Sinica, 20.

Libiszewski, S. and Bächler, G. (1997) 'Conflicts in mountain areas – a predicament for sustainable development', in B. Messerli and J.D. Ives (eds), *Mountains of the World: a global priority*, London and New York: Parthenon, pp. 103–30.

Liu, D. and Sun, H. (eds) (1981) *Proceedings on Symposium on Qinghai-Xizang (Tibet) Plateau*, Beijing: Science Press and New York: Gordon and Breach, 2 vols.

Low, B.L. (ed.) (1968) *Mountains and Rivers of India*, Prepared for 21st International Geographical Congress, India, New Delhi.

Lu, X.X., Ashmore, P. and Wang, J.F. (2003) 'Seasonal water discharge and sediment load changes in the upper Yangtze, China', *Mountain Research and Development*, 23(1): 56–64.

Lucas, P.H.C., Hardie, N.D. and Hodder, R.A.C. (1974) Report of the New Zealand Mission on Sagarmatha (Mount Everest) National Park, Department of National Parks and Wildlife Conservation, Kathmandu.

Lumley, J. (1997) *Joanna Lumley in the Kingdom of the Thunder Dragon*, London: BBC Books.

MacDonald, K.I. (1996) 'Population change in the upper Braldu Valley, Baltistan, 1900–1990: all is not as it seems', *Mountain Research and Development*, 16(4): 351–66.

MacDonald, K.I. and Butz, D. (1998) 'Investigating portering relations as a locus for transcultural interaction in the Karakorum region, northern Pakistan', *Mountain Research and Development*, 18(4): 333–43.

Mackinder, H.J. (1919) *Democratic Ideals and Reality*, reprinted, 1944, Harmondsworth: Penguin.

Mahat, T.B.S., Griffin, D.M. and Shepherd, K.R. (1986a) 'Human impact on some forests of the Middle Hills of Nepal: 1, forestry in the context of the traditional resources of the state', *Mountain Research and Development*, 6(3): 223–32.

—— (1986b) 'Human impact on some forests of the Middle Hills of Nepal: 2, some major human impacts before 1950 on the forests of Sindhu Palchok and Kabhre Palanchok', *Mountain Research and Development*, 6(4): 325–34.

—— (1987a) 'Human impact on some forests of the Middle Hills of Nepal: 3, forests in the subsistence economy of Sindhu Palchok and Kabhre Palanchok', *Mountain Research and Development*, 7(1): 53–70.

—— (1987b) 'Human impact on some forests of the Middle Hills of Nepal: 4, a detailed study in southeast Sindhu Palchok', *Mountain Research and Development*, 7(2): 111–34.

Malla, Y.B. and Griffin, D.M. (1999) 'The changing role of the forest resource in the hills of Nepal', Department of Forestry, Australian National University: Canberra, unpub. MS.

Mason, K. (1935) 'The study of threatening glaciers', *Geographical Journal*, 85(1): 24–35.

Mauch, S.P. (1976) 'The energy situation in the hills: imperative for development strategies?', *Mountain Environment and Development*, A collection of papers for the Swiss Association for Technical Assistance in Nepal (SATA), Kathmandu.

Mawdsley, E. (1999) 'A new Himalayan state in India: popular perceptions of regionalism, politics, and development', *Mountain Research and Development*, 19(2): 101–12.

McConnell, R.M. (1991) 'Solving environmental problems caused by adventure travel in developing countries: the Everest Environmental Expedition', *Mountain Research and Development*, 11(4): 359–66.

McKinnon, J. (1983) 'A highlander's geography of the highlands: mythology, process, and fact', *Mountain Research and Development*, 3(4): 313–17.

Mehra, B.S. and Mathur, P.K. (2001) 'Livestock grazing in the Great Himalayan National Park Conservation Area – A landscape level assessment', *Himalayan Research Bulletin*, XXI (2): 89–96.

Merzliakova, I.A. (1998) 'Environmental hazards (Khait earthquake) and administrative pressure for resource utilization (Pamiro-Alai case study)', in I. Stellrecht (ed.), *Karakorum—Hindukush—Himalaya: Dynamics of Change*, Köln: Rüdiger Köppe Verlag, pp. 269–86.

Messerli, B. and Ives, J.D. (1984) 'Gongga Shan (7556 m) and Yulongxue Shan (5596 m): geoecological observations in the Hengduan Mountains of southwestern China', in W. Lauer (ed.) *Natural Environment and Man in Tropical Mountain Ecosystems*. Stuttgart: Franz Steiner Verlag, Wiesbaden GMBH, pp. 55–77.

Messerli, B. and Ives, J.D. (eds) (1997) *Mountains of the World: A Global Priority*, London and New York: Parthenon.

Messerschmidt, D.A. (1990) 'Indigenous environmental management and adaptation: an introduction to four case studies from Nepal', *Mountain Research and Development*, 10(1): 5–6.

Metz, J.J. (1990) 'Conservation practices at an upper-elevation village of west Nepal', *Mountain Research and Development*, 10(1): 7–15.

—— (1991) 'A reassessment of the causes and severity of Nepal's environmental crisis', *World Development*, 19(7): 805–20.

—— (1997) 'Vegetation dynamics of several little disturbed temperate forests in east-central Nepal', *Mountain Research and Development*, 17(4): 333–51.

Miehe, S. and Miehe, G. (1998) 'Vegetation patterns as indicators of climatic humidity in the western Karakorum', in I. Stellrecht (ed.) *Karakorum–Hindukush–Himalaya: Dynamics of Change*, Pt. 1, Köln: Rüdiger Köppe Verlag, 101–26.

Mishra, H.R. (1973) 'Conservation in Khumbu: the proposed Mt. Everest National Park', preliminary report, Department of National Parks and Wildlife Conservation, Kathmandu.

Mool, P.K., Bajracharya, S.R. and Joshi, S.P. (2001a) *Inventory of Glaciers, Glacial Lakes and Glacial Lake Outburst Floods, Nepal*, Kathmandu: ICIMOD.

Mool, P.K., Wangda, D., Bajracharya, S.R., Kunzang, K., Gurung, D.R. and Joshi, S.P. (2001b) *Inventory of Glaciers, Glacial Lakes and Glacial Lake Outburst Floods, Bhutan*, Kathmandu: ICIMOD.

MRD (1987) Mohonk Mountain Conference Resolution 3, *Mountain Research and Development*, 7(3): 185.

MRD (2003) 'A Siachen Peace Park: The Solution to a Half-Century of International Conflict', letters to an author, *Mountain Research and Development*, 23(2): 208.

Mukerji, A.B. (1993) 'The greening of the Siwalik Himalayas', in B. Messerli, T. Hofer and S. Wymann (eds), *Himalayan Environment: Pressure – Problems – Processes: Twelve Years of Research*, Switzerland: Institute of Geography, University of Berne, pp. 63–77.

Müller, F. (1959) 'Eight months of glacier and soil research in the Everest region', *The Mountain World 1958/59*, London: Allen and Unwin, pp. 191–208.

Müller-Böker, U. and Kollmair, M. (2000) 'Livelihood strategies and local perceptions of a new nature conservation project in Nepal: the Kanchenjunga Conservation Area Project', *Mountain Research and Development*, 20(4): 324–31.

Müller-Hohenstein, K. (ed.) (1974) *International Workshop on the Development of Mountain Environment*, Munich: German Foundation for International Development.

Myers, N. (1986) 'Environmental repercussions of deforestation in the Himalayas', *Journal of World Forest Resources*, 2: 63–72.

Narayana, D.V.V. (1987) 'Downstream impacts of soil conservation in the Himalayan region', *Mountain Research and Development*, 7(3): 287–98.

Narayana, D.V.V. and Rambabu, I. (1983) 'Estimation of soil erosion in India, *Journal of Irrigation and Drainage Engineering*, 109(4): 409–34.

Naylor, R. (1970) Unpublished report, New Zealand Forest Service.

Negi, A.K., Bhatt, B.P., Todaria, N.P. and Saklani, A. (1997) 'The effects of colonialism on forests and the local people in the Garhwal Himalaya, India', *Mountain Research and Development*, 17(2): 159–68.

Nepal, S.K. (1999) 'Tourism induced environmental changes in the Nepalese Himalaya: A comparative analysis of the Everest, Annapurna and Mustang regions', unpub. PhD dissertation, University of Berne, Switzerland.

Nepal, S.K., Kohler, T. and Banzhaf, B.R. (2002) *Great Himalaya: tourism and the dynamics of change in Nepal*, Zurich: Swiss Foundation for Alpine Research.

Ngo, J.M. (1958) 'The glaciation of Yulungshan in China', *Erdkunde*, 12: 308–13.

Nicholas, C. and Singh, R. (1996) *Indigenous Peoples of Asia: Many peoples, one struggle*, Bangkok: Asia Indigenous Peoples Pact.

NPC (1993) *An Outline on Poverty Alleviation Policies and Programmes*, Kathmandu: National Planning Commission.

Numata, M. (ed.) (1983) *Structure and Dynamics of Vegetation in Eastern Nepal*, Laboratory for Ecology, Chiba University, Japan.

Nusser, M. and Clemens, J. (1996) 'Impacts on mixed mountain agriculture in the Rupal Valley, Nanga Parbat, northern Pakistan', *Mountain Research and Development*, 16(2): 117–133.

Ohsawa, M. (1983) 'Distribution, structure and regeneration of forest communities in eastern Nepal', in M. Numata (ed.) *Structure and Dynamics of Vegetation in Eastern Nepal*, Laboratory for Ecology, Chiba University, Japan.

Ortner, S.B. (1996) 'Making gender: toward a feminist, minority, postcolonial, subaltern, etc., theory of practice', in S.B. Ortner (ed.) *Making Gender: The Politics and Erotics of Culture*, Boston: Beacon Press.

——(2000) *Life and Death on Mt. Everest: Sherpas and Himalayan Mountaineering*, New Delhi: Oxford University Press.

Parish, R. (2002) *Mountain Environments*, London and New York: Prentice Hall.

Paul, B.K. (2003) 'Relief assistance to 1998 flood victims: A comparison of the performance of the government and NGOs', *Geographical Journal*, 169 (1): 75–89.

Pearce, F. (1991) 'The rivers that won't be tamed', *New Scientist*, 13 April, 130(1764): 38–41.

—— (1999) 'Hell and high water', *New Scientist*, 19 June 1999 p. 4.

—— (2002) 'Meltdown!'. *New Scientist*, 2 November, 176(2367): 44–8.

Pimbert, M.P. and Pretty, J.N. (1997) 'Parks, people and professionals: putting "participation" into protected area management', in K.B. Ghimire and M.P. Pimbert (eds) *Social Change and Conservation: Environmental Politics and Impacts of National Parks and Protected Areas*, London: Earthscan, pp. 297–330.

Poffenberger, M. (1980) *Patterns of Change in the Nepal Himalaya*, Delhi: Macmillan.

Price, M.F., Moss, L.A.G. and Williams, P.W. (1997) 'Tourism and amenity migration', in B. Messerli and J.D. Ives (eds) *Mountains of the World: a global priority*, London and New York: Parthenon, pp. 249–80.

Ramsay, W.J.H. (1985) 'Erosion in the Middle Himalaya, Nepal, with a case study of the Phewa Valley', unpublished thesis, University of British Columbia.

Rana, B., Shrestha, A.B., Reynolds, J.M., Aryal, R., Pokhrel, A.P. and Budhathoki, K.P. (2000) 'Hazard assessment of the Tsho Rolpa Glacier Lake and ongoing remediation measures', *Journal of Nepal Geological Society*, 22: 563–70.

Rawat, D.S. and Sharma, S. (1997) 'The development of a road network and its impact on the growth of infrastructure: a study of Almora District in the Central Himalaya', *Mountain Research and Development*, 17(2): 117–26.

Rawat, J.S. and Rawat, M.S. (1994a) 'Accelerated erosion in the Nana Kosi watershed, Central Himalaya, India. Part I: sediment load', *Mountain Research and Development*, 14(1): 25–38.

—— (1994b) 'Accelerated erosion in the Nana Kosi watershed, Central Himalaya, India. Part II: human impacts on stream runoff', *Mountain Research and Development*, 14(3): 255–60.

Renaud, F. (1997) ' Financial cost-benefit analysis of soil conservation practices in northern Thailand', *Mountain Research and Development*, 17(1): 11–18.

Renaud, F., Bechstedt, H.-D. and Nakorn, U.N. (1998) 'Farming systems and soil-conservation practices in a study area of northern Thailand', *Mountain Research and Development*, 18(4): 345–56.

Rerkasem, B. (ed.) (1996) *Montane Mainland Southeast Asia in Transition*, Chiang Mai University, Thailand.

Rhoades, R.E. (1997) *Pathways Towards a Sustainable Mountain Agriculture for the 21st Century: The Hindu Kush – Himalayan Experience*, Kathmandu: ICIMOD.

Richards, J.F. (1987) 'Environmental changes in Dehra Dun Valley, India: 1880–1980', *Mountain Research and Development*, 7(3): 299–304.

Rieck, A. (1997) 'From mountain refuge to "Model Area": Transformation of Shi'i communities in northern Pakistan', in I. Stellrecht and M. Winiger (eds), *Perspectives on History and Change in the Karakorum, Hindukush, and Himalaya*, Köln: Rüdiger Köppe Verlag, pp. 215–31.

Rieger, H.C. (1981) 'Man versus mountain: the destruction of the Himalayan ecosystem', in J.S. Lall and A.D. Moddie (eds) *The Himalaya: Aspects of Change*, Delhi and Calcutta: Oxford University Press, pp. 351–76.

Rock, J.F. (1924) 'Banishing the devil of disease among the Nashi: weird ceremonies performed by aboriginal tribe in the heart of Yunnan Province, China', *National Geographic Magazine*, 46: 473–99.

—— (1947) *The Ancient Na-Khi Kingdom of Southwest China*, 2 vols. Harvard-Yenching Monography Series VIII, Cambridge, MA: Harvard University Press.

Roder, W. (1997) 'Slash-and-burn rice systems in transition: challenges for agricultural development in the hills of northern Laos', *Mountain Research and Development*, 17(1): 1–10.

Roder, W., Gratzer, G. and Wangdi, K. (2002) 'Cattle grazing in the conifer forests of Bhutan', *Mountain Research and Development*, 22(4): 368–74.

Rogers, P., Lydon, P., and Seckler, D. (1989) *Eastern Waters Study: Strategies to Manage Flood and Drought in the Ganges-Brahmaputra Basin*. ISPAN, Irrigation Support Project for Asia and the Near East, Arlington, Virginia, USAID.

Rose, L.E. (1977) *The Politics of Bhutan*, Ithaca, New York: Cornell University Press.

Rowell, G. (1980) *Many People Come Looking, Looking*. Seattle: The Mountaineers.

Roy, R.D. (2002) *Land and Forest Rights in the Chittagong Hill Tracts, Bangladesh*, Kathmandu: ICIMOD.

Sabawal, V.K. (1999) *Pastoral Politics: Shepherds, Bureaucrats and Conservation in the Western Himalaya*, Delhi: Oxford University Press.

Saberwal, V.K. and Chhatre, A. (2001) 'The Parvarti and the Tragopan: conservation and development in the Great Himalayan National Park', *Himalayan Research Bulletin*, XXI (2): 79–88.

Savage, M. (1994) 'Land-use change and the structural dynamics of *Pinus kesiya* in a hill evergreen forest in Thailand', *Mountain Research and Development*, 14(3): 245–50.

Schickhoff, U. (1993) 'Interrelations between ecological and socioeconomic change: the case of the high altitude forests in the Northern Areas of Pakistan', *Pakistan Journal of Geography*, 3: 59–70.

—— (1995) 'Himalayan forest cover changes in historical perspective: a case study in the Kaghan Valley, northern Pakistan', *Mountain Research and Development*, 15(1): 1–18.

—— (1997) 'Ecological change as a consequence of recent road building: The case of the high altitude forests of the Karakorum', in I. Stellrecht and M. Winiger (eds) *Perspectives on History and Change in the Karakorum, Hindukush, and Himalaya*, Köln: Rüdiger Köppe Verlag, pp. 277–86.

—— (1998) 'Socio-economic background and ecological effects of forest destruction in northern Pakistan', in I. Stellrecht (ed.), *Karakorum–Hindukush–Himalaya: Dynamics of Change*, Köln: Rüdiger Köppe Verlag, pp. 287–302.

—— (2004) 'The forests of Hunza Valley: scarce resources under threat', in H. Kreutzmann (ed.), *Karakoram in Transition: The Hunza Valley*, Oxford: Oxford University Press (in press).

Schmidt-Vogt, D. (1998) 'Defining degradation: the impacts of swidden on forests in northern Thailand', *Mountain Research and Development*, 18(2): 135–49.

Schreier, H., Brown, S., Shah, P.B., Shrestha, B. and Merz, J. (2002) *Jhikhu Khola Watershed, Nepal*, 9 CD-Roms, Institute for Resources and Environment, University of British Columbia, Canada <http://www.ire.ubc.ca>

Schreier, H. and Wymann von Dach, S. (1996) 'Understanding Himalayan processes: shedding light on the dilemma', in H. Hurni, H. Kienholz, H. Wanner and U. Wiesmann (eds) *Umwelt Mensch Gebirge*, Festschrift for Bruno Messerli, Jahrbuch der Geographischen Gesellschaft Bern, Bd. 59/1994–1996, Berne, Switzerland, pp. 75–83.

Schuster, R.L. (1995) 'Landslide dams – a worldwide phenomenon', *Journal of the Japan Landslide Society*, 31(4): 38–49.

Schweinfurth, U. (1957) 'Die horizontale und vertikale Verbreitung der Vegetation im Himalaya', *Bonner Geogr. Abh.* 20, Bonn.

Sella, G.F., Dixon, T.H. and Mao, A. (2002) 'A model for recent plate velocities from space geodesy', *Journal of Geophysical Research*, 107(10): 1029–2000.

Shariff, K.M. (ed.) (2001–2002) *Pakistan Almanac*, Karachi: Royal Book Company.

Sharma, C.K. (1983) *Water and Energy Resources of the Himalayan Block: Pakistan, Nepal, Bhutan, Bangladesh and India*, Kathmandu: Mrs Sangeeta Sharma.

Sharma, P.D. and Minas, R.S. (1993) 'Land use and the biophysical environment of Kinnaur District, Himachal Pradesh, India', *Mountain Research and Development*, 13(1): 41–60.

Sharma, S., Koponen, J., Gyawali, D. and Dixit, A. (eds) (2004) *Aid Under Stress: Water, forests and Finnish support in Nepal*, Kathmandu: Himal Books for Institute of Development Studies, University of Helsinki and Interdisciplinary Analysts, Kathmandu.

Shelley, M.R. (2000) 'Socioeconomic status and development of Chittagong Hill Tracts (CHT) of Bangladesh: an overview', in M. Banskota, T.S. Papola and J. Richter (eds), *Growth, Poverty Alleviation and Sustainable Resource Management in the Mountain Areas of South Asia*, Feldafing, Germany: Deutsche Stiftung für internationale Entwicklung, pp. 107–35.

Sherpa, L.N., Peniston, B., Lama W. and Richard, C. (2003) *Hands Around Everest: Transboundary Cooperation for Conservation and Sustainable Livelihoods*, Kathmandu: ICIMOD and The Mountain Institute.

Shiva, V. (1988) *Staying Alive: Women, Ecology and Development*, London: Zed Books.

—— (1993) 'Monocultures of the mind: understanding the threat to biological and cultural diversity', unpublished report, International Development Research Centre, Ottawa, Canada.

Shiva, V. and Bandyopadhyay, J. (1986) 'The evolution, structure, and impact of the Chipko movement', *Mountain Research and Development*, 6(2): 133–42.

Shrestha, B. and Brown, S. (1995) 'Land-use dynamics and intensification', in H. Shreier, P.B. Shah and S. Brown (eds), *Challenge in Mountain Resource Management in Nepal: Process, Trend and Dynamics in Middle Mountain Watersheds*, Proceedings of a workshop, Kathmandu, Nepal, pp. 141–54.

Shroder, J.F. (ed.) (1998) Mass movement in the Himalayas, *Geomorphology*, special issue, 26: 1–222.

Shukla, J. (1987) 'Interannual variability of monsoon', in J.S. Fein and P.L. Stephens (eds) *Monsoons*, New York: John Wiley and Sons.

Sicroff, S. (1998) 'Approaching the Jade Dragon: Tourism in Lijiang County, Yunnan Province, China', unpublished Masters thesis, University of California, Davis.

Sicroff, S. and Alos Alabajos, E. (2001) 'Biodiversity and tourism in the sacred valley', in T. Watanabe, S. Sicroff, N.R. Khanal and M.P. Gautam (eds), Proceedings of the International Symposium on the Himalayan Environments, Kathmandu, 24–26 November, 2000, pp. 52–63.

Sicroff, S. and Ives, J.D. (2001) 'Tourism in Lijiang and the Jade Dragon Mountains, northwestern Yunnan, China: benign development or a pact with the devil?', in T. Watanabe, S. Sicroff, N.R. Khanal and M.P. Gautam (eds), Proceedings of the International Symposium on the Himalayan Environments, Kathmandu, 24–26 November, 2000, pp. 40–51.

Sinclair, A.J. (2003) 'Assessing the impacts of micro-hydro development in the Kullu District, Himachal Pradesh, India', *Mountain Research and Development*, 23(1): 11–13.

Singh, J.S. (ed.) (1985) *Environmental Regeneration in Himalaya: concepts and strategies*, Central Himalayan Environment Association: Nainital, U.P. India.

Singh, J.S. and Singh, S.P. (1987) 'Forest vegetation of the Himalaya', *Botanical Review*, 53(1): 80–192.

Singh, T.V. (1989) *The Kulu Valley: Impact of Tourism Development in the Mountain Areas*, New Delhi: Himalayan Books.

Singh, T.V. and Kaur, J. (1985) *Integrated Mountain Development*, New Delhi: Himalayan Books.

Slim, W.J. (Viscount) (1981) *Defeat into Victory*, Indian edition, Dehra Dun: Natraj Publ.

Smadja, J. (1992) 'Studies on climatic and human impacts and their relationship on a mountain slope above Salme in the Himalayan Middle Mountains, Nepal', *Mountain Research and Development*, 12(1): 1–28.

—— (ed.) (2003) *Histoire et Devenir des Paysages en Himalaya*, Paris: CNRS Publications.

Smadja, J. and Fort, M. (1998) 'Research on diversity, origin, and evolution of Himalayan Nepalese landscapes', in I. Stellrecht (ed.) *Karakorum–Hindukush–Himalaya: aspects of change*, Köln: Rüdiger Köppe Verlag, 387–413.

Speechley, H.T. (1978) Proposals for forest management in Sagarmatha (Mt. Everest) National Park, Department of National Parks and Wildlife Conservation, Kathmandu.

Stainton, J.D.A. (1972) *Forests of Nepal*, New York: Hafner.

Starkel, L. (1972a) 'The role of catastrophic rainfall in the shaping of the relief of the Lower Himalaya (Darjeeling Hills)', *Geographica Polonica*, 21: 103–47.

—— (1972b) 'The modelling of monsoon areas of India as related to catastrophic rainfall', *Geographica Polonica*, 23: 151–73.

Starkel, L. and Basu, S. (2000) *Rains, Landslides and Floods in the Darjeeling Himalaya*, Indian National Science Academy: New Delhi.

Starr, S.F. (2001) 'Altitude sickness: poverty and violence in the mountains', *The National Interest*, 65 (Fall, 2001): 90–100.

Stellrecht. I. (ed.) (1997) *The Past in the Present: Horizons of Remembering in the Pakistan Himalaya*, Köln: Rüdiger Köppe Verlag.

Stellrecht, I. (ed.) (1998) *Karakorum – Hindukush – Himalaya: Dynamics of Change*, Köln: Rüdiger Köppe Verlag.

Sterling, C. (1976) 'Nepal', *Atlantic Monthly* (Oct), 238(4): 14–25.

Stevens, S.F. (1993) *Claiming the High Ground: Sherpas, Subsistence and Environmental Change in the Highest Himalaya*, Berkeley: University of California Press.

—— (1997) 'Consultation, co-management, and conflict in the Sagarmatha (Mount Everest) National Park, Nepal', in S.F. Stevens (ed.) *Conservation Through Cultural Survival: Indigenous Peoples and Protected Areas*, Washington DC: Island Press.

—— (2003) 'Tourism and deforestation in the Mt Everest region of Nepal', *Geographical Journal*, 169(3): 255–77.

Stewart, K. (1989) 'Post flood: assessment and nutritional status of children in Matlab, Bangladesh', unpublished paper presented in *Regional and Global Environmental Perspectives*, Dhaka, 4–6.

Stone, L. (1990) 'Conservation and human resources: comments on four case studies from Nepal', *Mountain Research and Development*, 10(1): 5–6.

Subba, B. (2001) *Himalayan Waters: Promise and Potential, Problems and Prospects*, Kathmandu: Panos, South Asia.

—— (2003) 'Water, Nepal and India', in K.M. Dixit and S. Ramachandaran (eds) *State of Nepal*, Kathmandu: Himal Books, pp. 235–52.

Suslov, S.P. (1961) *Physical Geography of Asiatic Russia*, San Francisco: W.H. Freeman and Co.

Swope, L.H. (1995) 'Factors influencing rates of deforestation in Lijiang County, Yunnan Province, China: a village study', unpublished Masters thesis, University of California, Davis.

Swope, L., Swain, M.B., Yang, F. and Ives, J.D. (1997) 'Uncommon property rights in southwest China: trees and tourists', in B.R. Johnston (ed.) *Life and Death Matters: Human Rights and the Environment at the End of the Millennium*, Walnut Creek, California: AltaMira Press, Sage Publications.

Tabei, J. (2001) 'Climbers' impact on the natural environment in mountain areas and environmental conservation: Present situation of waste left by climbers in the Mount Everest region', in T. Watanabe, S. Sicroff, N.R. Khanal, and M.P. Gautam (eds) *Proceedings of an International Symposium on the Himalayan Environments: Mountain Science and Ecotourism/Biodiversity*, 24–26 November 2000, Kathmandu.

Tamrakar, R.M. (1995) A comparative study of land use changes in the Mahabharat Lekh area of Kabhre Palanchok District between 1978–1992, unpublished consultancy report to the Nepal-Australian Community Forestry Project, Kathmandu.

Tejwani, K.G. (1984a) 'Biophysical and socio-economic causes of land degradation and strategy to foster watershed reclamation in the Himalayas', in IUFRO symposium on The Effects of Forest Land Use on Erosion and Slope Stability, East–West Center, Honolulu.

—— (1984b) 'Reservoir sedimentation in India – its causes, control and future course of action', *Water International*, 9(4): 150–4.

—— (1987) 'Sedimentation of reservoirs in the Himalayan region, India', *Mountain Research and Development*, 7(3): 323–27.

Thapa, D. (2003) 'The Maobadi of Nepal', in K.M. Dixit and S. Ramachandaran (eds), *The State of Nepal*, Lalitpur, Nepal: Himal Books.

Thapa, D. and Sijapati, B. (2003) *A Kingdom Under Siege: Nepal's Maoist Insurgency, 1996–2002*, Kathmandu: Printhouse.

Thapa, G.B. and Paudel, G.S. (2002) 'Farmland degradation in the mountains of Nepal: a study of watersheds "with" and "without" external intervention', *Land Degradation and Development*, 13(6): 479–93.

Thapa, P.J. (1997) 'Water-led development in Nepal: myths, limitations and rational concerns', *Water Nepal*, 5(1): 35–57.

Thompson, M. (1995) 'Policy-making in the face of uncertainty: the Himalayas as unknowns', in G.P. Chapman and M. Thomson (eds), *Water and the Quest for Sustainable Development in the Ganges Valley*, London: Mansell, pp. 25–38.

Thompson, M. and Warburton, M. (1985a) 'Uncertainty on a Himalayan scale', *Mountain Research and Development*, 5(2): 115–35.

—— (1985b) 'Knowing where to hit it: a conceptual framework for the sustainable development of the Himalaya', *Mountain Research and Development*, 5(3): 203–20.

Thompson, M., Warburton, M. and Hatley, T. (1986) *Uncertainty on a Himalayan Scale*, London: Ethnographia.

Thorarinsson, S. (1960) 'Glaciological knowledge in Iceland before 1800', *Jökull*, 10: 1–14.

Tilman, H.W. (1952) *Nepal Himalaya*, Cambridge: Cambridge University Press, pp. 126–7.

Timsina, N.P. (2003) 'Promoting social justice and conserving montane forest environments: a case study of Nepal's community forestry programme', *Geographical Journal*, 169(3): 236–42.

Tiwari, C.K. (2002) Published lecture presented to Williams College, Massachusetts, 1 March 2002.

Troll, C. (1972) 'The upper limit of aridity and the arid core of Central Asia', in C. Troll (ed.) *Geoecology of the High-Mountain Regions of Eurasia*, Wiesbaden: Franz Steiner Verlag GMBH, 237–43.

Tucker, R.P. (1983) 'The British colonial system and the forests of the Western Himalayas, 1815–1914', in R.P. Tucker and J.F. Richards (eds), *Global Deforestation and the Nineteenth Century World Economy*, Durham, NC, USA: Duke University Press.

—— (1987) 'Dimensions of deforestation in the Himalaya: the historic setting', *Mountain Research and Development*, 7(3): 328–31.

Uhlig, H. (1995) 'Persistence and change in high mountain agricultural systems', *Mountain Research and Development*, 15(3): 199–212.

United Nations (2000) *Usoi Landslide Dam and Lake Sarez: An Assessment of Hazard and Risk in the Pamir Mountains, Tajikistan*, D. Alford and R. Shuster (eds), United Nations Publ., No. E.OO.III.M.1, International Strategy for Disaster Reduction (ISDR), Geneva, Switzerland.

UNEP (2001) *State of the Environment*, United Nations Environment Programme, Nepal.

UNICEF (1996) *Atlas of South Asian Children and Women*, Kathmandu: UNICEF Regional Office.

Valdiya, K.S. (1985) 'Accelerated erosion and landslide-prone zones in the Himalayan region', in J.S. Singh (ed.), *Environmental Regeneration in Himalayas: Concepts and Strategies*, Central Himalayan Environment Association., Nainital, UP, India.

—— (1987) *Environmental Geology: Indian Context*, New Delhi: Tata-McGraw-Hill.

—— (1992) 'Must we have high dams in the geodynamically active Himalayan domain?', *Current Science*, 63: 280–96.

—— (1993a) 'High dams in the Himalaya: environmental implication', *Proceedings, Indian National Science Academy*, 59A(1): 133–55.

—— (1993b) 'Tectonics of the Himalaya and recent crustal movements: an overview', in V.K. Gaur (ed.), *Earthquake Hazard and Large Dams in the Himalaya*, New Delhi: Indian National Trust for Art and Cultural Heritage (INTACH), pp. 1–34.

—— (1998) *Dynamic Himalaya*, Hyderabad: Universities Press (India).

Valdiya, K.S. and Bartarya, S.K. (1991) 'Hydrogeological studies of springs in the catchment of the Gaula River, Kumaun Lesser Himalaya, India', *Mountain Research and Development*, 11(3): 239–58.

Virgo, K.J. and Subba, K.J. (1994) 'Land-use change between 1978 and 1990 in Dhankuta District, Koshi Hills, Eastern Nepal', *Mountain Research and Development*, 14(2): 159–70.

Vuichard, D. and Zimmermann, M. (1986) 'The Langmoche flash-flood, Khumbu Himal, Nepal', *Mountain Research and Development*, 6(1): 90–4.

—— (1987) 'The catastrophic drainage of a moraine-dammed lake, Khumbu Himal, Nepal: cause and consequences', *Mountain Research and Development*, 7(2): 91–110.

Wagner, A. (1981) 'Rock structure and slope stability study of Walling area, central west Nepal', *Journal Nepal Geol. Soc.*, 1(2): 37–43.

—— (1983) 'The principal geological factors leading to landslides in the foothills of Nepal: a statistical study of 100 landslides; steps for mapping the risks of landslides', Helvetas/Swiss Technical Cooperation/ITECHO, Kathmandu.

Ward, F. Kingdon (1973) *Land of the Blue Poppy: Travels of a Naturalist in Eastern Tibet*, London: Minerva Press.

Watanabe, T, Ives, J.D. and Hammond, J.E. (1994) 'Rapid growth of a glacial lake in Khumbu Himal, Nepal: prospects for a catastrophic flood', *Mountain Research and Development*, 14(4): 329–40.

Watanabe, T., Kameyama, S. and Sato, T. (1995) 'Imja Glacier dead-ice melt rates and changes in a supra-glacial lake, 1989–1994, Khumbu Himal, Nepal: danger of lake drainage', *Mountain Research and Development*, 15(4): 293–300.

Watanabe, T. and Rothacher, D. (1996) 'The 1994 Lugge Tsho glacier lake outburst flood, Bhutan Himalaya', *Mountain Research and Development*, 16(1): 77–81.

Webster, P.J. (1987) 'The variable and interactive monsoon', in J.S. Fein and P.L. Stephens (eds), *Monsoons*, New York: John Wiley and Sons, pp. 269–330.

White, J.C. (1990) *Sikkim and Bhutan*, New Delhi: A. Sagar Book House.

Winkley, B.R., Lesleighter, E.J. and Cooney, J.R. (1994) 'Instability problems of the Arial Khan River, Bangladesh', in S.A. Schumm and B.R. Winkley, (eds), *The Variability of Large Alluvial Rivers*, New York: American Society of Civil Engineering Press, pp. 269–84.

Wohl, E.E. (ed.) (2000) *Inland Flood Hazards: Human, Riparian and Aquatic Communities*, Cambridge, UK: Cambridge University Press.

World Bank (1979) *Nepal: Development Performance and Prospects*, A World Bank Country Study, South Asia Regional Office, World Bank, Washington, DC.

World Resources Institute (1985) 'Tropical forests: A call to action', Report of an international task force convened by the World Resources Institute, the World Bank and the United Nations Development Programme, Washington DC: WRI.

WTTC (1999) *Travel and Tourism's Economic Perspective*, Brussels: World Travel and Tourism Council.

Wu, K. and Thornes, J.B. (1995) 'Terrace irrigation of mountainous hill slopes in the Middle Hills of Nepal: stability and instability', in G.P. Chapman and M. Thompson (eds) *Water and the Quest for Sustainable Development in the Ganges Valley*, London and New York: Mansell, pp. 41–63.

Wymann, S. (1991) 'Landnutzungsintensivierung und Bodenfruchtbarkeit im nepalischen Hugelgebiet Unveroff', unpublished doctoral thesis, University of Berne, Switzerland.

Xu, D. (1985) 'Characteristics of debris flows caused by outburst of glacial lake in Boqu River, in Xizang, China, 1981', unpublished manuscript, Institute of Glaciology and Cryopedology, Academia Sinica: Lanzhou.

Yablokov, A. (2001) 'The tragedy of Khait: a natural disaster in Tajikistan', *Mountain Research and Development*, 21(1): 91–3.

Yamada, T. (1998) 'Glacier lake and its outburst flood in the Nepal Himalaya', Tokyo, Japanese Society of Snow and Ice, Monograph No. 1.

Yoshino, M.M. (1980) 'Local climatological differences between highlands and lowlands in Thailand', in J.D. Ives, S. Sabhasri and P. Voraurai (eds), *Conservation and Development in Northern Thailand*, NRTS-3/UNUP-77: 63–74, Tokyo: United Nations University Press.

Zeitler, P.K., Johnson, N.M., Naeser, C.W. and Tahirkhedi, R.A.K. (1982) 'Fission-track evidence for Quaternary uplift of the Nanga Parbat region, Pakistan', *Nature*, 298: 255–7.

Zimmermann, M., Bichsel, M. and Kienholz, H. (1986) 'Mountain hazards mapping in the Khumbu Himal, Nepal', *Mountain Research and Development*, 6(1): 29–40.

Zinke, P.J., Sabhasri, S. and Kunstadter, P. (1978) 'Soil fertility aspects of a Lua forest fallow system of shifting cultivation', in P. Kunstadter, E.C. Chapman and S. Sabhasri (eds), *Farmers in the Forest*, Honolulu: University Press of Hawaii, pp. 134–59.

Zomer, R., Ustin, S. and Carpenter, C. (2001) 'Land use/cover change along tropical and subtropical riparian corridors within the Makalu Barun National Park and Conservation Area, Nepal', *Mountain Research and Development*, 21(2): 175–83.

Zurick, D.N. (1990) 'Traditional knowledge and conservation as a basis for development in a West Nepal village', *Mountain Research and Development*, 10(1): 23–33.

—— (1995) *Errant Journeys: Adventure Travel in a Modern Age*, Austin, Texas: University of Texas Press.

Zurick, D. and Karan, P.P. (1999) *Himalaya: Life on the Edge of the World*, Baltimore, USA: Johns Hopkins University Press.

Author index

Subject index